Innovative Techniques for Citrus Cultivation

Innovative Techniques for Citrus Cultivation

Guest Editors

Katsuya Ichinose
Bo Xiong

MDPI

Basel • Beijing • Wuhan • Barcelona • Belgrade • Novi Sad • Cluj • Manchester

Guest Editors
Katsuya Ichinose
Strategic Planning
Headquarters
National Agricultural
Research Organization
Tsukuba
Japan

Bo Xiong
College of Horticulture
Sichuan Agricultural
University
Chengdu
China

Editorial Office
MDPI AG
Grosspeteranlage 5
4052 Basel, Switzerland

This is a reprint of the Special Issue, published open access by the journal *Plants* (ISSN 2223-7747), freely accessible at: https://www.mdpi.com/journal/plants/special_issues/B77NBDWC3Z.

For citation purposes, cite each article independently as indicated on the article page online and as indicated below:

Lastname, A.A.; Lastname, B.B. Article Title. *Journal Name* **Year**, *Volume Number*, Page Range.

ISBN 978-3-7258-5621-3 (Hbk)
ISBN 978-3-7258-5622-0 (PDF)
https://doi.org/10.3390/books978-3-7258-5622-0

Cover image courtesy of Bo Xiong

Contents

About the Editors

Katsuya Ichinose

Katsuya Ichinose is a researcher at the Strategic Planning Headquarters, National Agricultural Research Organization (NARO), located in Tsukuba, Japan. His principal research area is entomology, in particular, the control of insect pests. He has been engaged in international projects for pest management, and one of his most focused research projects is on citrus greening disease, also known as huanglongbing. His interest in this disease extends widely from studies on ecology and management of its vectors to disease detection with optical measures. Following the recent trends in public concerns about human health and environmental conservation, he is shifting his research with pesticides to naturally derived materials and the integration of plant's physiological responses into pest infestation. Receiving support from researchers in various study areas, he is still conducting research projects for pest management.

Bo Xiong

Bo Xiong is a researcher at the College of Horticulture, Sichuan Agricultural University, China. He has served as both a project leader and a core researcher in more than 20 research initiatives, including projects under the National Key Research and Development Program, the Ministry of Agriculture and Rural Affairs' Industrial Cluster Program, the Ministry of Science and Technology's Key Research and Development Program, the Sichuan Provincial Science and Technology Support Program, and various international collaborative efforts. His research primarily focuses on citrus and plum species, including the mechanisms underlying fruit quality formation, plant stress physiology, and the regulation of secondary metabolites.

Preface

Citrus greening disease, also known as huanglongbing, is the most destructive disease in citrus fruits worldwide. Although it has been about 100 years since its first scientific report, practical measures for the control or management of this disease still remain to be developed. According to media sources, the turmoil of the global citrus market, especially oranges, in recent years can be attributed to the expansion of this disease, as well as global climate change. It seems to take significant time to stabilize the turmoil of the citrus market, which necessitates an increase in the number of studies from molecular to ambient levels. This Reprint consists of research papers published in the Special Issue, entitled "Innovative Techniques for Citrus Cultivation" in *Plants*. The Editorial Board Members planned this Special Issue to encourage these studies by providing knowledge on citrus fruits across a broad research area. The Editorial Board Members hope the papers in this Reprint will contribute to the development of citrus cultivation across the world.

Katsuya Ichinose and Bo Xiong
Guest Editors

plants

MDPI

Article

Physiological Mechanisms of Citrus Fruit Cracking: Study on Cell Wall Components, Osmoregulatory Substances, and Antioxidant Enzyme Activities

Shengjia Huang [1,2,†], Xinxia Yang [1,†], Tie Wang [1,†], Hang Li [1], Lijun Deng [1], Xiaoyi Bi [1], Juan Hu [1], Yan Gong [1], Yunjie Li [1], Zeyu Qin [1], Yuan Yao [1], Guochao Sun [3], Ling Liao [1], Mingfei Zhang [1], Siya He [3], Lingping Jiang [4] and Zhihui Wang [1,3,*]

[1] College of Horticulture, Sichuan Agricultural University, Chengdu 611130, China
[2] Institute of Economic Forest Research, Sichuan Academy of Forestry, Chengdu 610081, China
[3] Institute of Pomology and Olericulture, Sichuan Agricultural University, Chengdu 611130, China
[4] Agricultural and Rural Bureau, Shimian, Ya'an 625400, China
* Correspondence: wangzhihui318@sicau.edu.cn
† These authors contributed equally to this work.

Abstract: Fruit cracking affects both the yield and economic efficiency of citrus; however, the underlying mechanism remains unclear. Therefore, this study focused on resistant and susceptible cultivars to identify the mechanisms underlying fruit cracking. The results showed that in 'Mingrijian', pectin morphological transformation and hemicellulose and lignin degradation in the pericarp were important contributing factors. During the critical fruit-cracking period (115–150 days after flowering), the water-soluble pectin, protopectin, and lignin contents in the pericarp of 'Daya' presented inverse changes relative to those in 'Mingrijian', thus enhancing the mechanical properties and resistance of pericarp. From 115 to 150 days after flowering, the soluble sugar content in the pulp of 'Mingrijian' increased rapidly by 97.35%, aiding in pulp water absorption and expansion. Moreover, the soluble protein content in the pericarp of 'Mingrijian' exhibited a declining trend and was lower than that of 'Daya', thus affecting the overall metabolism. The superoxide dismutase (SOD) activity in the pericarp of 'Mingrijian' gradually decreased from 115 to 180 days after flowering, while the peroxidase (POD) activity remained at a low level, resulting in weaker antioxidant capacity and lower environmental resistance. This study provides valuable insights into the mechanisms of citrus fruit cracking, laying the foundation for preventive and control strategies.

Keywords: 'Mingrijian'; citrus; fruit cracking; cell wall component; osmoregulation; antioxidant enzyme

1. Introduction

Citrus belongs to the subfamily Aurantioideae and family Rutaceae and represents the world's most widely grown and most productive fruit genus [1,2]. Fruit cracking is a physiological disorder in citrus fruit development that occurs in various citrus varieties, such as mandarins (*Citrus reticulata*), navel oranges (*Citrus sinensis*), and mandarin hybrids (*Citrus reticulata* × *Citrus sinensis* or *Citrus reticulata* × *Citrus paradise*), with some citrus varieties having a fruit cracking rate of up to 50% or more, resulting in serious economic losses [3–6]. 'Mingrijian' (M), also referred to as 'Asumi', originated in Japan. It is a mandarin hybrid resulting from the crossbreeding of 'Harumi' (*Citrus reticulata* × (*Citrus reticulata* × *Citrus sinensis*)) and 'Sweet spring' (*Citrus reticulata* cv. Unshiu × *Citrus Hassaku* Hort. ex Tanaka). This hybrid is known for its thin easily peeled skin and high soluble solids content, thus it has gained popularity among consumers and has been recognized as a promising new cultivar in the citrus industry. However, the fruit is susceptible to cracking during its expanding period, which severely constrains the development of this cultivar.

Sudden heavy rainfall following an extended drought can increase the susceptibility of citrus fruits to cracking, a phenomenon that is often observed in the rapid expanding period of fruits. This phenomenon mainly stems from the uncoordinated growth between the pulp and pericarp. After rainfall, the fruit pulp rapidly absorbs water and expands. If the internal expansion pressure exceeds the pericarp strength, fruit cracking will occur.

The mechanical properties of the pericarp play a major role in resisting internal expanding pressure and controlling fruit cracking. As the fruit ripens, cell wall components change and degrade, leading to alterations in the cell wall structure and consequently affecting the mechanical properties of the pericarp [7–9]. During this process, the fruit is more susceptible to cracking, highlighting the significant role of cell wall components in the fruit-cracking process. The primary components of the cell wall include polysaccharides, such as pectin, cellulose, and hemicellulose, as well as lignin, proteins, and minerals. Previous research has revealed that during the maturation of jujube fruits, the content of protopectin and cellulose in the peel gradually decreased, while the content of water-soluble pectin (WSP) increased, which could potentially contribute to the increasing rate of fruit cracking [10]. In a comparative study on cracking-susceptible and cracking-resistant cultivars, the susceptible litchi cultivar 'Nuomici' had lower protopectin and cellulose contents in the cell wall of the pericarp than the resistant cultivar 'Huaizhi' [11]. Similarly, in tomato, the resistant cultivar 'LA1698' had higher protopectin and hemicellulose contents in the fruit than the susceptible cultivar 'LA2683' [12].

The susceptibility to cracking is closely linked to moisture absorption by the pulp, which becomes the main source of pressure on the pericarp. Osmoregulation serves as a critical regulatory mechanism in plants to maintain cellular osmotic pressure, and it subsequently influences levels of moisture absorption [13]. Soluble sugar (SS) is one of the main osmoregulatory substances in fruit. Previous studies have shown that exogenous nordihydroguaiaretic acid (NDGA) treatment could significantly reduce the contents of SS and soluble solids in grapefruits, thereby decreasing osmotic pressure, water uptake, and peel expansion and inhibiting fruit cracking [14]. According to Richardson et al. [15], sweet cherry fruits containing high SS levels typically experienced elevated osmotic pressure, which affected both moisture absorption and movement, ultimately leading to fruit cracking. Additionally, certain osmoregulatory substances can also enhance the stress resistance of plants. For example, plants protect themselves from the oxidative damage caused by dehydration by proactively accumulating soluble protein (SP) and respond to external stress by increasing reactive oxygen species (ROS), which inflict damage upon the cellular membrane system. Excessive accumulation of harmful substances leads to the occurrence of fruit cracking [16]. The response of the antioxidant enzyme system can reduce membrane damage and enhance stress resistance [17]. Antioxidant enzymes include superoxide dismutase (SOD), catalase (CAT), and peroxidase (POD), among others [18].

While preliminary research progress has been made on citrus fruit cracking, previous studies mainly focused on certain varieties, fruit morphologies (e.g., peel thickness and anatomical structure), cell wall components, and mineral elements [4,19]; thus, the underlying mechanism remains unclear. In 'Mingrijian', fruit cracking mainly happens at the top of the fruit, where the exocarp and mesocarp rapidly crack within a few minutes, leading to juice cell exposure and eventual fruit abscission (Figure 1). Comparative analyses between cracking and noncracking pericarps within a single citrus cultivar are difficult to perform because uncertain factors that arise after the occurrence of cracking, such as emergency response, oxidative stress, and pathogenic invasion, would interfere with the measured results. Local cultivation experience and preliminary laboratory surveys revealed that the citrus cultivar 'Daya' (D) hardly ever cracked, while 'Mingrijian' (M) cracked at a rate exceeding 60% in regions experiencing severe cracking. Both 'Daya' and 'Mingrijian' share certain traits, including a relatively thin peel, high soluble solids content in the pulp, and similar maturation periods in suitable planting areas. Therefore, this study utilized the cracking-susceptible cultivar 'Mingrijian' and the cracking-resistant cultivar 'Daya' as materials to conduct a comparative analysis of biochemical and metabolic changes primarily related to cell wall components, osmoregulatory substances, and antioxidant enzymes. The objective was to explore the contributions of these factors to fruit cracking and reveal the important elements that influence the pericarp mechanical properties and pulp water absorption capacity. This research contributes to elucidating the mechanism of citrus fruit cracking, thereby establishing a foundation for identifying effective prevention and control strategies.

Figure 1. Diagram of fruit cracking of 'Mingrijian'.

2. Materials and Methods

2.1. Materials

The samples were collected from Renshou County, Meishan City, Sichuan Province, China. A total of 15 healthy trees from both the 'Mingrijian' and 'Daya' citrus cultivars with uniform growth were selected, and the planting density was 3 m × 5 m. Each group of five trees formed a designated plot and replicate, thus ensuring uniform cultivation practices across all units. Four representative time points were selected based on preliminary laboratory statistics of fruit cracking rates for 'Mingrijian': before fruit cracking (75 days after flowering), initial stage of fruit cracking (115 days after flowering), peak stage of fruit cracking (150 days after flowering), and end stage of fruit cracking (180 days after flowering). Samples were collected separately from 'Mingrijian' and 'Daya' at each designated time point. From the periphery of the canopy, a total of thirty disease-free fruits of consistent size were carefully selected from each cultivar, covering all four cardinal directions: east, south, west, and north. After returning the fruits to the laboratory, they were cleaned, and the pericarp and pulp were separated, with a portion stored at −80 °C for measurements of osmoregulatory substances and enzyme activities, and the remaining portion underwent a drying process at 80 °C for 72 h for analysis of cell wall components.

2.2. Extraction and Determination of Cell Wall Material

Cell wall material (CWM) was extracted according to Huber et al. [20] with modifications. The dried pericarp tissue was crushed and sieved. A sample (1 g) was weighed and placed into a centrifuge tube. Then, 30 mL of 80% ethanol solution was added, and the tube was placed in a boiling water bath for 25 min. After cooling and vacuum filtration, the filter residue was rinsed three times with 30 mL of 80% ethanol solution and then collected. The collected residue was soaked in 30 mL of 90% dimethylsulfoxide solution, left overnight, and then vacuum-extracted again. The residue was rinsed with 30 mL acetone three times, dried to obtain CWM, and then weighed. The amount of CWM obtained per gram of sample was calculated. Three replicates were set up for each sample.

2.3. Determination of Cell Wall Polysaccharides

The separation of CWM was conducted according to Siddiqui et al. [21] with modifications. A CWM sample (0.01 g) was accurately weighed, and the protopectin content was determined using the protopectin test kit (Product No. G0703F; Suzhou Grace Biotechnology Co., Ltd., Suzhou, China). Additionally, 0.05 g of CWM was mixed with 5 mL of 50 mmol·L^{-1} sodium acetate buffer (pH 6.5), shaken for 6 h, and centrifuged at 9500 rpm for 10 min to obtain the WSP supernatant. Separately, 5 mL of 50 mmol·L^{-1} sodium acetate buffer (containing 50 mmol·L^{-1} EDTA, pH 6.5) was added to the sediment, shaken for 6 h, and then centrifuged at 9500 rpm for 10 min to obtain the ionic bound pectin (ISP) supernatant; and 5 mL of 50 mmol·L^{-1} Na_2CO_3 solution (containing 2 mmol·L^{-1} EDTA) was added to the sediment, shaken for 6 h, and then centrifuged at 9500 rpm for 10 min to obtain the supernatant containing covalently bound pectin (CSP). The contents of pectin (WSP, ISP, CSP) were determined by the carbazole colorimetric method [22].

After adding 5 mL of 4 mmol·L^{-1} NaOH solution (including 100 mmol·L^{-1} $NaBH_4$) to the sediment, the mixture was shaken for 6 h and then centrifuged at 9500 rpm for 10 min to obtain the supernatant containing hemicellulose. A 2 mL portion of the supernatant was collected, combined with 3 mL of 2 mol·L^{-1} sulfuric acid, and hydrolyzed at 100 °C in a boiling water bath for 5 h. The hydrolyzed reducing sugars were determined using the anthrone method. The sediment obtained from the previous step was cellulose, which was mixed with 1.5 mL of 80% sulfuric acid, left for 2 h, then mixed with 3 mL of water and hydrolyzed at 100 °C in a boiling water bath for 5 h. The hydrolyzed reducing sugars were determined using the anthrone method [23].

2.4. Determination of Lignin

The lignin content was determined by the acetylation method using a lignin test kit (Product No. G0708W; Suzhou Grace Biotechnology Co., Ltd., Suzhou, China).

2.5. Determination of Osmoregulatory Substances

The SS content was determined by anthrone colorimetry [24], and the SP content was determined by the Coomassie brilliant blue method [25].

2.6. Determination of Antioxidant Enzyme Activity

SOD activity (U/g FW) was determined by the NBT photoreduction method [26], where the amount required to inhibit the photochemical reduction of NBT to 50% of the control was defined as 1 unit of enzyme activity (U). POD activity (U/g FW) was determined by the guaiacol method [27], where a change of 0.01 in A470 within 1 min was defined as 1 unit of enzyme activity (U). CAT activity (U/g FW) was determined by UV spectrophotometry [27], where a decrease of 0.1 in A240 within 1 min was defined as 1 unit of enzyme activity (U).

2.7. Statistical Analysis

The experimental data underwent analysis employing an independent samples *t*-test through IBM SPSS Statistics 23.0. Duncan's multiple-range test was utilized to assess differences between the samples, with statistical significance set at $p < 0.05$. Furthermore, a correlation analysis was conducted using SPSS 23.0 and R 3.5.3. Graphs were generated using Excel 2016, R 3.5.3, and Adobe Illustrator 2020.

3. Results

3.1. Changes in the Cell Wall Components and Contents in the Pericarp during Fruit Development in 'Mingrijian' and 'Daya'

3.1.1. Changes in Cell Wall Material Content

Figure 2 shows that from 75 to 180 days after flowering, the content of CWM in the pericarp of 'Mingrijian' initially increased, reached its maximum at the peak cracking stage (150 days after flowering), and then decreased. The CWM content in the pericarp of 'Daya' gradually decreased from 0.70 g/g dry weight (DW) to 0.52 g/g DW. The results of the variance analysis indicated that the CWM content did not significantly differ between the pericarps of 'Daya' and 'Mingrijian' at 75 days after flowering. However, the CWM content in the pericarp of 'Daya' consistently remained lower than that of 'Mingrijian' during the period from 115 to 180 days after flowering.

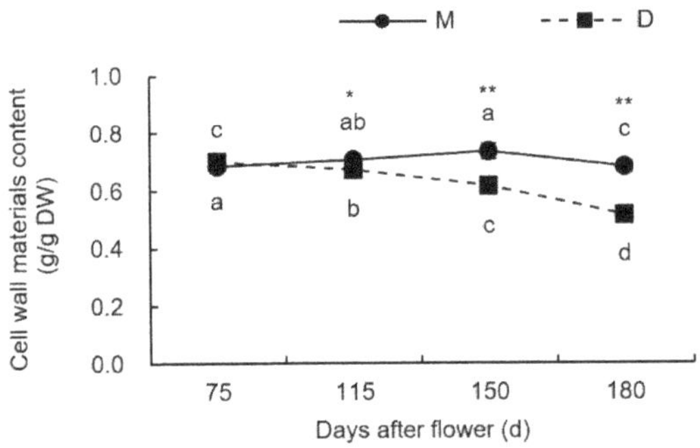

Figure 2. Changes in the cell wall material in the pericarp of 'Mingrijian' and 'Daya' during fruit development. Different lowercase letters represent significant differences in the different growth and development stages at the 0.05 level ($p < 0.05$). * Significant difference between two varieties ($p < 0.05$); ** extremely significant difference between two varieties ($p < 0.01$).

3.1.2. Changes in Pectin Content

As shown in Figure 3a, before the occurrence of cracking (75–115 days after flowering), the WSP content in the pericarp of both 'Mingrijian' and 'Daya' showed an increasing trend at 35.46% and 40.43%, respectively. During the period of fruit cracking from the initial to peak stage (115–150 days after flowering), the WSP content in the pericarp of 'Mingrijian' increased from 105. 96 mg/g DW to 121.77 mg/g DW as the fruit cracking rate increased, whereas that of 'Daya' decreased from 64.71 mg/g DW to 58.34 mg/g DW. From the peak to end stage of fruit cracking (150–180 days after flowering), the WSP content in the pericarp of both 'Mingrijian' and 'Daya' showed a decreasing trend, with similar rates of decrease. As shown in Figure 3b, the ISP content in the pericarp of 'Mingrijian' and 'Daya' exhibited fluctuating trends from 75 to 180 days after flowering, characterized by a 'decreasing-increasing-decreasing' pattern for 'Mingrijian' and an 'increasing-decreasing-increasing' pattern for 'Daya'. As shown in Figure 3c, the CSP and WSP contents in the pericarp of 'Mingrijian' exhibited contrasting patterns of change during the same period. From the initial to peak stage of fruit cracking (115–150 days after flowering), the CSP content in the pericarp of 'Mingrijian' decreased by 31.41% as the cracking rate increased. As shown in Figure 3d, from the pre-cracking to peak stage (75–150 days after flowering), the protopectin content in the pericarp of 'Mingrijian' continuously declined, reaching a minimum value of 122.03 mg/g DW at 150 days after flowering, while that in 'Daya' generally increased by 38.52%. As shown in Figure 3e, from 75 to 180 days after flowering, the total pectin content in the pericarp of both 'Mingrijian' and 'Daya' showed an increasing trend. Throughout this period, the contents of total pectin, WSP, and CSP in the pericarp of 'Daya' were significantly lower than those in 'Mingrijian'. However, after the occurrence of cracking (115–180 days after flowering), the protopectin content in the pericarp of 'Daya' exceeded that of 'Mingrijian'. The results indicated a mutual transformation between protopectin (predominantly CSP) and WSP in the pericarp of 'Mingrijian' in response to fruit cracking.

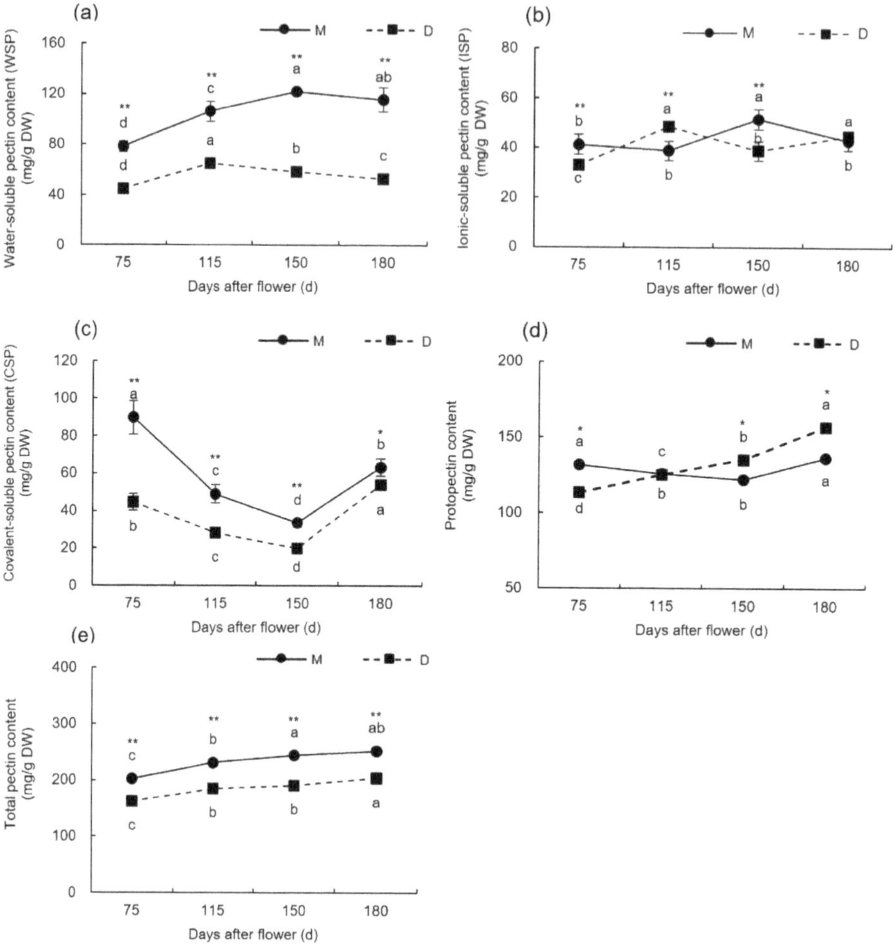

Figure 3. Changes in pectin contents in the pericarp of 'Mingrijian' and 'Daya' during fruit development. (**a**) Water-soluble content; (**b**) ionic-soluble pectin content; (**c**) covalent-soluble pectin content; (**d**) protopectin content; and (**e**) total pectin content. Different lowercase letters represent significant differences in the different growth and development stages at the 0.05 level ($p < 0.05$). * Significant difference between two varieties ($p < 0.05$); ** extremely significant difference between two varieties ($p < 0.01$).

3.1.3. Changes in Cellulose and Hemicellulose Contents

As shown in Figure 4a, the hemicellulose content in the pericarp of 'Mingrijian' decreased by 9.21% before the occurrence of fruit cracking (75–115 days after flowering), while that in 'Daya' significantly increased by 24.53%. This substantial increase in hemicellulose content in the pericarp of 'Daya' caused its hemicellulose levels to surpass those of 'Mingrijian' from 115 to 180 days after flowering. Notably, at 115 days after flowering, 'Daya' exhibited significantly higher hemicellulose content than 'Mingrijian'. From the initial to end stage of fruit cracking (115–180 days after flowering), both 'Mingrijian' and 'Daya' showed a decreasing trend in hemicellulose content. As shown in Figure 4b, no significant differences in cellulose content were observed in the pericarp of 'Mingrijian' with the occurrence of fruit cracking.

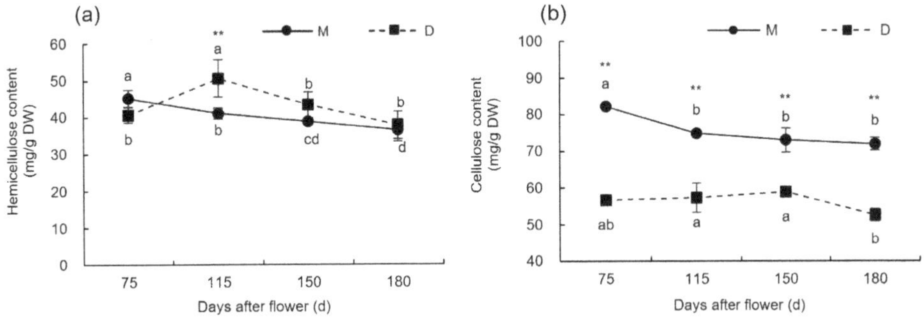

Figure 4. Changes in cellulose and hemicellulose contents in the pericarp of 'Mingrijian' and 'Daya' during fruit development. (**a**) Hemicellulose content; and (**b**) cellulose content. Different lowercase letters represent significant differences in the different growth and development stages at the 0.05 level ($p < 0.05$). ** Extremely significant difference between two varieties ($p < 0.01$).

3.1.4. Changes in Lignin Content

As illustrated in Figure 5, before the occurrence of cracking (75–115 days after flowering), the lignin content in the pericarp of both 'Mingrijian' and 'Daya' decreased. From the initial to peak stage of fruit cracking (115–150 days after flowering), the lignin content in the pericarp exhibited contrasting trends in 'Mingrijian' and 'Daya'. Specifically, 'Mingrijian' displayed a significant reduction from 11.63 mg/g DW to 7.72 mg/g DW, representing a decrease of 33.64%, whereas 'Daya' exhibited a notable increase from 5.03 mg/g DW to 5.92 mg/g DW, representing a rise of 17.63%. During the peak to end stage of fruit cracking (150–180 days after flowering), the lignin content in the pericarp showed no significant change in 'Mingrijian' but decreased to 3.82 mg/g DW in 'Daya'. These results indicated that with the occurrence of fruit cracking, lignin gradually degraded in the pericarp of 'Mingrijian'.

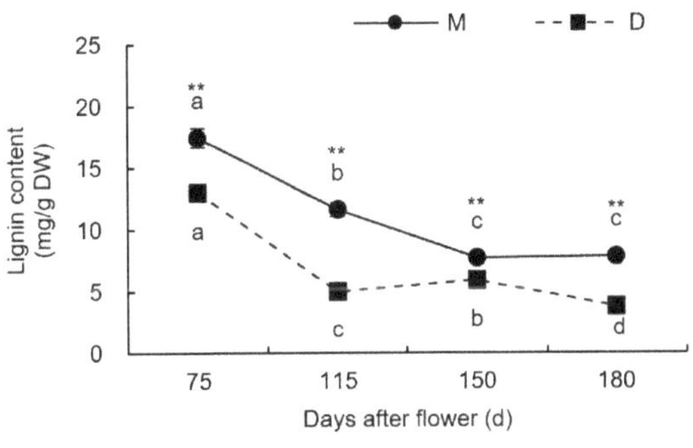

Figure 5. Changes in lignin content in the pericarp of 'Mingrijian' and 'Daya' during fruit development. Different lowercase letters represent significant differences in the different growth and development stages at the 0.05 level ($p < 0.05$). ** Extremely significant difference between two varieties ($p < 0.01$).

3.1.5. Correlation Analysis between Cell Wall Components and Fruit Cracking Rate

The cell wall components and contents were determined at 75, 95, 115, 135, 150, 165, and 180 days after flowering in the pericarp of 'Mingrijian'. Subsequently, a correlation analysis was performed between these measurements and the cracking rate data. The results are presented in Figure 6. Significance was evaluated using the *t*-test (* $p < 0.05$). A significant negative correlation was observed between CSP and WSP ($R^2 = -0.819$ *), while a significant positive correlation was found between CSP and protopectin ($R^2 = 0.872$ *), suggesting a mutual transformation of pectin forms within the pericarp, particularly involving protopectin (mainly CSP) and WSP. The rate of fruit cracking exhibited either a significant or highly significant negative correlation with hemicellulose and lignin ($R^2 \leq -0.888$ *), indicating that the occurrence of fruit cracking in 'Mingrijian' was associated with the degradation of hemicellulose and lignin.

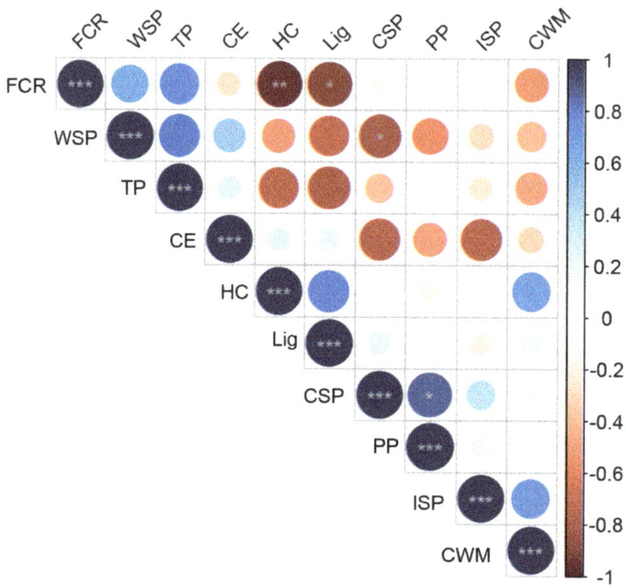

Figure 6. Heat map of the correlations between cell wall components and fruit cracking rate in 'Mingrijian'. Significance was evaluated using the *t*-test (* $p < 0.05$, ** $p < 0.01$, and *** $p < 0.001$); color depth represents the strength of the correlation; blue indicates the positive correlation, and red indicates the negative correlation. FCR: fruit cracking rate, WSP: water-soluble pectin, TP: total pectin, CE: cellulose, HC: hemicellulose, Lig: lignin, CSP: covalently bound pectin, PP: protopectin, ISP: ionic bound pectin, CWM: cell wall material.

3.2. *Variations in Osmoregulatory Substances during Fruit Development of 'Mingrijian' and 'Daya'*

3.2.1. Variations in Soluble Sugar Content in the Fruit

As shown in Figure 7, before the occurrence of fruit cracking (75–115 days after flowering), the SS content did not significantly vary in the pulp of 'Mingrijian'. However, from the initial to peak stage of fruit cracking (115–150 days after flowering), there was a significant rise in the SS content in the pulp of 'Mingrijian' from 2.26% to 4.46%, which represented a substantial increase of 96.97%. This value was significantly higher than that of 'Daya' at 150 days after flowering. From the peak to end stage of fruit cracking (150–180 days after flowering), both 'Mingrijian' and 'Daya' exhibited a continuous rise in SS content in their pulp. As for the pericarp, the SS content in 'Daya' gradually increased from 0.74% to 2.89% from 75 to 180 days after flowering, which was similar to the variation observed in the pulp. However, the SS content in 'Mingrijian' remained at a comparatively low level from 75 to 180 days after flowering, ranging from 0.44% to 1.90%. These results indicated a notable rise in the SS content in the pulp of 'Mingrijian' as the rate of fruit cracking rapidly increased (115–150 days after flowering).

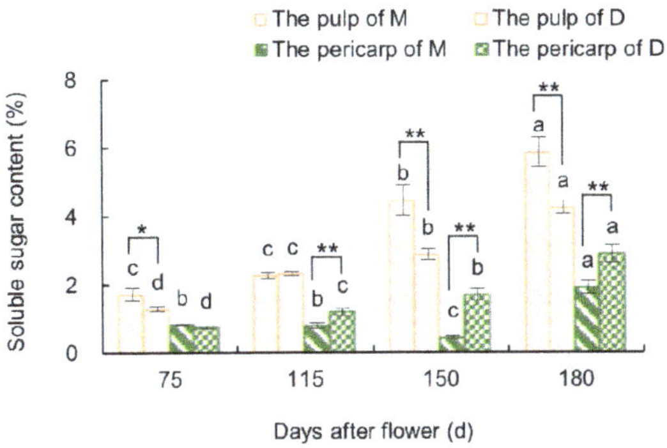

Figure 7. Changes in soluble sugar content in the fruit of 'Mingrijian' and 'Daya' during fruit development. Bars represent the SE (±), and asterisks indicate a significant difference; * significant difference between two varieties ($p < 0.05$); ** extremely significant difference between two varieties ($p < 0.01$). Different lowercase letters represent significant differences in the different growth and development stages at the 0.05 level ($p < 0.05$).

3.2.2. Variations in Soluble Protein Content in the Fruit

As shown in Figure 8, from the initial to peak stage of fruit cracking (115–150 days after flowering), the SP content in the pulp of 'Mingrijian' was significantly decreased by 28.63%, while that in the pulp of 'Daya' was significantly increased by 113.07%. Moreover, significant reductions in SP content were observed in the pericarp in both 'Mingrijian' and 'Daya' from 115 to 180 days after flowering. Following fruit cracking (115–180 days after flowering), the SP content in the pericarp of 'Mingrijian' was 16.15–30.85% lower than that of 'Daya'. The results showed that there was a significant and noticeable decrease in both the pulp and pericarp SP content in 'Mingrijian' as the rate of fruit cracking rapidly increased (115–150 days after flowering).

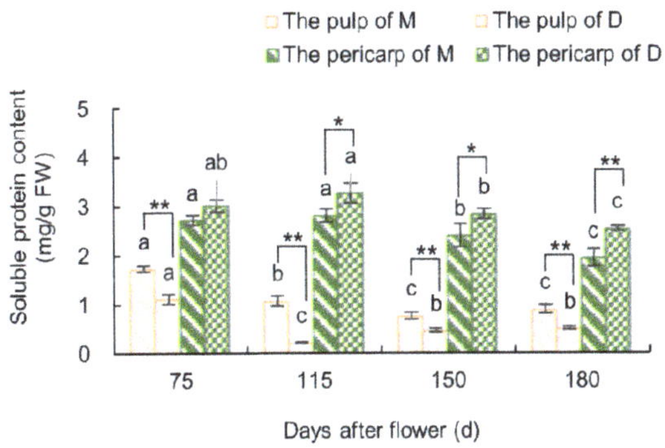

Figure 8. Changes in soluble protein content in the fruit of 'Mingrijian' and 'Daya' during fruit development. Bars represent the SE (±), and asterisks indicate a significant difference; * significant difference between two varieties ($p < 0.05$); ** extremely significant difference between two varieties ($p < 0.01$). Different lowercase letters represent significant differences in the different growth and development stages at the 0.05 level ($p < 0.05$).

3.3. Changes in Antioxidant Enzyme Activities in the Pericarp during Fruit Development in 'Mingrijian' and 'Daya'

3.3.1. Changes in Superoxide Dismutase Activity

As shown in Figure 9, the pericarp of both 'Mingrijian' and 'Daya' exhibited distinct patterns in SOD activity from 75 to 180 days after flowering. 'Mingrijian' displayed an initial increase followed by a subsequent decrease, while 'Daya' showed a fluctuating 'decreasing-increasing-decreasing' trend. Notably, before the occurrence of fruit cracking (75–115 days after flowering), the SOD activity in the pericarp of 'Mingrijian' significantly exceeded that of 'Daya'. From the initial to peak cracking stage (115–150 days after flowering), the SOD activity in the pericarp of 'Mingrijian' decreased significantly from 305.19 U/g FW to 231.71 U/g FW, representing a reduction of 24.08%, whereas that in 'Daya' showed a significant increase from 225.3 U/g FW to 267.07 U/g FW, representing an increase of 18.49%. Due to these changes, the SOD activity in the pericarp of 'Mingrijian' was notably lower than that in 'Daya' at 150 days after flowering. From the peak to end stage of fruit cracking (150–180 days after flowering), a distinct decline in SOD activity was observed in the pericarp of both 'Mingrijian' and 'Daya', with reductions of 26.77% and 34.16%, respectively. These findings indicated a significant reduction in SOD activity in the pericarp of 'Mingrijian' as the rate of fruit cracking rapidly increased (115–150 days after flowering).

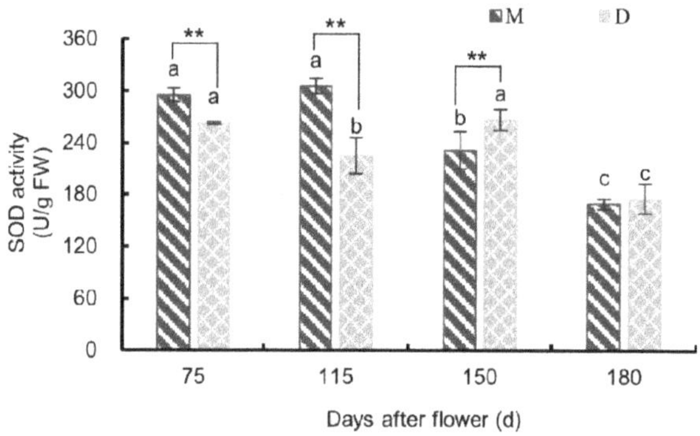

Figure 9. Changes in superoxide dismutase (SOD) activity in the pericarp of 'Mingrijian' and 'Daya' during fruit development. Bars represent the SE (±), and asterisks indicate a significant difference; ** extremely significant difference between two varieties ($p < 0.01$). Different lowercase letters represent significant differences in the different growth and development stages at the 0.05 level ($p < 0.05$).

3.3.2. Changes in Peroxidase Activity

As shown in Figure 10, the POD activity in the pericarp of both 'Mingrijian' and 'Daya' exhibited distinctive patterns from 75 to 180 days after flowering. 'Mingrijian' exhibited a 'decreasing-increasing-decreasing' trend, while 'Daya' displayed an initial increase followed by a decreasing trend. Notably, before fruit cracking (75 days after flowering), the POD activity in the pericarp of 'Mingrijian' was 100,235.28 U/g FW, which was significantly higher (2.02 times) than that of 'Daya'. From 75 to 115 days after flowering, the POD activity in the pericarp of 'Mingrijian' decreased rapidly to 36,449.17 U/g FW, representing a decrease of 63.64%, whereas that in 'Daya' increased to 64,107.54 U/g FW, representing an increase of 28.89%. From the initial to peak stage of fruit cracking (115–150 days after flowering), the POD activity in the pericarp of both 'Mingrijian' and 'Daya' increased significantly, with 'Daya' reaching its peak value of 72,671.78 U/g FW at 150 days after flowering. From 115 to 150 days after flowering, the POD activity in the pericarp of 'Daya' was significantly higher than that in 'Mingrijian'. These results indicated a marked decrease in POD activity in the pericarp of 'Mingrijian' before the occurrence of fruit cracking (75–115 days after flowering).

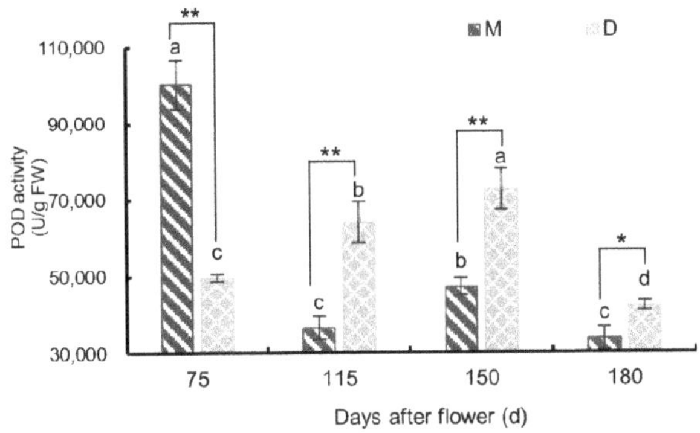

Figure 10. Changes in peroxidase (POD) activity in the pericarp of 'Mingrijian' and 'Daya' during fruit development. Bars represent the SE (±), and asterisks indicate a significant difference; * significant difference between two varieties ($p < 0.05$); ** extremely significant difference between two varieties ($p < 0.01$). Different lowercase letters represent significant differences in the different growth and development stages at the 0.05 level ($p < 0.05$).

3.3.3. Changes in Catalase Activity

As shown in Figure 11, the CAT activity in the pericarp of both 'Mingrijian' and 'Daya' showed a distinct pattern, with an initial increase followed by a decrease from 75 to 180 days after flowering. Notably, both cultivars exhibited maximum activity at 150 days after flowering, with values of 126.31 U/g FW for 'Mingrijian' and 185.09 U/g FW for 'Daya'. From 75 to 180 days after flowering, the CAT activity in the pericarp of 'Mingrijian' was significantly lower than that of 'Daya'. These results indicated that with the rapid escalation of the fruit cracking rate, CAT activity in the pericarp increased.

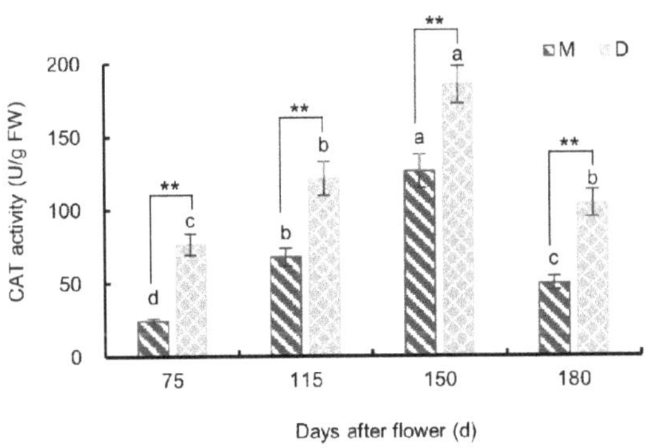

Figure 11. Changes in catalase (CAT) activity in the pericarp of 'Mingrijian' and 'Daya' during fruit development. Bars represent the SE (±), and asterisks indicate a significant difference; ** extremely significant difference between two varieties ($p < 0.01$). Different lowercase letters represent significant differences in the indexes among different growth and development stages at 0.05 level ($p < 0.05$).

3.4. Correlation Analysis of Osmoregulatory Substances, Antioxidant Enzymes, and Fruit Cracking Rates

The cell wall components and contents were determined at 75, 95, 115, 135, 150, 165, and 180 days after flowering for the pericarp of 'Mingrijian'. Subsequently, a correlation analysis was performed between these measurements and the cracking rate data. The results are presented in Figure 12. As shown in Figure 12, significance was evaluated using the *t*-test (* $p < 0.05$, and ** $p < 0.01$). A significant positive correlation was observed between the fruit cracking rate and SS content in both the pulp ($R^2 = 0.960$ **) and pericarp

($R^2 = 0.791$ *), indicating a close relationship between the SS content and fruit cracking. Notably, the rapid increase in pulp SS content had a significant effect on fruit cracking. Conversely, a remarkably significant negative correlation was observed between the fruit cracking rate and the SP content and SOD activity ($R^2 \leq -0.895$ **), emphasizing the substantial impact of SP content and SOD activity on fruit cracking.

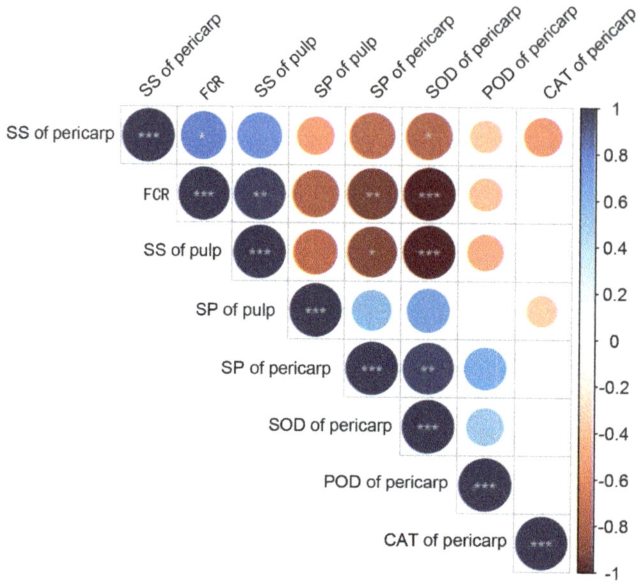

Figure 12. Heat map of the correlation between the fruit cracking rate and osmoregulatory substances and antioxidant enzyme in the fruit of 'Mingrijian'. Significance was evaluated using the *t*-test (* $p < 0.05$, ** $p < 0.01$, and *** $p < 0.001$); color depth represents the strength of the correlation; blue indicates the positive correlation, and red indicates the negative correlation. SS: soluble sugar, FCR: fruit cracking rate, SP: soluble protein, SOD: superoxide dismutase, POD: peroxidase, CAT: catalase.

4. Discussion

4.1. Relationship between Cell Wall Material and Fruit Cracking

The components of the cell wall play a crucial role in determining the strength, toughness, and structural integrity of the cell wall and represent key factors in maintaining the mechanical properties of fruit peel [11,12]. Studies on jujube have shown that a reduction in protopectin and cellulose content, coupled with an increase in WSP content, could lead to an increasing rate of fruit cracking [10]. Research on citrus fruits has revealed that cracked fruit exhibited lower levels of protopectin and higher levels of WSP than normal fruit [19]. In studies on grapes, immersing 'Xiangfei' grapefruits in a calcium solution effectively suppressed the production of WSP, delayed the degradation of protopectin, stabilized the cell wall structure, and improved the mechanical properties of the peel [28]. In the present study, as the fruit cracking rate rapidly increased (from 115 to 150 days after flowering), the protopectin (mainly CSP) in the pericarp of 'Mingrijian' progressively transformed into WSP, while hemicellulose and lignin gradually degraded. These changes probably affected the strength and structural integrity of the cell wall and were correlated closely with the occurrence of fruit cracking. Notably, the rate of fruit cracking exhibited a significant and highly negative correlation with the hemicellulose and lignin content ($R^2 \leq -0.888$ *). Presently, a greater number of research studies have focused on the impact of pectin morphological transformations and cellulose degradation on fruit cracking, whereas few investigations have addressed the connection between lignin and fruit cracking [12,29].

During the growth and development of fruit, the cell wall components of different cultivars exhibited distinct patterns of change. Previous research on tomatoes has shown that the crack-resistant tomato genotype 'LA1698' exhibited higher levels of protopectin (SSP and CSP) and cellulose than the cracking-susceptible genotype 'LA2683' [12]. In this study, from the initial to peak stage of fruit cracking (from 115 to 150 days after flowering), the cracking-resistant cultivar 'Daya' exhibited higher levels of protopectin and hemicellulose in its pericarp than 'Mingrijian'. Furthermore, a gradual accumulation of lignin was observed in 'Daya', which reinforced its cell wall and enhanced the mechanical strength of the pericarp. This adaptation enabled the cultivar to withstand the internal expansion stress generated during the fruit's growth process. From 75 to 180 days after flowering, the cracking-resistant cultivar 'Daya' exhibited markedly lower levels of total pectin, WSP, CSP, cellulose and lignin in its pericarp than the cracking-susceptible cultivar 'Mingrijian'. However, these results were inconsistent with previous studies on citrus [19], tomato [12], and lychee [11], potentially because of the comparatively lower levels of CWM in the pericarp of 'Daya'.

4.2. Relationship between Osmoregulatory Substances and Fruit Cracking

SS is one of the major osmoregulatory substances in fruits. This study revealed a rapid increase in SS content in the pulp of 'Mingrijian' during the critical period of fruit cracking and a highly significant positive correlation between the SS content and fruit cracking rate. The phenomenon is likely attributed to the climatic conditions of the growing area, which is marked by prolonged drought during the rapid expansion phase of fruit. In such circumstances, cells proactively accumulate osmoregulatory substances (solutes) to enhance the plant's water-holding capacity. The rapid accumulation of SS in 'Mingrijian' fruits would lead to an increased osmotic pressure. Consequently, if a sudden rainfall event occurred, the fruit would rapidly absorb a substantial amount of water, ultimately causing fruit cracking. During 115 to 180 days after flowering, it is under the climate condition characterized by alternating high-temperature drought and heavy rainfall. This result is consistent with previous research on fruit, such as sweet cherry [15], wax apple [30], grape [14,31], and tomato [12], which identified elevated SS contents and reduced osmotic potential as contributing factors to fruit cracking. In conclusion, the rapid increase in SS content in the pulp appears to be a crucial factor underlying fruit cracking.

The majority of SPs in plants are enzymes involved in various metabolic processes and serve as crucial indicators of overall plant metabolism [32]. In this study, the SP content in the pericarp of the cracking-susceptible cultivar 'Mingrijian' was found to be lower than that of the cracking-resistant cultivar 'Daya'. Furthermore, during the critical period of fruit cracking (115–150 days after flowering), the SP content in the fruit of 'Mingrijian' exhibited a decreasing trend. The decrease in SP content could be attributed to severe damage to the photosynthetic apparatus and cell membranes, which hindered SP synthesis. The other reason could be that the plants accelerated the decomposition of energy substances to supply energy to the organism [33].

4.3. Relationship between Antioxidant Enzymes and Fruit Cracking

SOD scavenges ROS generated during the ageing of cells, tissues, or organs, thereby protecting cell membranes and maintaining cellular metabolic balance [10]. In this study, the SOD activity in the pericarp of the cracking-susceptible cultivar 'Mingrijian' decreased as the fruit cracking rate increased (115–150 days after flowering), while that in the cracking-resistant cultivar 'Daya' increased. This result is inconsistent with previous research [12,34]. The decrease in SOD activity in the pericarp of 'Mingrijian' could potentially be related to ROS-induced enzyme inactivation or reduced enzyme synthesis, which leads to an increase in membrane lipid peroxidation in the pericarp and ultimately triggers fruit cracking. In this study, at the peak stage of fruit cracking (150 days after flowering), SOD activity was lower in the pericarp of the cracking-susceptible cultivar 'Mingrijian' than in the cracking-resistant cultivar 'Daya'. This suggests that cracking-resistant cultivars exhibited a greater ability to scavenge oxygen radicals than cracking-susceptible cultivars. Furthermore, cellular senescence adapts more slowly to changes in external factors, such as moisture and light [12]. Previously, the SOD activity in the peel of a cracking-resistant variety of tomatoes was found to be higher than that of the cracking-susceptible variety [35]. Therefore, SOD plays a role in preventing the occurrence of fruit cracking.

POD is an oxidoreductase prevalent in fruit trees and plays a dual role in scavenging ROS and cross-linking the phenolic groups, which affects peel extensibility [35,36]. In this study, the POD activity in the pericarp of 'Mingrijian' decreased sharply before the occurrence of fruit cracking (75–115 days after flowering). Additionally, the POD activity in the pericarp of the cracking-susceptible cultivar 'Mingrijian' was significantly lower than that of the cracking-resistant cultivar 'Daya' from the initial to peak stage of fruit cracking (115–150 days after flowering). These results are consistent with those of a previous study on *Akebia trifoliata* [34]. Higher POD activity in the pericarp could contribute to ROS scavenging and improve the resistance to cracking. However, this result is inconsistent with that of other studies on several other fruits, which indicated that POD activity was significantly higher in cracking-susceptible cultivars than in cracking-resistant cultivars and higher in cracked fruits than in normal fruits [12,35]. This may be attributed to the fact that in the previously studied fruits, POD primarily engages in cross-linking with cell wall phenolic components, resulting in increased POD activity and subsequent cell wall stiffening [36,37], thereby diminishing the mechanical properties of the fruit peel.

CAT can reduce oxidative damage to tissues inflicted by H_2O_2 to some extent [38]. In this study, the CAT activity in the pericarp of both cracking-susceptible 'Mingrijian' and cracking-resistant 'Daya' showed an increasing and then decreasing trend from 75 to 180 days after flowering, with the highest activity at the peak stage (150 days after flowering). This suggests that during the critical period of fruit cracking, the intracellular activity of CAT increases to ease oxidative damage by ROS [18,39]. Moreover, the CAT activity was significantly higher in the cracking-susceptible tomato cultivar 'LA1698' than in the same tissues of the cracking-resistant cultivar 'LA2683' and was higher at the site where cracking occurred than in the fruit that did not crack [35]. The increase in CAT activity could be a defensive response against oxidative stress. However, further research is needed to clarify the relationship between fruit cracking and CAT.

5. Conclusions

Pectin morphological transformations and hemicellulose and lignin degradation in the cell wall of 'Mingrijian' fruit pericarp are essential factors influencing fruit cracking. With the occurrence of fruit cracking in 'Mingrijian', covalently bound pectin transformed into WSP, leading to a notable decrease in protopectin content, which was concomitant with the degradation of both hemicellulose and lignin. From the initial to end stage of fruit cracking (115–180 days after flowering), the cracking-resistant cultivar 'Daya' exhibited higher contents of protopectin and hemicellulose in the pericarp than the cracking-susceptible cultivar 'Mingrijian'. Furthermore, at 115–150 days after flowering, a marked increase in lignin content was observed in the pericarp of 'Daya'. Such compositional variations in the cell wall contribute to enhancing the mechanical properties of the pericarp.

The SS content in the pulp, SP content in the pericarp, and SOD and POD activities in the pericarp play a crucial role in fruit cracking susceptibility in 'Mingrijian'. During the rapid expansion phase of the 'Mingrijian' fruit, there was a rapid accumulation of SS in the pulp. This alternation in fruit solutes could facilitate significant water uptake by the pulp, especially in response to sudden rainfall. Consequently, this could lead to peel rupture and fruit cracking. From 115 to 180 days after flowering, a significant decrease in SP content and SOD activity and consistently lower levels of POD activity were observed in the pericarp of 'Mingrijian'. These changes resulted in a weakened metabolic level, lower cellular antioxidant capacity, and reduced resistance to the external environment, thereby increasing the susceptibility to fruit cracking.

Author Contributions: Conceptualization, S.H. (Shengjia Huang), Z.W. and L.L.; methodology, S.H. (Shengjia Huang); software, S.H. (Shengjia Huang); validation, S.H. (Shengjia Huang), L.D., T.W. and H.L.; formal analysis, S.H. (Shengjia Huang); investigation, S.H. (Shengjia Huang), X.Y., X.B., J.H., Y.G., Y.L., Z.Q. and Y.Y.; resources, G.S., L.L., L.J. and S.H. (Shengjia Huang); data curation, S.H. (Shengjia Huang); writing—original draft preparation, S.H. (Shengjia Huang); writing—review and editing, X.Y. and S.H. (Shengjia Huang); visualization, S.H. (Shengjia Huang) and T.W.; supervision, Z.W. and M.Z.; project administration, G.S. and S.H. (Siya He); funding acquisition, Z.W. All authors have read and agreed to the published version of the manuscript.

Funding: This work was supported by the Ministry of Science and Technology of the People's Republic of China (Grant number 2021YFD1600802-02), and the Science and Technology Department of Sichuan Province (Grant number 2021YFYZ0023-14).

Data Availability Statement: Data are contained within the article.

Conflicts of Interest: The authors declare no conflict of interest.

References

1. Chen, Y.Z.; Shi, M.M.; Duan, T.T.; Zhang, D.X. Isolation and characterization of ten microsatellite loci for wild *Citrus japonica* (Rutaceae). *J. Genet.* **2014**, *93*, 41–43. [CrossRef]
2. Biswas, M.K.; Chai, L.J.; Amar, M.H.; Zhang, X.L.; Deng, X.X. Comparative analysis of genetic diversity in Citrus germplasm collection using AFLP, SSAP, SAMPL and SSR markers. *Sci. Hortic.* **2011**, *129*, 798–803. [CrossRef]
3. Odemis, B.; Turhan, S.; Buyuktas, D. The effects of irrigation and fertilizer applications on yield, pomological characteristics and fruit cracking in Nova mandarin. *Agric. Water Manag.* **2014**, *135*, 54–60. [CrossRef]
4. Khadivi-Khub, A. Physiological and genetic factors influencing fruit cracking. *Acta Physiol. Plant.* **2014**, *37*, 1718. [CrossRef]
5. Cronje, P.J.; Stander, O.P.; Theron, K.I. Fruit splitting in citrus. *Hortic. Rev.* **2013**, *41*, 177–200. [CrossRef]
6. Krajewski, A.; Ebert, T.; Schumann, A.; Waldo, L. Pre-Harvest Fruit Splitting of Citrus. *Agronomy* **2022**, *12*, 1505. [CrossRef]
7. Brüggenwirth, M.; Knoche, M. Cell wall swelling, fracture mode, and the mechanical properties of cherry fruit skins are closely related. *Planta* **2017**, *245*, 765–777. [CrossRef] [PubMed]
8. Vicens, A.; Fournand, D.; Williams, P.; Sidhoum, L.; Moutounet, M.; Doco, T. Changes in polysaccharide and protein composition of cell walls in grape berry skin (Cv. Shiraz) during ripening and over-ripening. *J. Agric. Food Chem.* **2009**, *57*, 2955–2960. [CrossRef]
9. Jiang, F.; Lopez, A.; Jeon, S.; de Freitas, S.T.; Yu, Q.; Wu, Z.; Labavitch, J.M.; Tian, S.; Powell, A.L.T.; Mitcham, E. Disassembly of the fruit cell wall by the ripening-associated polygalacturonase and expansin influences tomato cracking. *Hortic. Res.* **2019**, *6*, 17. [CrossRef]
10. Wang, B.; Ding, G.; Wang, X.; Fu, C.; Qin, G.; Yang, J.; Cang, G.; Wen, P. Changes of histological structure and water potential of huping jujube fruit cracking. *Sci. Agric. Sin.* **2013**, *46*, 4558–4568.
11. Huang, X.M.; Wang, H.C.; Gao, F.F.; Huang, H.B. A comparative study of the pericarp of litchi cultivars susceptible and resistant to fruit cracking. *J. Hortic. Sci. Biotechnol.* **1999**, *74*, 351–354. [CrossRef]
12. Yang, Z.; Wu, Z.; Zhang, C.; Hu, E.; Zhou, R.; Jiang, F. The composition of pericarp, cell aging, and changes in water absorption in two tomato genotypes: Mechanism, factors, and potential role in fruit cracking. *Acta Physiol. Plant.* **2016**, *38*, 215. [CrossRef]
13. Huo, J.; Shi, Y.; Zhang, H.; Hu, R.; Huang, L.; Zhao, Y.; Zhang, Z. More sensitive to drought of young tissues with weak water potential adjustment capacity in two desert shrubs. *Sci. Total Environ.* **2021**, *790*, 148103. [CrossRef]
14. Yu, J.; Zhu, M.T.; Wang, M.J.; Tang, W.Y.; Wu, S.; Zhang, K.; Yang, G.S. Effect of nordihydroguaiaretic acid on grape berry cracking. *Sci. Hortic.* **2020**, *261*, 108979. [CrossRef]
15. Richardson, D.G. Rain-cracking of 'Royal Ann' sweet cherries: Fruit physiological relationships, water temperature, orchard treatments, and cracking index. *Acta Hortic.* **1998**, *468*, 677–682. [CrossRef]
16. Liu, J.; Liang, L.; Jiang, Y.M.; Chen, J.J. Changes in metabolisms of antioxidant and cell wall in three pummelo cultivars during postharvest storage. *Biomolecules* **2019**, *9*, 319. [CrossRef]
17. Xu, H.; Gao, X.; Yu, C. Physiological and transcriptomic analysis of Pinus massoniana seedling response to osmotic stress. *Biol. Plantarum.* **2021**, *65*, 145–156. [CrossRef]
18. Yang, Y.; Han, X.; Liang, Y.; Ghosh, A.; Chen, J.; Tang, M. The combined effects of arbuscular mycorrhizal fungi (AMF) and lead (Pb) stress on Pb accumulation, plant growth parameters, photosynthesis, and antioxidant enzymes in *Robinia pseudoacacia* L. *PLoS ONE* **2015**, *10*, e0145726. [CrossRef]
19. Li, J.; Chen, J.Z. Citrus fruit-cracking: Causes and occurrence. *Hortic. Plant J.* **2017**, *3*, 255–260. [CrossRef]
20. Huber, D.J. Polyuronide degradation and hemicellulose modifications in ripening tomato fruit. *J. Am. Soc. Hortic. Sci.* **1983**, *108*, 405–409. [CrossRef]
21. Siddiqui, S.; Brackmann, A.; Streif, J.; Bangerth, F. Controlled atmosphere storage of apples: Cell wall composition and fruit softening. *J. Hortic. Sci.* **1996**, *71*, 613–620. [CrossRef]

22. Iii, P.K.K.; Buren, J.P.V. Carbohydrate interference and its correction in pectin analysis using the m-hydroxydiphenyl method. *J. Food Ence* **2010**, *47*, 756–759.
23. Wen-qian, L.; Ming-ming, H.; Dang-wei, P.; Jin, C.; Yuan-yuan, W.; He-he, D.; Yong-lan, C.; Min, J.; Yong-li, L.; Yong, L.; et al. Characteristics of lodging resistance of high-yield winter wheat as affected by nitrogen rate and irrigation managements. *J. Integr. Agric.* **2022**, *21*, 1290–1309.
24. Buysse, J.; Merckx, R. An improved colorimetric method to quantify sugar content of plant-tissue. *J. Exp. Bot.* **1993**, *44*, 1627–1629. [CrossRef]
25. Bradford, M.M. A rapid and sensitive method for the quantitation of microgram quantities of protein utilizing the principle of protein-dye binding. *Anal. Biochem.* **1976**, *72*, 248–254. [CrossRef]
26. Stewart, R.R.C.; Bewley, J.D. Lipid-peroxidation associated with accelerated aging of soybean axes. *Plant Physiol.* **1980**, *65*, 245–248. [CrossRef]
27. Yang, Z.F.; Zheng, Y.H.; Cao, S.F. Effect of high oxygen atmosphere storage on quality, antioxidant enzymes, and DPPH-radical scavenging activity of chinese bayberry fruit. *J. Agric. Food Chem.* **2009**, *57*, 176–181. [CrossRef]
28. Yu, J.; Zhu, M.; Bai, M.; Xu, Y.; Fan, S.; Yang, G. Effect of calcium on relieving berry cracking in grape (*Vitis vinifera* L.) 'Xiangfei'. *PeerJ* **2020**, *8*, e9896. [CrossRef]
29. Ng, J.K.T.; Schroder, R.; Sutherland, P.W.; Hallett, I.C.; Hall, M.I.; Prakash, R.; Smith, B.G.; Melton, L.D.; Johnston, J.W. Cell wall structures leading to cultivar differences in softening rates develop early during apple (*Malus x domestica*) fruit growth. *BMC Plant Biol.* **2013**, *13*, 183. [CrossRef]
30. Lu, P.-L.; Lin, C.-H. Physiology of fruit cracking in wax apple (*Syzygium samarangense*). *Bot. Orient. J. Plant Sci.* **2012**, *8*, 70–76. [CrossRef]
31. Considine, J.A. Physical aspects of fruit growth: Cuticular fracture and fracture patterns in relation to fruit structure in vitis vinifera. *J. Hortic. Sci.* **1982**, *57*, 79–91. [CrossRef]
32. Cai, H.M.; Zhou, Y.; Xiao, J.H.; Li, X.H.; Zhang, Q.F.; Lian, X.M. Overexpressed glutamine synthetase gene modifies nitrogen metabolism and abiotic stress responses in rice. *Plant Cell Rep.* **2009**, *28*, 527–537. [CrossRef] [PubMed]
33. Yang, X.; Lu, M.; Wang, Y.; Wang, Y.; Liu, Z.; Chen, S. Response mechanism of plants to drought stress. *Horticulturae* **2021**, *7*, 50. [CrossRef]
34. Jiang, Y.; Yin, H.; Wang, D.; Zhong, Y.; Deng, Y. Exploring the mechanism of *Akebia trifoliata* fruit cracking based on cell-wall metabolism. *Food Res. Int.* **2022**, *157*, 111219. [CrossRef] [PubMed]
35. Zhang, C.; Zhao, Y.J.; Jiang, F.L.; Wu, Z.; Cui, S.Y.; Lv, H.M.; Yu, L. Differences of reactive oxygen species metabolism in top, middle and bottom part of epicarp and mesocarp influence tomato fruit cracking. *J. Hortic. Sci. Biotechnol.* **2020**, *95*, 746–756. [CrossRef]
36. Zhu, M.T.; Yu, J.; Zhao, M.; Wang, M.J.; Yang, G.S. Transcriptome analysis of metabolisms related to fruit cracking during ripening of a cracking-susceptible grape berry cv. Xiangfei (*Vitis vinifera* L.). *Genes Genom.* **2020**, *42*, 639–650. [CrossRef]
37. Elstner, E.F. Oxygen activation and oxygen toxicity. *Annu. Rev. Plant Physiol.* **1982**, *33*, 73–96. [CrossRef]
38. Karkonen, A.; Kuchitsu, K. Reactive oxygen species in cell wall metabolism and development in plants. *Phytochemistry* **2015**, *112*, 22–32. [CrossRef]
39. Zahedi, S.M.; Hosseini, M.S.; Meybodi, N.D.H.; Abadia, J.; Germ, M.; Gholami, R.; Abdelrahman, M. Evaluation of drought tolerance in three commercial pomegranate cultivars using photosynthetic pigments, yield parameters and biochemical traits as biomarkers. *Agric. Water Manag.* **2022**, *261*, 107357. [CrossRef]

plants

MDPI

Article

The Effects of a New Citrus Rootstock *Citrus junos* cv. Shuzhen No. 1 on Performances of Ten Hybrid Citrus Cultivars

Wen He [1,2,†], Jiufeng Chai [1,†], Rui Xie [1], Yang Wu [1], Hao Wang [1], Yan Wang [1,2], Qing Chen [1,2], Zhiwei Wu [1], Mengyao Li [1], Yuanxiu Lin [1], Yunting Zhang [1], Ya Luo [1], Yong Zhang [1], Haoru Tang [1] and Xiaorong Wang [1,2,*]

[1] College of Horticulture, Sichuan Agricultural University, Chengdu 611130, China; hewen0724@gmail.com (W.H.); wangyanwxy@sicau.edu.cn (Y.W.); supnovel@gmail.com (Q.C.); 71444@sicau.edu.cn (Z.W.); limy@sicau.edu.cn (M.L.); linyx@sicau.edu.cn (Y.L.); asyunting@sicau.edu.cn (Y.Z.); luoya945@sicau.edu.cn (Y.L.); zhyong@sicau.edu.cn (Y.Z.); htang@sicau.edu.cn (H.T.)

[2] Key Laboratory of Agricultural Bioinformatics, Ministry of Education, Chengdu 611130, China

* Correspondence: wangxr@sicau.edu.cn

† These authors contributed equally to this work.

Abstract: The importance of rootstock in citrus production lies in its crucial role in determining tree growth, environmental stress tolerance, and fruit quality. *Citrus junos* Siebold ex Tanaka cv. Shuzhen No. 1, a recently developed rootstock, demonstrates excellent graft compatibility and abiotic stress tolerance. The objective of this study was to assess ten hybrid citrus cultivars grafted onto two *C. junos* rootstock selections, with the aim of determining the potential for industrial utilization of the new citrus rootstock. All graft junctions are mature and well established. Vigorous growth characterized all ten citrus cultivars on Shuzhen No. 1, with the largest tree's height reaching 280.33 cm (Wogan scion) and the widest scion's diameter being 67.52 cm (Chunjian scion). However, the scion-to-rootstock diameter ratio was the lowest at 0.62 (Chunxiang scion). *C. junos* rootstock selections significantly affected fruit weight (five of ten scions) and fruit color (seven of ten scions) but had negligible impact on peel thickness (nine of ten scions). Furthermore, rootstock type had a significant influence on fruit quality. In conclusion, our findings indicate strong graft compatibility between all scions and *C. junos* rootstocks, which can impact overall size and fruit quality. Based on these results, Shuzhen No. 1 is recommended as a valuable citrus rootstock.

Keywords: fragrant citrus; fruit quality; grafting; rootstock; tree vigor

1. Introduction

The genus *Citrus* encompasses a broad range of cultivated fruit crops of high global value [1]. Grafting, a commonly employed technique, involves merging a desired citrus scion with a compatible rootstock to propagate desired traits [2,3]. Rootstocks play a pivotal role in determining graft compatibility and scion vigor by influencing nutrient uptake, water absorption, disease resistance, and stress tolerance [4]. Selecting appropriate rootstocks is crucial, accounting for distinct environmental conditions [5,6]. Numerous studies have explored rootstock diversity, domestication, and their influence on plant vigor and stress responses [7–9].

Currently, rootstock selection for citrus production prioritizes compatibility and resistance; limited information, however, addresses rootstock impacts on performance of scion cultivars [10–12]. Despite the wide citrus grafting use, certain extensively employed rootstocks might still display graft incompatibility in the orchard, with manifestation taking years [13]. For instance, swingle citrumelo (*Citrus paradisi* × *Citrus trifoliata*), noted for biotic stress tolerance and enhancing scion fruit quality, exhibits incompatibility with specific sweet orange clones (*Citrus sinensis*) [14]. Trifoliate orange (*C. trifoliata*) serves as a commonly employed citrus rootstock due to its high resistance against various abiotic and biotic stresses [15,16]. However, it exhibits incompatibility with certain lemon (*Citrus limon*) and pummelo (*Citrus maxima*) cultivars [17,18]. With the dynamic shift in citrus cultivars, the need for rootstocks boasting compatibility and tolerance to multiple stresses hinders citrus industry growth.

Recently, hybrid citrus cultivars have been produced in southwest China, where the extensive calcareous purple soil type and the complex and variable climate pose huge challenges to citrus production [19]. The application of appropriate rootstocks in this region significantly impacts citrus scion resistance and fruit quality [20,21]. *Citrus junos* Siebold ex Tanaka cv. Ziyang Xiangcheng (CjZy), originating from southwest China, has gained widespread recognition as an iron-deficiency-, alkaline-, cold-, and acid-tolerant rootstock [22–24]. Nonetheless, for some specific hybrid citrus cultivars, CjZy cannot achieve entirely satisfactory indicators of fruit quality. For instance, in terms of the impact of rootstocks on the sugar content of Qingjian fruits, the effectiveness of CjZy was inferior to trifoliate orange [25]. Further, to obtain cultivars that perform better in the environment than the ones currently used, our previous citrus rootstock breeding program introduced a novel rootstock, *C. junos* Siebold ex Tanaka cv. Shuzhen No. 1 (CjSz), renowned for its tolerance to multiple stresses, such as flooding, alkaline, and freezing stress [26–28]. Meanwhile, it performs well as a rootstock for pummelo [29]. However, there is currently a lack of systematic research on the application of CjSz in practical production and its impact on tree growth and fruit quality.

Currently, there is rapid progress in the updating and selection of scion cultivars. However, the development of appropriate rootstocks is lagging behind [30]. In this study, ten cultivars of hybrid citrus, appreciated by consumers and commonly planted in southwestern China, were selected as scions. This study investigates the performance of the hybrid citrus cultivars grafted onto two *C. junos* rootstock selections, offering insights for rootstock selection and hybrid citrus production.

2. Materials and Methods

2.1. Plant Materials

The trial was conducted in Sichuan Agricultural University's orchard located in Chengdu, China (latitude 30°56′ N, longitude 103°65′ E, altitude 518 m), where water and fertilizer were used appropriately, and pest control was standardized. The climate is subtropical humid monsoon, featuring an annual mean temperature of 15.9 °C and 1012.4 mm precipitation. In January 2019, ten promising hybrid citrus cultivars with diverse genetic backgrounds were grafted onto one-year-old rootstock seedlings: *Citrus junos* Siebold ex Tanaka cv. Shuzhen No. 1 (CjSz) and *C. junos* Siebold ex Tanaka cv. Ziyang Xiangcheng (CjZy). The graft combinations in this study were prepared using the bud-grafting method. Table 1 illustrates the ripening time and origin of these hybrid citrus cultivars in the Sichuan region. The grafted seedlings were planted in purple soil with a pH range of 6.8 to 7.4, spaced at 2.0 × 2.7 m intervals.

Table 1. Background information about scions used in this study.

Genotype	Parents	Maturity Stage in Sichuan Area
Ai Yuan 38	*Citrus reticulata* Blanco × *C. reticulata* Blanco	October to December
Chunxiang	*Citrus tamurana* hort. ex Tanaka × *C.* Spp. hort. ex	December to January
Chunjian	*C. reticulata* Blanco × *C. reticulata* Blanco	December to January
Shougan	*C. reticulata* Blanco × *Citrus sinensis* (L.) Osbeck	December to February
Mingrijian	*C. reticulata* Blanco × *C. reticulata* Blanco	Late December to middle February
Ganping	*C. reticulata* Blanco × *C. reticulata* Blanco	Early January to early March
Wogan	*C. reticulata* Blanco × *C. reticulata* Blanco	Late January to early March

Table 1. *Cont.*

Genotype	Parents	Maturity Stage in Sichuan Area
Buzhihuo	*C. reticulata* Blanco × *C. reticulata* Blanco	February to April
Laihujian	*C. reticulata* Blanco × *C. reticulata* Blanco	Middle February to April
Qingjian	*Citrus unshiu* Macf × *C. sinensis* (L.) Osbeck	March to April

2.2. Vegetative Growth Parameter of the Tree

Three healthy trees per graft combination, exhibiting consistent growth potential, were chosen for assessing tree growth parameters. Plant height, rootstock, and scion diameter, along with the scion-to-rootstock diameter ratio, were measured in December 2022. To investigate the sprouting in 2022, ten newly developed shoots were selected from each tree. Their length and diameter were measured during the cessation of growth for spring, summer, and autumn shoots. The scion and rootstock diameters were measured five centimeters above and below the grafting junction, respectively. The length of the new shoot was assessed from branch base to top bud. The new shoot diameter was determined at 3 cm above its base.

2.3. Relative Chlorophyll Contents Determination

Relative chlorophyll content was measured by SPAD-502 Chlorophyll Meter Model (Konica Minolta, Kyoto, Japan). During the cessation of growth for spring, summer, and autumn shoots, 20 leaves from different new shoot types of each tree were selected to measure SPAD. The SPAD value of a leaf was determined by calculating the average SPAD values of its upper, middle, and lower parts.

2.4. Photosynthetic Rate Measurement

The leaf photosynthetic rates of different new shoot types were measured using a LI-6400 portable photosynthesis system (Li-Cor, Inc., Lincoln, NE, USA) on sunny mornings after growth cessation of spring, summer, and autumn shoots, respectively. Each tree contributed three biological replicates, with each replicate consisting of three healthy, intact, sun-exposed, disease-free functional leaves selected randomly.

2.5. Fruit Collection and Quality Parameters Assessment

Between December 2022 and March 2023, mature fruits were harvested from various crown directions to assess fruit quality based on GB/T 8210-2011 standards [31] (Figure 1). Each biological replicate consisted of ten fruits per tree, and three trees were selected for each period of the biological repeat. External quality was observed in five fruits from each tree, while the remaining fruits were stored at −20 °C for internal quality assessment. Fruit weight (FW) was measured. The horizontal and vertical diameters of fruits and equatorial peel thickness (PT) were measured directly using a Digital Vernier scale (Deli, Ningbo, China). The ratio of horizontal to vertical diameter defined the fruit shape index (FSI). The fruit's color was quantified on opposing equatorial sides at room temperature using a Konica Minolta Hunter Lab colorimeter by the CIE (The Commission Internationale de l'Eclairage) system, yielding L*, a*, b*, C* and H0 values. The citrus color index ($CCI = 1000 \times a^*/L^* \times b^*$) gauged surface color variance. L* value represents the brightness of the color. a* value implies its location between green and red, with positive and negative numbers indicating red and green, respectively. b* value is an indicator for measuring whether it leans towards yellow or blue, with positive and negative numbers indicating yellow and blue, respectively. C* value is a chromaticity indicator that refers to the ratio of colored components to white components, where 0 represents no color and 100 represents a very bright color. H0 value is the hue angle, and different angles represent different colors.

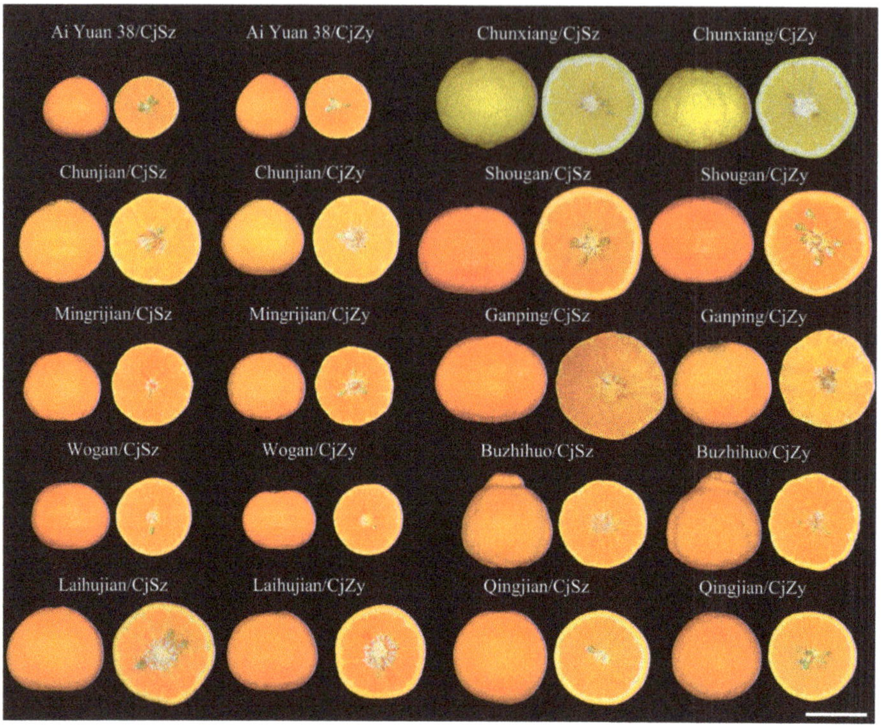

Figure 1. The appearance of fruits of ten hybrid citrus cultivars grafted onto two *C. junos* rootstocks. CjSz: *Citrus junos* Sieb. Tanaka cv. Shuzhen No. 1, CjZy: *C. junos* Sieb. Tanaka cv. Ziyang Xiangcheng. Scale bar = 5 cm.

2.6. Fruit Inner Quality Parameters Assessment

The total soluble solids (TSS) and titratable acidity (TA) of the juice were measured using a digital refractometer (ATAGO, Tokyo, Japan) expressed as percentages. The filtered fruit juice was used for the determination of the ascorbic acid content and total sugar. Ascorbic acid (Vc) content was determined via titration with 2, 6-dichlorophenolindophenol sodium salt 0.08 $g \cdot L^{-1}$; 2, 6-dichlorophenolindophenol sodium salt was used to titrate a mixture of 0.5 mL fruit juice and 4.5 mL 1% oxalic acid. Total sugar (TS) was ascertained using the sulphate–anthrone method. Diluted fruit juice and sulphate–anthrone solution were placed in the test tube and cooled naturally to room temperature, after keeping them in a thermostatic water bath at 95 °C for 10 min. Finally, the absorbance of the reaction solution was measured at the wavelength of 620 nm.

2.7. Data Processing and Analysis

Data were processed using Microsoft Excel 2020. The significance of differences among the data were assessed by using Software SPSS v.22.0 (SPSS-IBM, Corp., Armonk, NY, USA), with the following metrics: independent samples, *t*-test, two-tailed test, and a statistical significance level of $p = 0.05$. Significance was tested for indicators of the same scion with different rootstocks. The correlation matrixes of the different variables were visualized and conducted by the Pearson method using the OmicShare tools v.1.0 (https://www.omicshare.com/tools (accessed on 10 October 2023).

3. Results

3.1. Tree Growth

Tree height, shoot length, and diameter, along with stem thickness above and below the graft joint were measured in 2022, when all grafts were approximately four years old (Figure 2). Remarkable tree vigor was displayed by the scion Wogan and the rootstock CjSz in comparison to other scion–rootstock combinations (Figure 2A). This particular combination not only exhibited the largest tree height at 280.33 cm but also showed the second largest scion diameter at 54.23 cm. The highest scion-to-rootstock diameter ratio, indicating successful grafting fusion and harmonious growth, was observed on Chunjian grafted onto CjSz and Buzhihuo grafted onto CjZy, both showcasing a ratio of 0.99 (Figure 2A). Conversely, when Chunxiang was grafted onto CjSz, the scion-to-rootstock diameter ratio reached its lowest value. Despite this minimum value of 0.69, the findings suggested successful grafting compatibility (Figure 2A).

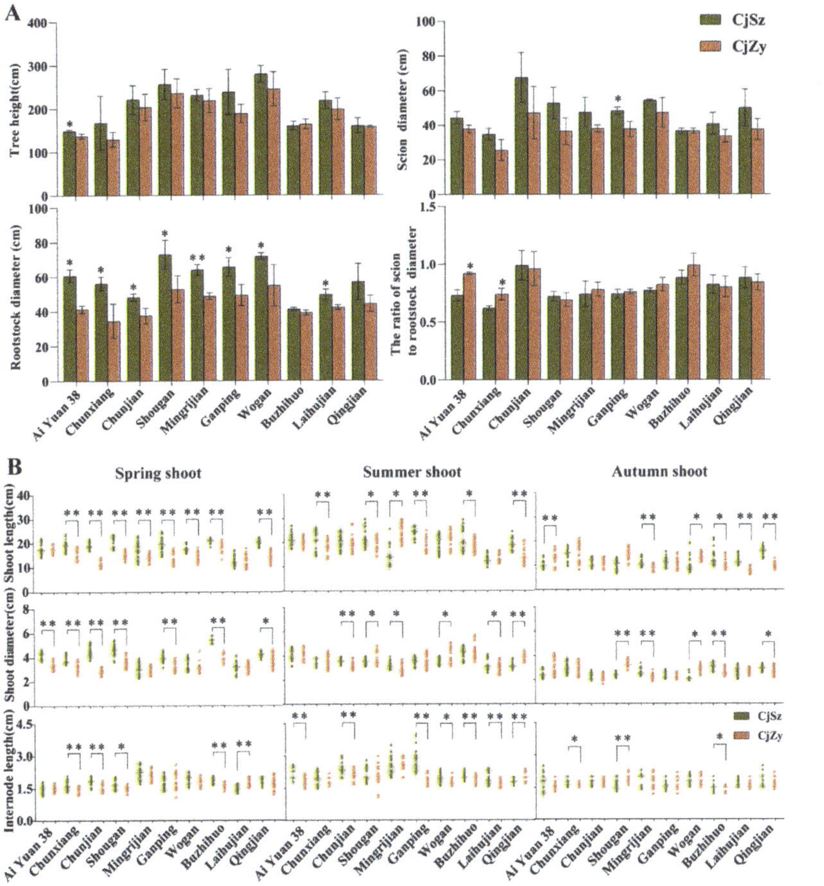

Figure 2. Effects of different rootstocks on tree growth of ten citrus scions. (**A**) Tree height, scion diameter, rootstock diameter, and the ratio of scion-to-rootstock diameter. Data were collected in December 2022. (**B**) Shoot length, diameter length, and internode length of spring, summer, and autumn shoots. Data were collected in 2022. The horizontal coordinates of the graph represent scions. The different colors represent scions grafted on different rootstocks. Significance was tested for indicators of the same scion with different rootstocks. A single asterisk (*) indicated significant differences at $p < 0.05$ and double asterisks (**) indicated significant differences at $p < 0.01$.

Compared to cultivars grafted on CjZy, most cultivars grafted on CjSz exhibited significantly higher growth potential in terms of stem length, stem thickness, and internode length (Figure 2B). Notably, trees of the cultivars AiYuan 38, Chunxiang, Chunjian, Shougan, Ganping, Buzhihuo, and Qingjian, when grafted on CjSz, had significantly larger spring shoot diameters than CjZy, with diameters of 4.23 cm, 3.83 cm, 4.49 cm, 4.61 cm, 3.94 cm, 5.39 cm, and 4.20 cm, respectively (Figure 2B). The use of CjSz rootstock also increased the spring shoot length of the cultivars Chunxiang, Chunjian, Shougan, Mingrijian, Ganping, Wogan, Buzhihuo, and Qingjian, compared to CjZy (Figure 2B). Regarding autumn shoot growth indicators, the diameters of Mingrijian, Buzhihuo, and Qingjian on the CjSz rootstock were significantly larger than that of CjZy. Additionally, the shoot lengths of Mingrijian, Laihujian, Buzhihuo, and Qingjian were significantly thicker than that of CjSz (Figure 2B).

3.2. Leaf Greenness and Photosynthetic Capacity

The highest SPAD value at 85.27, an indicator of leaf greenness, emerged in leaves obtained from the summer shoot, which were grafted on CjZy, specifically Buzhihuo and Chunxiang. Conversely, leaves grafted with Chunjian onto CjZy displayed the lowest SPAD value at 59.91 (Table 2). The use of CjSz rootstock positively influenced the leaf SPAD value of summer shoot of Chunjian and Ganping, with values of 73.25 and 80.31 (Table 2). Based on the leaf performance during the autumn shoot cessation, there was a significant difference in leaf greenness of scions grafted on CjSz rootstock compared to CjZy, including Chunjian, Mingrijian, Ganping, Wogan, and Qingjian (Table 2). Overall, when the CjSz was used as a rootstock, the relative chlorophyll content of the leaves of the cultivars Mingrijian, Wogan, Chunjian, Buzhihuo, Qingjian, AiYuan 38, and Ganping exhibited a significant decrease. The rootstock choice significantly affected the net photosynthetic rates (P_n) of the leaves of the grafted cultivars (Table 2). In spring shoots, CjSz rootstock led to elevated P_n in six hybrid citrus cultivars—Mingrijian, Laihujian, Wogan, Ganping, Chunjian, and Buzhihuo—compared to CjZy. This underscored CjSz's role in bolstering photosynthetic efficiency for these citrus cultivars in spring. Throughout summer shoots, all citrus cultivars displayed similar photosynthetic capacities, regardless of their rootstock. For autumn shoots, the P_n value was notably higher in Ganping grafted onto CjZy than when grafted onto CjSz, and higher in AiYuan 38 grafted onto CjSz than when grafted onto CjZy (Table 2). In autumn shoots, Ganping cultivar's P_n value significantly exceeded when grafted onto CjSz, compared to when grafted onto CjZy. Similarly, AiYuan 38 cultivar exhibited higher P_n value when grafted onto CjSz than onto CjZy (Table 2).

Table 2. Leaf greenness and net photosynthetic rate among different graft combinations.

Graft Combinations	SPAD Value			Photosynthetic Rate ($\mu mol \cdot m^{-2} \cdot s^{-1}$)		
	Spring Shoot	Summer Shoot	Autumn Shoot	Spring Shoot	Summer Shoot	Autumn Shoot
AiYuan 38/CjSz	69.72 ± 2.99 a	82.74 ± 2.29 b	80.84 ± 6.87 a	8.52 ± 0.19 a	17.62 ± 1.29 a	16.73 ± 1.66 a
AiYuan 38/CjZy	62.85 ± 2.45 b	84.77 ± 2.42 a	80.76 ± 2.27 a	7.81 ± 0.67 a	16.47 ± 1.96 a	13.06 ± 2.31 b
Chunxiang/CjSz	74.45 ± 2.39 a	84.90 ± 2.89 a	75.98 ± 1.30 b	8.36 ± 0.96 a	16.14 ± 1.33 a	15.61 ± 2.80 a
Chunxiang/CjZy	66.78 ± 2.97 b	85.19 ± 2.70 a	78.00 ± 1.71 a	8.02 ± 0.71 a	15.51 ± 2.08 a	11.94 ± 2.22 a
Chunjian/CjSz	66.57 ± 2.36 a	73.25 ± 4.11 a	61.06 ± 6.41 a	8.57 ± 0.13 a	17.76 ± 1.46 a	15.60 ± 2.98 a
Chunjian/CjZy	59.88 ± 2.21 b	59.91 ± 4.57 b	51.27 ± 4.75 b	7.00 ± 0.84 b	16.31 ± 2.32 a	16.43 ± 1.56 a
Shougan/CjSz	67.35 ± 1.34 a	75.54 ± 2.23 a	72.98 ± 3.84 a	10.05 ± 1.13 a	17.97 ± 2.02 a	14.63 ± 342 a
Shougan/CjZy	64.97 ± 3.20 b	77.24 ± 3.47 a	74.74 ± 1.67 a	7.19 ± 1.01 a	18.03 ± 3.42 a	13.46 ± 2.99 a
Mingrijian/CjSz	62.70 ± 1.07 a	78.71 ± 3.34 a	78.94 ± 3.28 a	7.70 ± 0.04 a	10.30 ± 1.31 a	9.30 ± 2.09 a
Mingrijian/CjZy	64.06 ± 2.45 a	77.91 ± 1.60 a	76.42 ± 3.16 b	6.28 ± 0.10 b	10.50 ± 0.59 a	9.57 ± 3.02 a
Ganping/CjSz	66.24 ± 1.47 a	80.31 ± 4.11 a	79.18 ± 3.79 a	10.45 ± 1.06 a	18.36 ± 2.89 a	11.47 ± 1.93 b
Ganping/CjZy	67.80 ± 1.82 a	75.91 ± 4.20 b	67.29 ± 5.41 b	9.10 ± 0.45 b	17.69 ± 3.02 a	15.24 ± 1.65 a
Wogan/CjSz	60.81 ± 3.30 a	74.72 ± 2.51 a	74.77 ± 2.75 b	9.16 ± 0.31 a	11.66 ± 2.34 a	9.50 ± 1.36 a
Wogan/CjZy	60.08 ± 2.45 a	73.37 ± 3.49 a	70.38 ± 3.05 a	7.45 ± 0.47 b	10.04 ± 1.87 a	11.87 ± 2.06 a
Buzhihuo/CjSz	75.18 ± 3.20 a	82.58 ± 1.89 a	75.64 ± 6.21 a	10.60 ± 0.49 a	16.03 ± 1.57 a	11.42 ± 3.54 a
Buzhihuo/CjZy	73.04 ± 2.10 b	85.27 ± 2.82 a	75.62 ± 2.91 a	8.38 ± 1.26 b	17.09 ± 1.16 a	14.80 ± 3.10 a
Laihujian/CjSz	69.24 ± 3.44 a	79.33 ± 0.77 a	74.04 ± 6.25 a	8.17 ± 0.35 a	13.09 ± 1.84 a	9.93 ± 2.10 a
Laihujian/CjZy	64.39 ± 3.49 b	79.60 ± 2.82 a	77.89 ± 3.17 a	7.24 ± 0.35 b	13.41 ± 0.54 a	12.39 ± 1.90 a

Table 2. *Cont.*

Graft Combinations	SPAD Value			Photosynthetic Rate (μmol·m^{-2}·s^{-1})		
	Spring Shoot	Summer Shoot	Autumn Shoot	Spring Shoot	Summer Shoot	Autumn Shoot
Qingjian/CjSz	72.41 ± 2.05 a	77.42 ± 3.86 a	79.44 ± 3.46 a	7.12 ± 1.06 a	17.31 ± 0.49 a	12.77 ± 1.58 a
Qingjian/CjZy	66.04 ± 3.11 b	76.72 ± 2.61 a	77.35 ± 2.47 b	7.85 ± 0.28 a	16.75 ± 0.96 a	13.19 ± 2.34 a

Note: The results were presented as mean ± SD. The performance of the same scion on two rootstocks was compared using the *t*-test method (comparison between two rootstock–scion combinations). Different lowercase letters indicate significant differences at $p < 0.05$.

3.3. Fruit Exterior Quality

Fruit weight characterized all ten citrus cultivars on CjSz, with the largest average individual fruit weight reaching 281.43 g (Chunxiang scion) and the smallest mean fruit weight at 102.72 g (AiYuan 38 scion). When Ganping, Buzhihuo, and AiYuan 38 were grafted onto the CjZy, there was a noteworthy increase in fruit weight compared to when they were grafted onto the CjSz (Table 3). Furthermore, the results showed that significant differences occurred primarily in terms of fruit vertical horizontal diameter, showing a significant impact on fruit size. It is noteworthy that the fruit horizontal and vertical diameters of Chunjian and Qingjian, grafted to CjSz, had significantly higher horizontal and vertical diameters when compared to CjZy, while AiYuan 38 showed the opposite trend (Table 3). Ganping grafted on both studied rootstocks only demonstrated differences in fruit vertical diameter (Table 3). With respect to peel thickness, there was no significant difference in the performance of the nine cultivars on the two studied rootstocks, except when Ganping grafted on CjSz, whose combination of the fruits increased peel thickness by 37.6%, compared to CjZy (Table 3).

Table 3. Effects of different rootstocks on appearance quality of ten citrus scions.

Graft Combination	Fruit Weight (g)	Horizontal Diameter (mm)	Vertical Diameter (mm)	Fruit Shape Index	Peel Thickness (mm)	CCI
AiYuan 38/CjSz	102.72 ± 7.74 b	58.25 ± 2.38 b	59.65 ± 0.53 b	1.02 ± 0.03 a	2.00 ± 0.32 a	10.40 ± 1.29 b
AiYuan 38/CjZy	177.44 ± 27.85 a	62.57 ± 2.53 a	63.07 ± 2.56 a	0.97 ± 0.03 b	2.56 ± 0.96 a	12.81 ± 1.25 a
Chunxiang/CjSz	281.43 ± 22.73 a	87.68 ± 3.57 a	72.14 ± 3.42 a	0.83 ± 0.04 a	5.64 ± 0.62 a	0.84 ± 0.34 a
Chunxiang/CjZy	267.92 ± 18.62 a	86.49 ± 3.35 a	71.88 ± 3.57 a	0.83 ± 0.05 a	5.55 ± 0.24 a	1.32 ± 0.83 b
Chunjian/CjSz	158.73 ± 28.28 a	72.61 ± 6.14 a	66.87 ± 2.48 a	0.93 ± 0.03 a	2.31 ± 0.70 a	7.61 ± 0.49 a
Chunjian/CjZy	120.27 ± 6.22 b	64.30 ± 1.17 b	58.49 ± 1.80 b	0.88 ± 0.02 b	2.15 ± 0.18 a	8.32 ± 1.51 a
Shougan/CjSz	266.44 ± 34.66 a	87.35 ± 4.23 a	70.90 ± 3.31 a	0.84 ± 0.04 a	4.60 ± 0.45 a	10.41 ± 1.04 b
Shougan/CjZy	254.40 ± 17.03 a	87.35 ± 2.36 a	69.25 ± 2.26 a	0.80 ± 0.03 b	4.58 ± 0.40 a	13.32 ± 1.18 a
Mingrijian/CjSz	137.73 ± 23.54 a	67.99 ± 4.10 a	57.18 ± 4.01 a	0.84 ± 0.01 a	2.23 ± 0.56 a	7.79 ± 0.57 b
Mingrijian/CjZy	124.02 ± 6.21 a	64.23 ± 1.60 a	54.85 ± 1.75 a	0.85 ± 0.01 a	2.30 ± 0.41 a	10.98 ± 1.55 a
Ganping/CjSz	247.59 ± 22.61 b	85.83 ± 2.73 a	64.13 ± 2.73 b	0.75 ± 0.03 a	2.45 ± 0.49 a	9.92 ± 0.71 b
Ganping/CjZy	290.97 ± 27.24 a	88.90 ± 3.73 a	66.98 ± 2.64 a	0.77 ± 0.05 a	1.78 ± 0.31 b	12.72 ± 1.27 a
Wogan/CjSz	105.30 ± 6.69 a	61.98 ± 1.94 a	50.03 ± 3.11 a	0.81 ± 0.01 a	2.50 ± 0.46 a	6.30 ± 1.00 a
Wogan/CjZy	103.85 ± 12.83 a	61.18 ± 2.93 a	49.39 ± 1.28 a	0.81 ± 0.02 a	2.63 ± 0.12 a	7.01 ± 1.31 a
Buzhihuo/CjSz	212.79 ± 15.82 b	76.30 ± 2.72 a	74.07 ± 5.03 a	1.01 ± 0.10 a	3.02 ± 0.23 a	5.88 ± 1.10 a
Buzhihuo/CjZy	259.12 ± 27.84 a	81.40 ± 4.24 a	78.02 ± 1.04 a	0.92 ± 0.02 a	3.40 ± 0.89 a	6.28 ± 0.90 a
Laihujian/CjSz	220.76 ± 26.79 a	82.87 ± 4.38 a	65.22 ± 2.84 a	0.82 ± 0.02 a	3.95 ± 0.48 a	7.13 ± 1.31 a
Laihujian/CjZy	218.16 ± 15.65 a	81.75 ± 2.18 a	64.54 ± 4.07 a	0.80 ± 0.03 a	3.67 ± 0.28 a	8.14 ± 0.71 a
Qingjian/CjSz	230.43 ± 27.97 a	80.92 ± 3.61 a	70.35 ± 3.73 a	0.90 ± 0.01 a	3.41 ± 0.23 a	3.83 ± 0.73 b
Qingjian/CjZy	179.37 ± 16.84 b	73.63 ± 2.54 b	66.57 ± 4.01 b	0.91 ± 0.04 a	3.20 ± 0.20 a	5.92 ± 0.65 a

Note: The results are presented as mean ± SD. The performance of the same scion on two rootstocks was compared using the *t*-test method (comparison between two rootstock–scion combinations). Different lowercase letters indicate significant differences at $p < 0.05$.

The color of the fruit peel of ten cultivars grafted on two *C. junos* rootstocks is shown in Table 3. The lightness (L* value) in the fruit peel ranged from 54.73 (Shougan grafted on CjZy) to 69.23 (Chunxiang grafted on CjSz). It is worth noting that the L* values of the ten types of scions using CjSz was significantly higher than that using CjZy as the rootstock, indicating that CjSz can significantly improve the brightness of the fruit. Likewise, compared with CjZy as rootstock, the b* values of eight hybrid citrus cultivars grafted on CjSz were remarkably higher except for Wogan and Qingjian. In terms of a* value and the citrus color index (CCI), apart from four graft combinations that do not show significant differences, the remaining graft combinations displayed noticeable variations in color. The range of C* values were from 35.87 to 76.91, and the values of the fruit of eight of the ten scions grafted on the CjSz was significantly higher than that of the fruit grafted on the CjZy. This indicated that compared to CjZy, CjSz produced the more vibrant and saturated color. About the H0 values of the peel color, no significant differences were observed in only three of the ten citrus cultivars.

3.4. Fruit Inner Quality

Variations in sugar and acid content are shown in Figure 3. Due to differences among cultivars, the range of titratable acids varied significantly, with a maximum value of 1.27% and a minimum value of 0.32%. The CjZy rootstock resulted in higher acid content for Laihujian, Chunjian, Buzhihuo, and Qingjian, compared to grafting onto the CjSz. Conversely, Ganping displayed elevated acid content when grafted on CjSz, as opposed to the CjZy. However, inconsistencies were observed in Vc content and TA content (Figure 3C,D). The Vc content of Buzhihuo and AiYuan 38 was higher when grafted on the CjSz rootstock, compared to CjZy (Figure 3D). Regarding TSS content comparison among different graft combinations, the positive impact was shown when CjSz rootstock was used, where there was an increase of 5.93%, 4.64%, and 6.91% in the TSS content of the fruits of the Wogan, Shougan, and Qingjian, respectively, compared to CjZy. In terms of TS content, except for AiYuan 38 and Chunjian, there was no significant difference between the other eight grafting combinations.

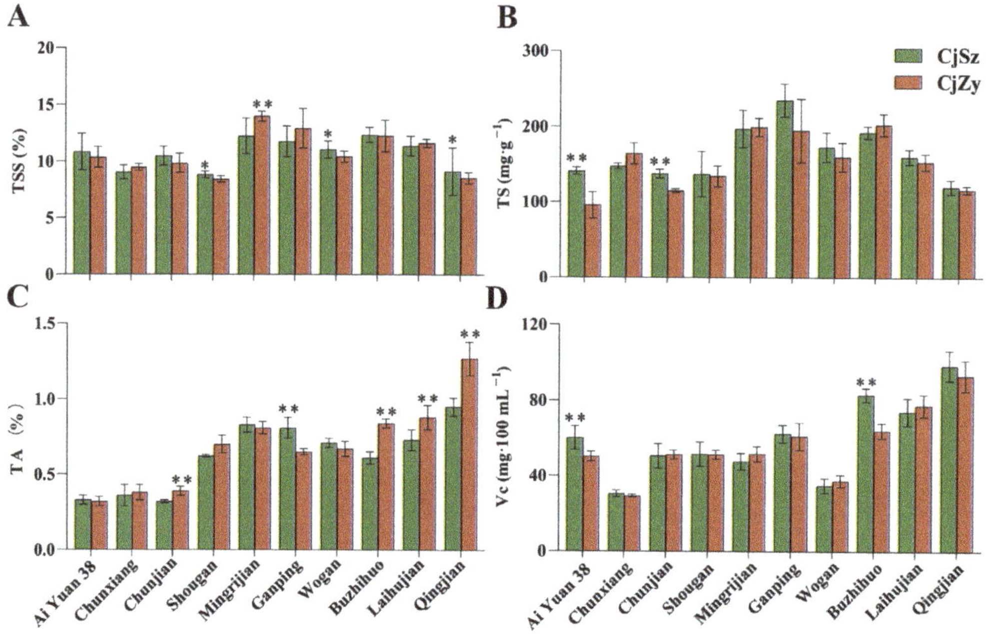

Figure 3. Effects of rootstocks on internal quality of ten citrus scions. (**A**) Juice total soluble solids (TSS); (**B**) total sugar (TS); (**C**) titratable acidity (TA); (**D**) ascorbic acid (Vc) content. The horizontal coordinates of the graph represent scions. The different colors represent scions grafted on different rootstocks. Significance was tested for indicators of the same scion with different rootstocks. A single asterisk (*) indicate significant differences at $p < 0.05$ and double asterisk (**) indicate significant differences at $p < 0.01$.

3.5. Correlation among the Parameters

In the context of the overall values of ten citrus cultivars grafted onto two *C. junos* rootstock selections, an extremely significant correlation of 0.74 between total sugar (TS) and total soluble solids (TSS). This correlation was also observed for individual scion cultivars, with CjSz at 0.78 and CjZy at 0.76, respectively (Figure 4). Likewise, a significant correlation was noted between scion diameter (SD) and SPAD value in all graft combinations, for CjSz and CjZy, with respective coefficients of −0.47, −0.88, and −0.75 (Figure 4).

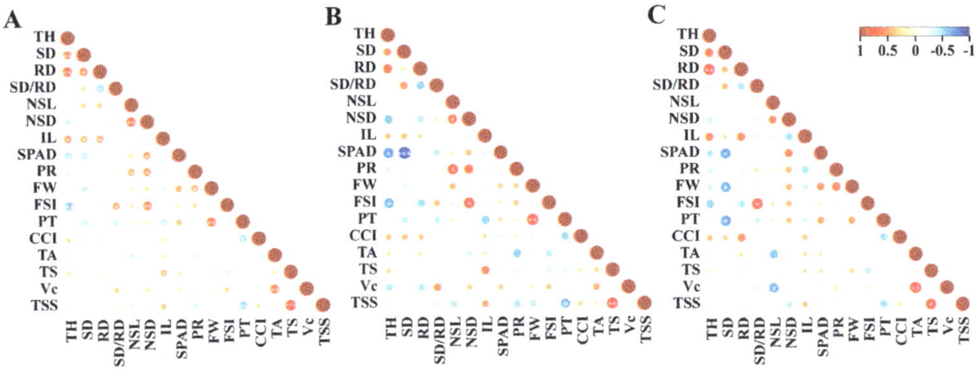

Figure 4. Correlation analysis of the parameters on horticultural characteristics. (**A**) Two rootstocks overall. (**B**) CjSz. (**C**) CjZy. TH, tree height; SD, scion diameter; RD, rootstock diameter; SD/RD, the ratio of scion-to-rootstock diameter; NSL, new shoot length; NSD, new shoot diameter; IL, internode length; SPAD, SPAD value; PR, photosynthetic rate; FW, fruit weight; FSI, fruit shape index; PT, peel thickness; CCI, citrus color index; TA, titratable acid; TS, total sugar; Vc, vitamin C; TSS, total soluble solids. A single asterisk (*) indicates statistically significant difference ($p < 0.05$), double asterisks (**) indicate highly statistically significant difference ($p < 0.01$) and three asterisks (***) indicate p value less than 0.001.

Tree height (TH) showed extremely significant correlation with SPAD value and fruit shape index (FSI) at −0.75, −0.69, −0.48, and −0.58 in CjSz and all graft combinations, respectively (Figure 4). Similarly, new shoot length (NSL) exhibited an extremely significant correlation with new shoot diameter (NSD) and photosynthetic rate (PR) in CjSz and all graft combinations (Figure 4). The thickness of fruit peels (PT) demonstrated a negative correlation with TSS, with an extremely significant coefficient of −0.66 in CjSz, and significant coefficients of −0.57 in overall values, but insignificance in CjZy at −0.51 (Figure 4).

4. Discussion

The significant influence of rootstock on plant growth, photosynthesis, and fruit quality is widely acknowledged [32,33]. Therefore, it is crucial to ascertain the interactions between different commercial cultivars and rootstocks in specific regional environments, providing a new perspective for rootstock selection and citrus production. *Citrus junos* Siebold ex Tanaka cv. Shuzhen No. 1 (CjSz), a new citrus rootstock, has proven reliable against various abiotic and biotic stresses [26,27,34].

Graft joint assessments reveal no noticeable stem swelling, indicating strong graft compatibility between the scions and *C. junos* rootstocks (Figure 2A), consistent with previous findings [19]. Compared to the use of CjZy as the rootstock, most combinations grafted on CjSz had excellent performance in the length, coarseness, and internode length of spring, summer, and autumn shoots (Figure 2B), consistent with previous findings that CjSz had stronger tree vigor as the rootstock [35]. Previous studies have shown that the influence of rootstocks on scions also affects the leaves, causing significant differences in the content of chlorophyll and nutrients in scion leaves, thereby affecting leaf photosynthesis [36]. In addition, the higher photosynthetic capacity of fruit trees is the foundation for excellent quality. In this study, the same scion grafted onto the two rootstocks exhibited varied chlorophyll contents and photosynthetic capacities in leaves (Table 2). Noticeably, the CjSz rootstock led to higher net photosynthetic rates (P_n) in spring shoots of six hybrid citrus cultivars compared to CjZy (Table 2), consistent with previous conclusion that CjSz rootstock caused an up-regulated photosynthetic capacity [29]. These differences may stem from varying root-borne resource supplies, such as water and minerals, to the shoots [37,38].

The rootstock has a significant impact on fruit external quality [39]. Notably, our results indicate higher fruit weights for Ganping, Buzhihuo, and AiYuan 38 grafted on CjZy compared to CjSz (Table 3). Similarly, compared to Hongju (*C. reticulata* Blanco) and trifoliate orange rootstocks, CjZy-based grafts yield a significantly higher average weight per fruit. Fruit size, an important quality characteristic of citrus fruits, determines their popularity in the fresh fruit market [40]. The influence of citrus rootstocks on the scion fruits size may stem from differences in various pathways that regulate fruit size. In the work of Liu et al. [41], it was proposed that the higher concentration of abscisic acid may inhibit the synthesis of growth-promoting hormones and hinder the growth and cell expansion of the Shatangju/trifoliate orange fruit. In our study, Chunjian and Qingjian grafted on CjSz exhibited significantly larger fruit horizontal and vertical diameters than those on the CjZy, while AiYuan 38 showed the opposite trend (Table 3). This provides a research basis for exploring the intrinsic mechanism of citrus rootstock influence on fruit size. Fruit color, a subjective attribute, to a certain extent determines consumers' purchasing desire and is frequently associated with the ripeness or taste of fruits, affecting commercial activities in the market [42]. In this study, observable color differences were identified between two *C. junos* when grafted with the same scion (Table 3), providing valuable insights into genetic variations and underlying physiological mechanisms driving citrus fruit development [43]. Moreover, it is worth noting that the brightness (L*) of all types of scions using CjSz was significantly higher than CjZy.

The quality and taste of citrus fruits, and subsequently consumer preferences, are heavily influenced by sugar and acid contents in juice [44]. Prior research demonstrated higher soluble solids content in citrus scions grafted onto trifoliate orange rootstock compared to those grafted on CjZy. Similar to previous results, we found that Shougan, Wogan, and Qingjian grafted on CjSz rootstocks had higher total soluble solid content than grafted on CjZy (Figure 3A). It is worth noting that rootstocks have more of an effect on TA than that of TSS, which was inconsistent with the results in pummelo 'Guanxi Miyou' [29]. This may be due to the difference between the hybrid citrus and pummelo scions used in the two studies. The titratable acidity is not only used as an indicator of citrus juice quality, but also as a reference standard for judging the proper harvest time in production implementation [45]. In this study, the fruit pulp titratable acid content of the Chunjian, Buzhihuo, Laihujian, and Qingjian grafted onto CjSz were notably lower than that onto CjZy (Figure 3C). However, further research will explore the quality comparison of the same scion grafted on CjSz and trifoliate orange.

To sum up, although rootstock can affect scion growth and fruit quality, there were some scions in this trial that did not differ significantly after being grafted on two rootstocks. This may be due to differences in scion–rootstock interaction caused by different genetic backgrounds of the scion. In this study, CjSz can enable scions to have stronger sprouting ability, stronger tree vigor, and more sufficient nutrition, which is beneficial for plants to better cope with various pressures and environmental changes, thus achieving high-quality production. Our results support that CjSz is a rootstock that can be used in citrus production.

5. Conclusions

Citrus rootstocks are the underground parts of grafting combinations that help crops adapt to both biological and abiotic conditions. However, the lagging status of citrus rootstock breeding affects the industry's development. *Citrus junos* Siebold ex Tanaka cv. Shuzhen No. 1 (CjSz), a new citrus rootstock introduced in our previous breeding program, is renowned for its tolerance to multiple stresses. Compared to the commonly used rootstock CjZy, our study highlights the notable benefits of using CjSz rootstocks, leading to increased tree sprouting ability, tree vigor, and photosynthetic activity. The fruit produced by scions grafted onto CjSz exhibited acceptable physiochemical quality. In summary, our study provides valuable insights into the potential value of CjSz as a substitute for currently used rootstocks for hybrid citrus cultivars. These findings have important value for both citrus growers and breeders, effectively facilitating optimized cultivation practices of citrus hybrid cultivars and offering perspective for further rootstock breeding.

Author Contributions: Conceptualization and supervision, X.W.; methodology, W.H. and Q.C.; investigation, J.C., Y.W. (Yang Wu), Y.W. (Yan Wang), Z.W., M.L., Y.L. (Yuanxiu Lin), Y.L. (Ya Luo), Y.Z. (Yong Zhang), Y.Z. (Yunting Zhang) and H.W.; bioinformatic analyses, W.H., J.C. and Y.W. (Yang Wu); data curation, W.H., R.X. and J.C.; manuscript preparation, W.H.; writing—review and editing, W.H., H.T. and X.W. All authors have read and agreed to the published version of the manuscript.

Funding: This research was founded by the Sichuan Provincial Postdoctoral Science Foundation, and the Shuangzhi Project Innovation Team of Sichuan Agricultural University (Grant No. P202107).

Data Availability Statement: Data are contained within the article.

Conflicts of Interest: The authors declare no conflict of interest.

References

1. Wu, G.A.; Terol, J.; Ibanez, V.; López-García, A.; Pérez-Román, E.; Borredá, C.; Domingo, C.; Tadeo, F.R.; Carbonell-Caballero, J.; Alonso, R.; et al. Genomics of the Origin and Evolution of Citrus. *Nature* **2018**, *554*, 311–316. [CrossRef] [PubMed]
2. Yang, L.; Xia, L.; Zeng, Y.; Han, Q.; Zhang, S. Grafting Enhances Plants Drought Resistance: Current Understanding, Mechanisms, and Future Perspectives. *Front. Plant Sci.* **2022**, *13*, 1015317. [CrossRef] [PubMed]
3. Feng, M.; Augstein, F.; Kareem, A.; Melnyk, C.W. Plant Grafting: Molecular Mechanisms and Applications. *Mol. Plant* **2023**, *17*, 75–91. [CrossRef] [PubMed]
4. Tworkoski, T.; Fazio, G. Hormone and Growth Interactions of Scions and Size-Controlling Rootstocks of Young Apple Trees. *Plant Growth Regul.* **2016**, *78*, 105–119. [CrossRef]
5. Warschefsky, E.J.; Klein, L.L.; Frank, M.H.; Chitwood, D.H.; Londo, J.P.; Von Wettberg, E.J.B.; Miller, A.J. Rootstocks: Diversity, Domestication, and Impacts on Shoot Phenotypes. *Trends Plant Sci.* **2016**, *21*, 418–437. [CrossRef]
6. De Carvalho, D.U.; Neves, C.S.V.J.; Da Cruz, M.A.; Colombo, R.C.; Yada, I.F.U.; Leite Junior, R.P.; Tazima, Z.H. Performance of 'Salustiana' Sweet Orange on Different Rootstocks under Brazilian Subtropical Conditions. *Sci. Hortic.* **2021**, *287*, 110226. [CrossRef]
7. Loupit, G.; Brocard, L.; Ollat, N.; Cookson, S.J. Grafting in Plants: Recent Discoveries and New Applications. *J. Exp. Bot.* **2023**, *74*, 2433–2447. [CrossRef]
8. Kehr, J.; Morris, R.J.; Kragler, F. Long-Distance Transported RNAs: From Identity to Function. *Annu. Rev. Plant Biol.* **2022**, *73*, 457–474. [CrossRef]
9. Li, H.; Testerink, C.; Zhang, Y. How Roots and Shoots Communicate through Stressful Times. *Trends Plant Sci.* **2021**, *26*, 940–952. [CrossRef]
10. Vives-Peris, V.; López-Climent, M.F.; Moliner-Sabater, M.; Gómez-Cadenas, A.; Pérez-Clemente, R.M. Morphological, Physiological, and Molecular Scion Traits Are Determinant for Salt-Stress Tolerance of Grafted Citrus Plants. *Front. Plant Sci.* **2023**, *14*, 1145625. [CrossRef]
11. Habibi, F.; Liu, T.; Folta, K.; Sarkhosh, A. Physiological, Biochemical, and Molecular Aspects of Grafting in Fruit Trees. *Hortic. Res.* **2022**, *9*, uhac032. [CrossRef]
12. Morales, J.; Bermejo, A.; Navarro, P.; Forner-Giner, M.Á.; Salvador, A. Rootstock Effect on Fruit Quality, Anthocyanins, Sugars, Hydroxycinnamic Acids and Flavanones Content during the Harvest of Blood Oranges 'Moro' and 'Tarocco Rosso' Grown in Spain. *Food Chem.* **2021**, *342*, 128305. [CrossRef]
13. Raiol-Junior, L.L.; De Carvalho, E.V.; Moreira, A.S.; Marques, J.P.R.; Stuchi, E.S.; Peña, L.; Girardi, E.A. Graft Compatibility Classification within Aurantioideae Based on Biometric Traits and the Anatomy of Graft Union. *Agriculture* **2022**, *12*, 76. [CrossRef]
14. Fadel, A.L.; Stuchi, E.S.; Silva, S.R.D.; Parolin, L.G.; Oliveira, C.R.D.; Müller, G.W.; Donadio, L.C. Compatibility and Horticultural Performance of Pera Sweet Orange Clones Grafted to Swingle Citrumelo Rootstock. *Bragantia* **2019**, *78*, 564–572. [CrossRef]
15. Dai, W.-S.; Peng, T.; Wang, M.; Liu, J.-H. Genome-Wide Identification and Comparative Expression Profiling of the WRKY Transcription Factor Family in Two Citrus Species with Different *Candidatus Liberibacter* Asiaticus Susceptibility. *BMC Plant Biol.* **2023**, *23*, 159. [CrossRef]

16. Zhang, Y.; Xiao, W.; Wang, M.; Khan, M.; Liu, J. A C2H2 -type Zinc Finger Protein ZAT12 of *Poncirus Trifoliata* Acts Downstream of CBF1 to Regulate Cold Tolerance. *Plant J.* **2023**, *117*, 1317–1329. [CrossRef] [PubMed]
17. He, W.; Xie, R.; Wang, Y.; Chen, Q.; Wang, H.; Yang, S.; Luo, Y.; Zhang, Y.; Tang, H.; Gmitter, F.G.; et al. Comparative Transcriptomic Analysis on Compatible/Incompatible Grafts in *Citrus*. *Hortic. Res.* **2022**, *9*, uhab072. [CrossRef] [PubMed]
18. Bevington, K. Development of Union Abnormalities in Grafts between Lemon (*Citrus Limon*) and *Poncirus Trifoliata*. *Aust. J. Agric. Res.* **1976**, *27*, 661. [CrossRef]
19. Zhu, S.; Huang, T.; Yu, X.; Hong, Q.; Xiang, J.; Zeng, A.; Gong, G.; Zhao, X. The Effects of Rootstocks on Performances of Three Late-Ripening Navel Orange Varieties. *J. Integr. Agric.* **2020**, *19*, 1802–1812. [CrossRef]
20. Carvalho, D.U.D.; Junior, R.P.L.; Yada, I.F.U.; Tazima, Z.H. Trifoliate Orange-Related Rootstocks Enhance the Horticultural Performance of 'Shamouti' Sweet Orange under Humid Subtropical Condition. *Agriculture* **2022**, *12*, 1782. [CrossRef]
21. Qureshi, M.A.; Shafqat, W.; Ashraf, E.; Albaayit, S.F.A. Rootstock Influence on Performance of Different Citrus Scion Cultivars: A Review. *J. Glob. Innov. Agric. Sci.* **2023**, *11*, 273–283. [CrossRef]
22. Wu, J.; Cao, J.; Su, M.; Feng, G.; Xu, Y.; Yi, H. Genome-Wide Comprehensive Analysis of Transcriptomes and Small RNAs Offers Insights into the Molecular Mechanism of Alkaline Stress Tolerance in a Citrus Rootstock. *Hortic. Res.* **2019**, *6*, 33. [CrossRef]
23. Zhu, S.; Nong, J.; Luo, G.; Li, Q.; Wang, F.; Jiang, D.; Zhao, X. Varied Tolerance and Different Responses of Five Citrus Rootstocks to Acid Stress by Principle Component Analysis and Orthogonal Analysis. *Sci. Hortic.* **2021**, *278*, 109853. [CrossRef]
24. Xu, C.; Cao, J.; Su, M.; Yan, X.; Yi, H.; Yang, H.; Wu, J. Comprehensive Analysis Provides Insights into Ziyang Xiangcheng (*Citrus junos* Sieb.) Tolerance of Alkalinity Stress. *Fruit Res.* **2024**, *4*, e001. [CrossRef]
25. Dong, T.; Xiong, B.; Huang, S.; Liao, L.; Qiu, X.; Sun, G.; He, Y.; Duan, C.; Wang, X.; Zhang, X.; et al. Investigation of the Cause of Reduced Sugar Content in Kiyomi Tangor Fruit of Ziyang Xiangcheng (*Citrus junos* Sieb. Ex Tanaka) Rootstock. *Sci. Rep.* **2019**, *9*, 19263. [CrossRef] [PubMed]
26. He, W.; Xie, R.; Chai, J.; Wang, H.; Wang, Y.; Chen, Q.; Wu, Z.; Li, M.; Lin, Y.; Zhang, Y.; et al. Comprehensive Evaluation of Abiotic Stress Tolerance and Graft Compatibility of *Citrus Junos* cv. *"Shuzhen No.1." Fruit Res.* **2023**, *11*, 273–283. [CrossRef]
27. He, W.; Luo, L.; Xie, R.; Chai, J.; Wang, H.; Wang, Y.; Chen, Q.; Wu, Z.; Yang, S.; Li, M.; et al. Transcriptome Sequencing Analyses Uncover Mechanisms of Citrus Rootstock Seedlings under Waterlogging Stress. *Front. Plant Sci.* **2023**, *14*, 1198930. [CrossRef] [PubMed]
28. He, W.; Luo, L.; Xie, R.; Chai, J.; Wang, H.; Wang, Y.; Chen, Q.; Wu, Z.; Yang, S.; Li, M.; et al. Genome-Wide Identification and Functional Analysis of the AP2/ERF Transcription Factor Family in Citrus Rootstock under Waterlogging Stress. *Int. J. Mol. Sci.* **2023**, *24*, 8989. [CrossRef]
29. Xie, R.; He, W.; Chai, J.; Luo, L.; Wang, Y.; Chen, Q.; Tang, H.; Wang, X. A Study of Scion Phenotypes in Pummelo Grafted onto a New Citrus Rootstock *Citrus Junos* 'Pujiang Xiangcheng'. *Horticulturae* **2022**, *8*, 1039. [CrossRef]
30. Deng, X. A Review and Perspective for Citrus Breeding in China during the Last Six Decades. *Acta Hortic. Sin.* **2022**, *49*, 2063–2074, (Chinese with English Abstract). [CrossRef]
31. *GB/T 8210-2011*; Method of inspection for fresh citrus fruit. AQSIQ and SAC. China National Standardization Administration: Beijing, China, 2011.
32. Nawaz, M.A.; Imtiaz, M.; Kong, Q.; Cheng, F.; Ahmed, W.; Huang, Y.; Bie, Z. Grafting: A Technique to Modify Ion Accumulation in Horticultural Crops. *Front. Plant Sci.* **2016**, *7*, e1457. [CrossRef]
33. Hu, Z.; Wang, F.; Yu, H.; Zhang, M.; Jiang, D.; Huang, T.; Xiang, J.; Zhu, S.; Zhao, X. Effects of Scion-Rootstock Interaction on Citrus Fruit Quality Related to Differentially Expressed Small RNAs. *Sci. Hortic.* **2022**, *298*, 110974. [CrossRef]
34. Chen, Q.; Min, A.; Luo, S.; He, J.; Wu, R.; Lin, X.; Wang, Y.; He, W.; Zhang, Y.; Lin, Y.; et al. Metabolomic Analysis Revealed Distinct Physiological Responses of Leaves and Roots to Huanglongbing in a Citrus Rootstock. *Int. J. Mol. Sci.* **2022**, *23*, 9242. [CrossRef]
35. Fu, X.; Huang, X.; Chen, T.; Zhang, J.; Wang, Y.; Chen, Q.; Lei, Q.; Tang, H.; Wang, X. A New Citrus Rootstock "Pujiang Xiangcheng" (*Citrus junos*). *J. Fruit Sci.* **2017**, *34*, 917–920, (Chinese with English Abstract). [CrossRef]
36. Liu, Y.-F.; Qi, H.-Y.; Bai, C.-M.; Qi, M.-F.; Xu, C.-Q.; Hao, J.-H.; Li, Y.; Li, T.-L. Grafting Helps Improve Photosynthesis and Carbohydrate Metabolism in Leaves of Muskmelon. *Int. J. Biol. Sci.* **2011**, *7*, 1161–1170. [CrossRef] [PubMed]
37. Romero, P.; Navarro, J.M.; Perez-Perez, J.; Garcia-Sanchez, F.; Gomez-Gomez, A.; Porras, I.; Martinez, V.; Botia, P. Deficit Irrigation and Rootstock: Their Effects on Water Relations, Vegetative Development, Yield, Fruit Quality and Mineral Nutrition of Clemenules Mandarin. *Tree Physiol.* **2006**, *26*, 1537–1548. [CrossRef] [PubMed]
38. Migicovsky, Z.; Harris, Z.N.; Klein, L.L.; Li, M.; McDermaid, A.; Chitwood, D.H.; Fennell, A.; Kovacs, L.G.; Kwasniewski, M.; Londo, J.P.; et al. Rootstock Effects on Scion Phenotypes in a 'Chambourcin' Experimental Vineyard. *Hortic. Res.* **2019**, *6*, 64. [CrossRef]
39. Kaleem, M.M.; Nawaz, M.A.; Alam, S.M.; Ding, X.; Cheng, J.; Bie, Z. Rootstock–Scion Interaction Mediated Impact on Fruit Quality Attributes of Thick-Skinned Melon during Storage under Different Temperature Regimes. *Sci. Hortic.* **2023**, *312*, 111823. [CrossRef]
40. Dubey, A.; Sharma, R.M. Effect of Rootstocks on Tree Growth, Yield, Quality and Leaf Mineral Composition of Lemon (*Citrus Limon* (L.) Burm.). *Sci. Hortic.* **2016**, *200*, 131–136. [CrossRef]
41. Liu, X.; Li, J.; Huang, M.; Chen, J. Mechanisms for the Influence of Citrus Rootstocks on Fruit Size. *J. Agric. Food Chem.* **2015**, *63*, 2618–2627. [CrossRef]

42. Agustin, C.; Javier, M.; José M., B.; Ignacio, P. Relation between Temperture and the Temperature and the Colour Coordinate. *Acta Hortic.* **2015**, 305–311. [CrossRef]
43. Emmanouilidou, M.G.; Kyriacou, M.C. Rootstock-Modulated Yield Performance, Fruit Maturation and Phytochemical Quality of 'Lane Late' and 'Delta' Sweet Orange. *Sci. Hortic.* **2017**, *225*, 112–121. [CrossRef]
44. Nasiruddin, M.; Hasan, M.M.; Roy, U.; Islam, A.R.; Islam, M.B. Quantitative Analysis of Juice, Citric Acid, Vitamin C Content, Sugar Levels and Sugar Acid Quantitative Relation in Some Cultivated Citrus Fruits. *Int. J. Food Sci. Nutr.* **2019**, *4*, 38–41.
45. Perez-Perez, J.G.; Castillo, I.P.; Garcia-Lidon, A.; Botia, P.; Garcia-Sanchez, F. Fino Lemon Clones Compared with the Lemon Varieties Eureka and Lisbon on Two Rootstocks in Murcia (Spain). *Sci. Hortic.* **2005**, *106*, 530–538. [CrossRef]

plants

MDPI

Article

Spatial Distribution Characteristics of Micronutrients and Their Deficiency Effect on the Root Morphology and Architecture in Citrus Rootstock

Gaofeng Zhou [1], Yiping Fu [1], Mei Yang [1], Yanhong Li [1] and Jing Zhang [2,*]

1 National Navel Orange Engineering Research Center, College of Navel Orange, Gannan Normal University, Ganzhou 341000, China; zhougaofeng428@163.com (G.Z.); fuyiping2024@163.com (Y.F.); yangmei208@163.com (M.Y.); liyanhong_208@163.com (Y.L.)

2 International Education School, Gannan Normal University, Ganzhou 341000, China

* Correspondence: jzhang0614@163.com

Abstract: Roots play essential roles in the acquisition of water and minerals from soils in higher plants. However, water or nutrient limitation can alter plant root morphology. To clarify the spatial distribution characteristics of essential nutrients in citrus roots and the influence mechanism of micronutrient deficiency on citrus root morphology and architecture, especially the effects on lateral root (LR) growth and development, two commonly used citrus rootstocks, trifoliate orange (*Poncirus trifoliata* L. Raf., Ptr) and red tangerine (*Citrus reticulata* Blanco, Cre), were employed here. The analysis of the mineral nutrient distribution characteristics in different root parts showed that, except for the P concentrations in Ptr, the last two LR levels (second and third LRs) had the highest macronutrient concentrations. All micronutrient concentrations in the second and third LRs of Ptr were higher than those of Cre, except for the Zn concentration in the second LR, which indicates that Ptr requires more micronutrients to maintain normal root system growth and development. Principal component analysis (PCA) showed that B and P were very close in terms of spatial distribution and that Mo, Mn, Cu, and Fe contributed significantly to PC1, while B, Cu, Mo, and Zn contributed significantly to PC2 in both rootstocks. These results suggest that micronutrients are major factors in citrus root growth and development. The analysis of root morphology under micronutrient deficiency showed that root growth was more significantly inhibited in Ptr and Cre under Fe deficiency (FeD) than under other micronutrient deficiencies, while Cre roots exhibited better performance than Ptr roots. From the perspective of micronutrient deficiency, FeD and B deficiency (BD) inhibited all root morphological traits in Ptr and Cre except the average root diameter, while Mn deficiency (MnD) and Zn deficiency (ZnD) had lesser impacts, as well as the morphology of the stem. The mineral nutrient concentrations in Ptr and Cre seedlings under micronutrient deficiency revealed that single micronutrient deficiencies affected both their own concentrations and the concentrations of other mineral nutrients, whether in the roots or in stems and leaves. Dynamic analysis of LR development revealed that there were no significant decreases in either the first or second LR number in Ptr seedlings under BD and ZnD stress. Moreover, the growth rates of first and second LRs in Ptr and Cre did not significantly decrease compared with the control under short-term (10 days) BD stress. Altogether, these results indicate that micronutrients play essential roles in citrus root growth and development. Moreover, citrus alters its root morphology and biological traits as a nutrient acquisition strategy to maintain maximal micronutrient acquisition and growth. The present work on the spatial distribution characteristics and micronutrient deficiency of citrus roots provides a theoretical basis for effective micronutrient fertilization and the diagnosis of micronutrient deficiency in citrus.

Keywords: citrus; lateral root; nutrient acquisition; nutrient deficiency; root architecture; root morphology

1. Introduction

Root systems play many essential roles in adaptive functions in higher plants, including water and nutrient uptake, anchorage to the soil, and the establishment of biotic interactions in the rhizosphere [1]. Roots serve as the interface between plants and the complex soil environment. The acquisition of water and minerals from the soil is the key function of plant roots, and well-developed root system architecture (RSA) is essential for the realization of this function [2]. However, the morphological traits of plant roots can change when water or nutrients are limited. According to the latest reviews on the roles of plant root traits in acquiring mineral nutrients, root traits influence the acquisition of mineral nutrients by plants through six main processes: (1) high-affinity/high-capacity transport systems; (2) the modification of the rhizosphere pH and the efflux of low-molecular-weight organic solutes and/or enzymes from roots; (3) the distribution of roots in the soil profile; (4) the relative biomass allocation to the roots (root/shoot biomass quotient) and the root growth rate; (5) the architectural and anatomical characteristics of the root system; and (6) interactions with microorganisms [3]. Therefore, advantageous root traits play a fundamental role in improving the efficiency of water and nutrient acquisition.

Many previous studies have focused on the role of nutrient availability in regulating root architecture. The ability of plants to respond appropriately to nutrient availability is of fundamental importance for their adaptation to the environment. Deficiencies of nutrients such as nitrogen (N) and phosphorus (P) act as signals that can be perceived. These signals trigger molecular and physiological mechanisms that modify cell division and cell differentiation processes within the root and have a profound impact on RSA [1]. The relationship between P deficiency and plant RSA has been elucidated in numerous plants by many authors. Low P availability increases lateral root (LR) elongation and the number of LRs, whereas taproot root growth is reduced [4–6]. Under low P concentrations, mitotic activity is relocated to the sites of LR formation, which leads to increased LR density [7,8]. With respect to N, low N promotes LR growth but has no significant effect on taproot growth [9,10]. The molecular mechanisms that control taproot and LR development and growth under P and N deficiency have mainly been explored separately [11–13]. Comparatively few studies have focused on the effects of micronutrient deficiency on root development and growth.

Citrus is one of the most important fruit crops produced in tropical and subtropical regions around the world. However, citrus production can be seriously affected by micronutrient limitations, such as boron (B) [14], zinc (Zn) [15], and iron (Fe) limitations [16]. Grafting is widely employed in citrus production, and good citrus production, therefore, depends mainly on the availability of suitable rootstocks, which can influence citrus growth, yield, fruit quality, and tolerance to different biotic and abiotic stresses [17,18]. Previous research has shown that micronutrient shortages, which can also affect the RSA, have become one of the major factors hindering citrus production around the world. Over the last decade, most studies in this area have examined the effects of B deficiency on citrus root [19–22]. In contrast, very little research has been conducted on other micronutrients in citrus rootstocks, and there is little information regarding how citrus root traits improve micronutrient acquisition and the necessity of different traits under micronutrient limitations.

It is well known that mineral nutrients are essential for citrus plant growth and development, fruit yield, and quality. Therefore, effective fertilization is one of the most important management measures in citrus cultivation. Fertilization recommendations are proposed based on the mobility and distribution of mineral nutrients, and the mobility and distribution of the same mineral nutrients differ significantly in different crops [23]. RSA and root system function are important for crop yield and stress tolerance. The root system is the main organ involved in mineral nutrient absorption. Therefore, studying the distribution characteristics of mineral nutrients in the root system holds reference significance for nutrient management in citrus. As detailed in previous research, root nutrient acquisition includes proliferation, transporter function, exudation, symbioses, and the delivery of dissolved nutrients from the bulk soil to the root surface via mass flow and diffusion [24]. Hence, shaping a good citrus RSA is a key factor in improving root nutrient acquisition. In this work, to explore the effects of micronutrient deficiency on citrus RSA and the differences in root traits under different micronutrient deficiency conditions, two of the most commonly used citrus rootstocks, trifoliate orange (Ptr) and red tangerine (Cre), were used to conduct the subsequent experiments. The aims of this study were to (1) examine the mineral nutrient distribution in different parts of the root system (the taproot and three levels of LRs); (2) determine the effects of micronutrient Fe, manganese (Mn), B, and Zn deficiency on the root traits of the two rootstocks, including root growth, root morphology, and root nutrient accumulation; and (3) clarify the relationships between micronutrients and the RSA of citrus, especially in regard to the effects on LR growth and development. It is well known that improved root nutrient acquisition can enhance fertilizer use efficiency and is an important factor in securing food production. Therefore, research on the effects of micronutrients on the roots of citrus rootstocks is vital to increase micronutrient fertilizer use efficiency and improve citrus fruit product quality.

2. Results

2.1. Mineral Nutrient Distribution in Different Root Parts

As shown in Figure 1A, the concentration of P was significantly higher in the root collar than in the taproot, but there was no significant difference between the concentrations of P in LRs at all levels. The K concentration in Ptr and Cre showed the same trend, with the second lateral root (second LR) and tertiary lateral root (third LR) having the highest K concentrations, significantly higher than those in the taproot and primary lateral root (first LR), while the root collar had the lowest K concentration (Figure 1B). For Ca, there was no significant difference between the Ca concentration of the taproot and that of the first LR, as well as between the second LR and third LR, but the former was significantly higher than the latter in both Ptr and Cre (Figure 1C). The Mg concentration decreased significantly in the order of third LR, second LR, and first LR in both Ptr and Cre (Figure 1D). Overall, except for the P concentrations in Ptr, the last two levels (second LR and third LR) had the highest concentrations of macronutrients.

Among the micronutrients, the Fe concentration was remarkably higher in both the second LR and third LR than in the other parts of the root system of Ptr (Figure 1E). This trend was also observed in Cre. The Fe concentration of the first LR was significantly higher than that of the root collar for both rootstocks. Similar trends were observed for Mn, Zn, Cu, and Mo, which exhibited significant increases in concentration in the order of root collar, taproot, first LR, second LR, and third LR in both Ptr and Cre (Figure 1F,H–J). For the B concentration in Ptr, while there were no significant differences between the B concentrations at all LR levels and the taproot, the B concentrations at all root levels were significantly higher than that in the root collar. In Cre, no significant differences were found between the B concentrations at all LR levels, but the taproot B concentration was significantly higher than that in the root collar and lower than the B concentrations of the second LR and third LR (Figure 1G). Taken as a whole, all micronutrient concentrations in the last two root levels (second LR and third LR) of Ptr were higher than those of Cre, except for the Zn concentration in the second LR. These results indicate that Ptr requires more micronutrients than Cre to maintain the normal growth and development of the roo system.

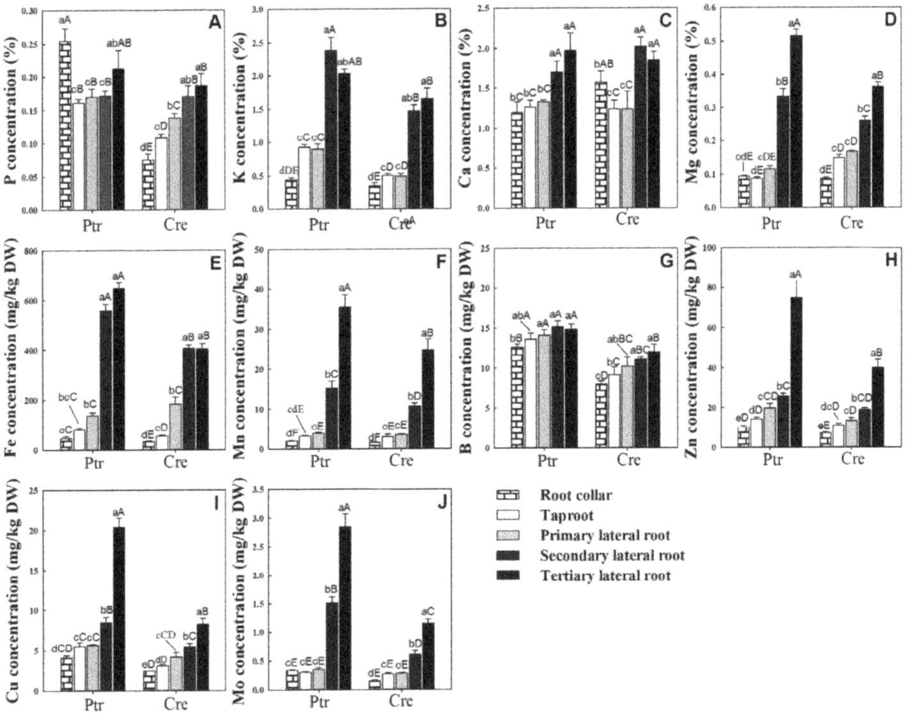

Figure 1. Nutrient distribution in different root parts of trifoliate orange (Ptr) and red tangerine (Cre) seedlings. **(A)** P concentration; **(B)** K concentration; **(C)** Ca concentration; **(D)** Mg concentration; **(E)** Fe concentration; **(F)** Mn concentration; **(G)** B concentration; **(H)** Zn concentration; **(I)** Cu concentration; **(J)** Mo concentration. Data are presented as the mean ± SE of six biological replicates. Different lowercase and uppercase letters above the bars indicate significant differences ($p < 0.05$) between the different root parts and the citrus rootstock species, respectively.

2.2. Principal Component Analysis of Nutrient Distribution

The principal component analysis (PCA) results showed that the first and second principal components could explain 70.39% and 12.39% of the total variance, respectively. The first component generally distinguished the different root parts, particularly the second and third LRs, while the second component separated the roots of Ptr from those of Cre (Figure 2A). As shown in Figure 2B, loading scatter plots of 10 elements in different root parts were analyzed. The results showed that Mo and Mn contributed more to PC1, while B, P, and Ca contributed more to PC2. Furthermore, B and P were very close in terms of spatial distribution. This result implies that P and B may play similar roles in root growth and development. Interestingly, no elements were distributed in the negative area of PC1.

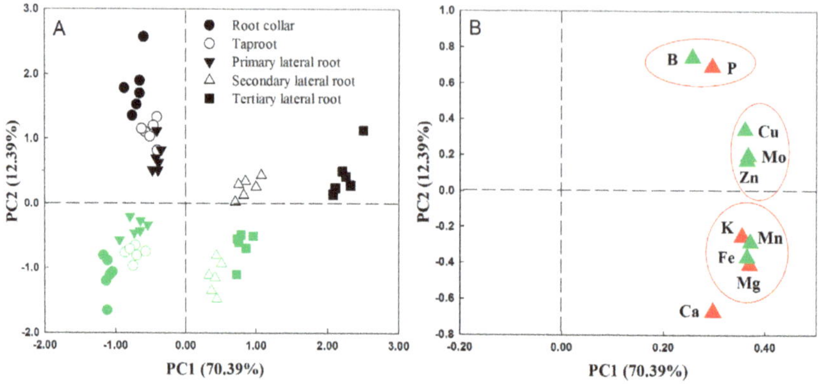

Figure 2. Principal component analysis (**A**) and loading scatter plot (**B**) of 10 elements in different root parts of trifoliate orange and red tangerine seedlings. Green symbols in (**A**) represent red tangerine, and black symbols represent trifoliate orange; green symbols in (**B**) represent micronutrients, and red symbols represent macronutrients. The red circle in subfigure (**B**) indicated that the correlation between nutrient elements was significant.

2.3. *Root Morphology Under Micronutrient Deficiency Conditions*

Visible symptoms and plant growth are depicted in Figure 3. Root growth was significantly inhibited in both Ptr and Cre under iron deficiency (FeD) conditions compared with other micronutrient deficiencies, but the Cre roots exhibited better performance than the Ptr roots under FeD. For manganese deficiency (MnD) treatments, the LRs number was obviously reduced and the taproot became thinner in diameter, but the reduction in taproot length was lower than that observed in other micronutrient deficiency treatments in both Ptr and Cre. In the zinc deficiency (ZnD) treatment, the root morphology was slightly affected compared with other treatments. There were no significant differences in root morphology between Ptr and Cre following the MnD and ZnD treatments. The most dramatic morphological difference was found between the roots of Ptr and Cre under the boron deficiency (BD) treatment. A large number of short second LRs were formed in the root system of Ptr under BD conditions, while this was not observed in the Cre root system. Effects of micronutrient deficiency stress on stem growth parameters in Ptr and Cre seedlings were also measured. Stem lengths (seedling height) were decreased significantly in both Ptr and Cre under micronutrient deficiency conditions, except for Ptr under ZnD. Stem diameter decreased significantly in Ptr under FeD and in Cre under FeD, MnD, and ZnD treatment but increased significantly in Ptr under BD. Stem and root dry weight were decreased significantly in Cre under all micronutrient deficiency treatments. For Ptr, the stem dry weights decreased significantly under FeD and MnD but increased significantly under BD and ZnD conditions (Table 1).

Table 1. Effects of micronutrient deficiency stress on stem and root growth parameters in trifoliate orange and red tangerine seedlings.

Species	Growth Parameters	CK	FeD	MnD	BD	ZnD
Ptr	Stem length (cm)	83.64 ± 3.37 b	29.21 ± 3.50 d	34.60 ± 6.35 d	74.50 ± 6.37 c	95.98 ± 4.47 a
	Stem diameter (mm)	4.57 ± 0.08 b	4.56 ± 0.12 b	3.12 ± 0.23 c	5.94 ± 0.11 a	5.13 ± 0.14 ab
	Stem dry weight (g)	2.39 ± 0.25 b	0.89 ± 0.10 c	0.50 ± 0.15 c	3.18 ± 0.50 a	3.32 ± 0.21 a
	Root dry weight (g)	1.58 ± 0.21 a	0.34 ± 0.10 c	1.07 ± 0.09 b	0.58 ± 0.04 c	1.40 ± 0.18 a

Table 1. *Cont.*

Species	Growth Parameters	CK	FeD	MnD	BD	ZnD
Cre	Stem length (cm)	76.92 ± 2.60 a	27.76 ± 2.08 c	50.32 ± 5.13 b	30.62 ± 5.38 c	30.54 ± 5.98 c
	Stem diameter (mm)	4.81 ± 0.17 a	3.26 ± 0.07 b	3.41 ± 0.21 b	4.15 ± 0.33 ab	3.46 ± 0.20 b
	Stem dry weight (g)	2.49 ± 0.20 a	0.52 ± 0.04 b	0.97 ± 0.16 b	0.87 ± 0.28 b	0.87 ± 0.20 b
	Root dry weight (g)	1.39 ± 0.06 a	0.57 ± 0.02 c	0.90 ± 0.08 b	0.52 ± 0.05 c	0.60 ± 0.07 c

Note: Trifoliate orange (Ptr) and red tangerine (Cre) seedlings were grown under different micronutrient deficiency conditions for 12 weeks. CK: control, FeD: iron deficiency, MnD: manganese deficiency, BD: boron deficiency, and ZnD: zinc deficiency. Different lowercase letters following the mean values indicate significant differences ($p < 0.05$) between different growth conditions.

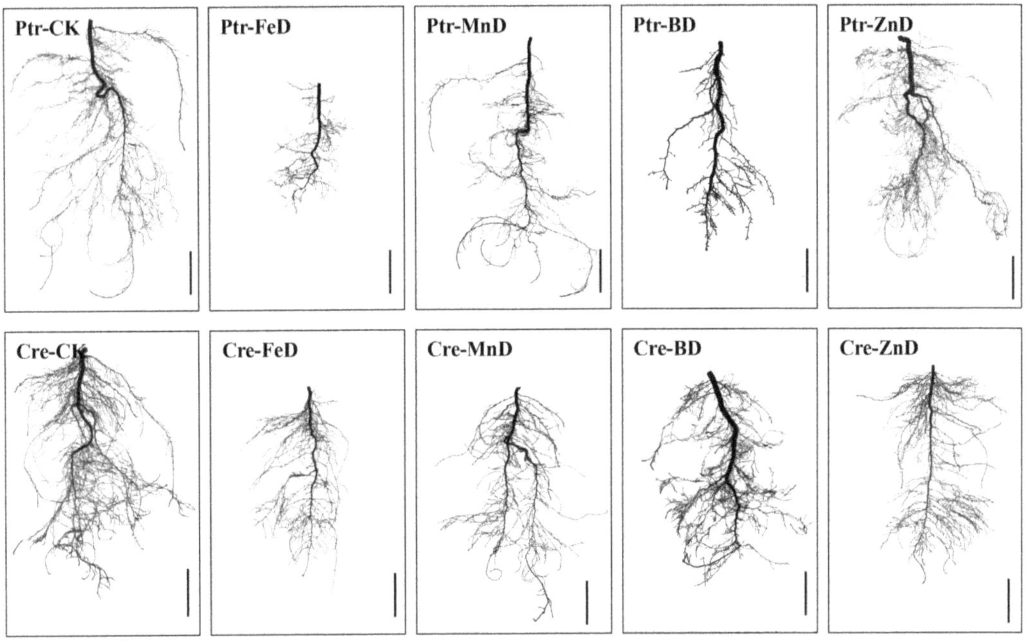

Figure 3. Scanned images of the root morphology of two types of citrus rootstocks under different micronutrient deficiency conditions. Ptr: trifoliate orange, Cre: red tangerine, CK: control, FeD: iron deficiency, MnD: manganese deficiency, BD: boron deficiency, and ZnD: zinc deficiency. Bar = 2 cm.

2.4. RSA Assay

Further quantitative root morphology analysis results showed that the taproot length was significantly decreased in the Ptr root system under all micronutrient deficiency conditions, while significantly decreased taproot length only occurred under the FeD and MnD treatments in Cre. The total root length, root surface area, and root volume exhibited similar varying tendencies under all micronutrient deficiency conditions in Ptr and Cre roots. In both Ptr and Cre roots, the total root length, root surface area, and root volume decreased significantly under all micronutrient deficiency conditions, except for the Ptr root surface under ZnD conditions. Under MnD conditions, the total root length, root surface area, and root volume of Ptr were significantly higher than the corresponding values in the FeD and BD treatments, while the same was not found for Cre roots. Compared with the control (CK), the average root diameter was significantly increased in the roots of Ptr under FeD and BD conditions but not for the roots of Cre. It is worth noting that the average root diameter was significantly higher under BD conditions than under MnD and ZnD conditions (Figure 4). These results demonstrate that the FeD and BD treatments inhibited all root morphology traits in both Ptr and Cre, except for the average root diameter. In addition, the influence of the MnD and ZnD treatments on root morphology was less than that of the FeD and BD treatments.

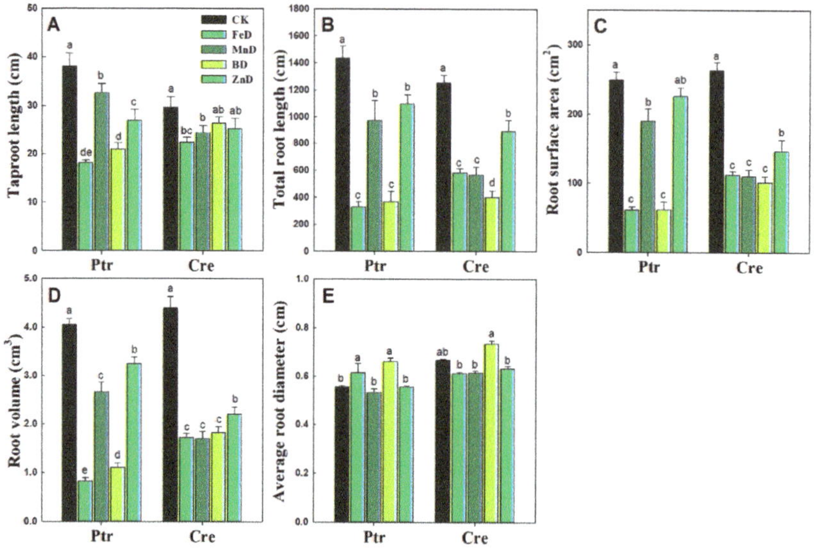

Figure 4. Root system architecture response to micronutrient deficiency stress in trifoliate orange (Ptr) and red tangerine (Cre) seedlings. (**A**) Taproot length; (**B**) Total root length; (**C**) Root surface area; (**D**) Root volume; (**E**) Average root diameter. Data are presented as the mean ± SE of six biological replicates. Different lowercase letters above the bars indicate significant differences ($p < 0.05$) between the different root parts and the citrus rootstock species, respectively. CK: control, FeD: iron deficiency, MnD: manganese deficiency, BD: boron deficiency, and ZnD: zinc deficiency.

Compared with the CK treatment, the root length of Ptr was significantly decreased under FeD and BD conditions at all root diameter classes, but for the MnD and ZnD treatments, the root length only decreased significantly when the root diameter was 0.0–0.5 mm. The distribution ratio of Ptr root length only decreased significantly at a root diameter of 0.0–0.5 mm and increased significantly at a root diameter of 0.5–1.0 mm under the ZnD treatment (Figure 5A). Different from Ptr, the root length of Cre was significantly decreased compared with CK under all micronutrient deficiency conditions and in all root diameter classes. The distribution ratio of Cre root length was significantly increased at a root diameter of 0.0–0.5 mm and significantly decreased at a root diameter of 0.5–0.1 mm under the FeD treatment. In contrast, under the BD treatment, the distribution ratio of Cre root length was significantly decreased at a root diameter of 0.5–0.1 mm and significantly increased at a root diameter of >1.5 mm (Figure 5B). The root surface area in Ptr and Cre was significantly inhibited by FeD, MnD, and BD stress in all root diameter classes, but the Ptr root surface area was not significantly reduced under ZnD conditions when the root diameter was >0.5 mm. The distribution ratio of the Ptr root surface area was significantly decreased at a root diameter of 0.0–0.5 mm but increased at root diameters of 0.5–1.0 mm and >1.5 mm under the ZnD treatment. For the FeD and BD treatments, the distribution ratio of the Ptr root surface area decreased significantly at root diameters of 0.5–1.0 and 1.0–1.5 mm but increased at a root diameter of >1.5 mm. Different from Ptr, the distribution ratio of the Cre root surface area was significantly increased at a root diameter of 0.0–0.5 mm and significantly decreased at a root diameter of 1.0–1.5 mm under the FeD and MnD treatments. For BD treatments, the distribution ratio of the Cre root surface area was decreased significantly at root diameters of 0.5–1.0 and 1.0–1.5 mm but increased at a root diameter of >1.5 mm (Figure 5C,D). The effects of micronutrient deficiency on the distribution and ratio of root volume in Ptr and Cre seedlings were also analyzed in this work (Figure 6). Compared with the CK treatment, the root volume of Ptr was decreased significantly under FeD and BD conditions at all root diameter classes, but there was no significant difference for the MnD and ZnD treatments. The root volume only increased significantly when the root diameter was >1.5 mm under the ZnD treatment. For Cre, the root volumes decreased significantly under FeD, MnD, and BD conditions at all root diameter classes, but the root volume only decreased significantly when the root diameter was 0.5–1.0 mm under the ZnD treatment. The distribution ratio of Ptr root volume decreased significantly at a root diameter of 0.5–1.5 mm and increased significantly at a root diameter of >1.5 mm under the FeD and BD treatment. But, no significant change was found in Ptr root volume under MnD treatment. For Cre, the distribution ratio of root volume decreased significantly at a root diameter of 0.0–1.5 mm and increased significantly at a root diameter of >1.5 mm under the BD treatment. The distribution ratios of root volume were significantly increased at a root diameter of 0.0–1.0 mm but decreased at root diameters of 1.0–1.5 mm under the FeD and MnD treatment and only decreased at root diameters of 0.5–1.5 mm for ZnD treatment.

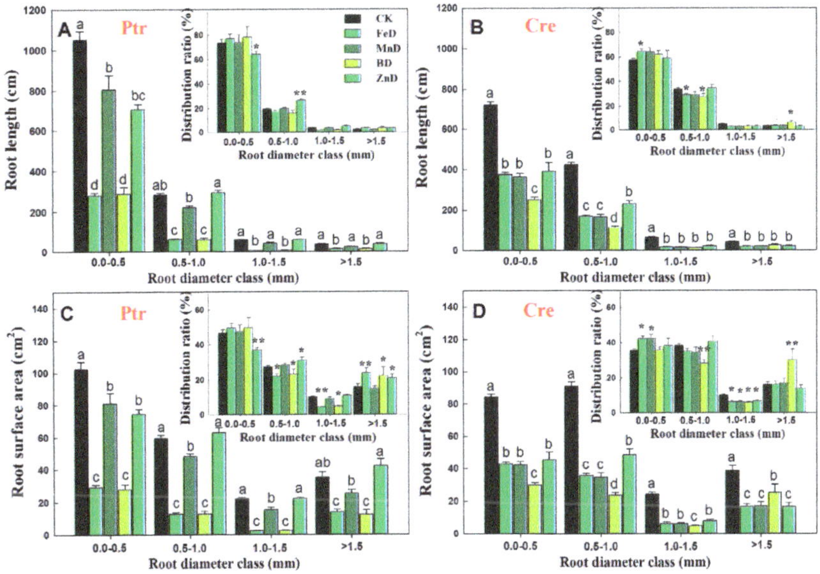

Figure 5. Effects of micronutrient deficiency on the distribution and ratio of root length and root surface area in trifoliate orange (Ptr, **A**,**C**) and red tangerine (Cre, **B**,**D**) seedlings. Data are presented as mean ± SE of six biological replicates. Different lowercase letters above the bars indicate significant differences ($p < 0.05$) between the different root parts and the citrus rootstock species, respectively. Significance of analysis of variance (ANOVA): * $p < 0.05$; ** $p < 0.01$. CK: control, FeD: iron deficiency, MnD: manganese deficiency, BD: boron deficiency, and ZnD: zinc deficiency.

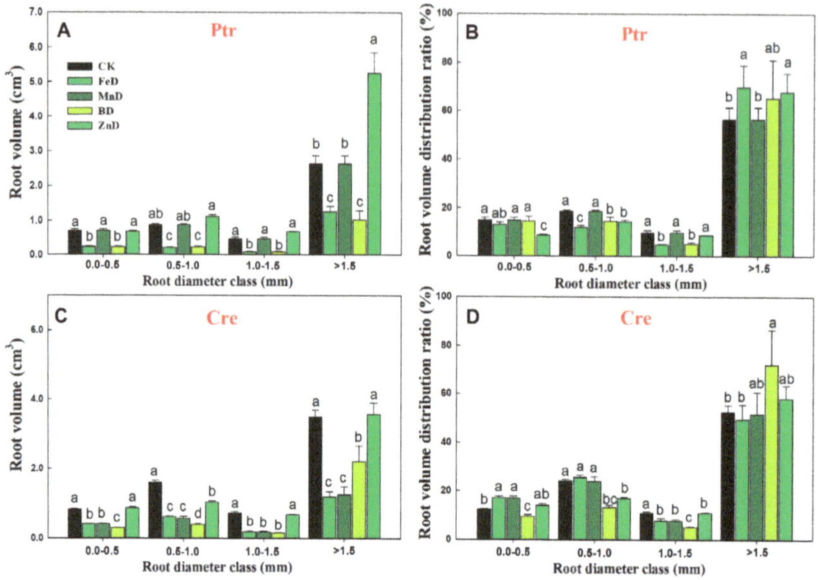

Figure 6. Effects of micronutrient deficiency on the distribution and ratio of root volume in trifoliate orange (Ptr) and red tangerine (Cre) seedlings. (**A**,**B**) Root volume distribution and their ratio of Ptr; (**C**,**D**) Root volume distribution and their ratio of Cre. Data are presented as mean ± SE of six biological replicates. Different lowercase letters above the bars indicate significant differences ($p < 0.05$) between the different root parts and the citrus rootstock species, respectively. CK: control, FeD: iron deficiency, MnD: manganese deficiency, BD: boron deficiency, and ZnD: zinc deficiency.

2.5. Mineral Nutrient Concentration Analysis

The concentrations of mineral macronutrients (P, K, Ca, and Mg) in the leaves and roots of Ptr and Cre seedlings are presented in Figure 7. MnD significantly increased the P concentrations in the leaves of Ptr and Cre, and a similar tendency was found in the leaves of Cre under FeD conditions. In contrast, the root P concentration was significantly reduced in both Ptr and Cre, except for in MnD- and ZnD-treated Ptr. For K, the MnD treatment significantly enhanced the leaf K concentration in Ptr, while the BD treatment significantly decreased the leaf K concentration in Cre. All treatments reduced the root K concentrations, except for MnD in Ptr and ZnD in Cre. The leaf Ca concentration was significantly elevated in Ptr under BD and in Cre under FeD, whereas no significant changes were detected in root Ca concentrations, except for in FeD-, BD-, and ZnD-treated Ptr seedlings. The Mg concentration in the leaves of Ptr and Cre was reduced under BD conditions, whereas it was increased by the FeD treatment. The root Mg concentration was also reduced under micronutrient deficiency conditions, except for in FeD- and MnD-treated Cre seedlings.

The concentrations of micronutrients (Fe, Mn, B, Zn, Cu, and Mo) in the leaves and roots of Ptr and Cre seedlings are presented in Figure 8. The Fe concentrations were significantly higher in the roots than in the leaves of Ptr and Cre seedlings, and the Fe concentrations were significantly reduced under FeD conditions in both the leaves and roots. The root Fe concentrations were also significantly reduced under BD and ZnD conditions (Figure 8A,B). The Mn concentrations declined under MnD conditions in both the leaves and roots of Ptr and Cre seedlings, and the Mn concentrations were also decreased in BD-treated Ptr. In contrast, the Mn concentrations were increased in the leaves and roots of Ptr under ZnD conditions (Figure 8C,D). The B concentrations were significantly higher in the leaves than in the roots of Ptr and Cre seedlings under CK conditions. The BD treatment significantly reduced the B concentrations in the leaves and roots of both Ptr and Cre seedlings, while the B concentration increased in the MnD-treated leaves of Cre seedlings (Figure 8E,F). The leaf concentrations of Zn were significantly reduced in both ZnD- and BD-treated Ptr and Cre seedlings, and the same trend was found in roots, except for in BD-treated Cre seedlings. Interestingly, elevated Zn concentrations were detected in the leaves and roots of both rootstock species under MnD treatment conditions (Figure 8G,H). In the case of Cu, the Cu concentration increased significantly under MnD treatment in both Ptr leaves and roots, and the same trend was also found in MnD- and ZnD-treated Cre seedlings (Figure 8I,J). The Mo concentration showed no significant change in the leaves of Ptr under micronutrient deficiency conditions, whereas the Mo concentration decreased significantly in Cre, except for under the ZnD treatment. In the roots, the Mo concentrations were reduced in FeD-treated Ptr and BD-treated Cre, whereas they were increased in MnD-treated Ptr and ZnD-treated Cre (Figure 8K,L). The concentrations of mineral nutrients in the stems of Ptr and Cre seedlings were also determined (Figure 9). There was no significant decrease in the concentrations of macronutrients (P, K, Ca, and Mg) in the stem of Ptr and Cre seedlings under micronutrient deficiency conditions. Especially in the treatment of FeD, the concentrations of macronutrients in stems increased significantly in both Ptr and Cre, except Ca in Ptr seedlings. The concentrations of Fe, Mn, B, and Zn in stems of Ptr and Cre decreased significantly in the corresponding treatments of FeD, MnD, BD, and ZnD. There was no significant decrease in the Cu and Mo concentrations in the stem of Ptr and Cre seedlings under micronutrient deficiency conditions. Moreover, MnD promoted the increase in stem Cu concentration, and FeD and BD promoted the increase in stem Mo concentration in both Ptr and Cre.

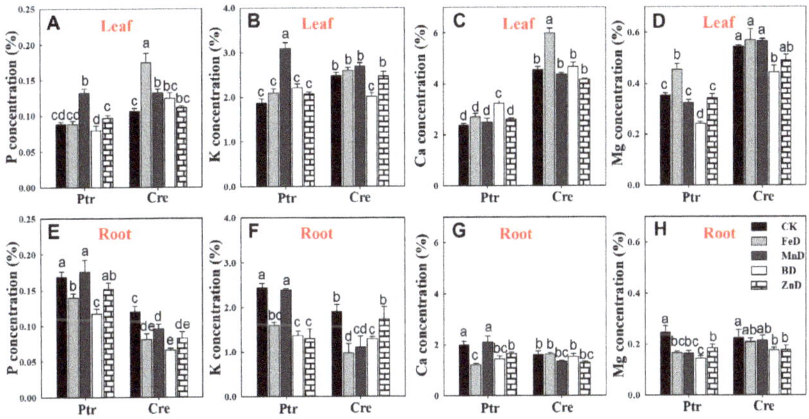

Figure 7. Effects of micronutrient deficiency on macronutrient concentrations (%) in the leaves and roots of trifoliate orange and red tangerine seedlings. P concentration in leaf (**A**) and root (**E**); K concentration in leaf (**B**) and root (**F**); Ca concentration in leaf (**C**) and root (**G**); Mg concentration in leaf (**D**) and root (**H**).Trifoliate orange (Ptr) and red tangerine (Cre) seedlings were grown under different micronutrient deficiency conditions for 12 weeks. Data are presented as means ± SE of nine replicates (n = 9, one plant for each replicate). Different lowercase letters above the bars indicate significant differences ($p < 0.05$) between different growth conditions. CK: control, FeD: iron deficiency, MnD: manganese deficiency, BD: boron deficiency, and ZnD: zinc deficiency.

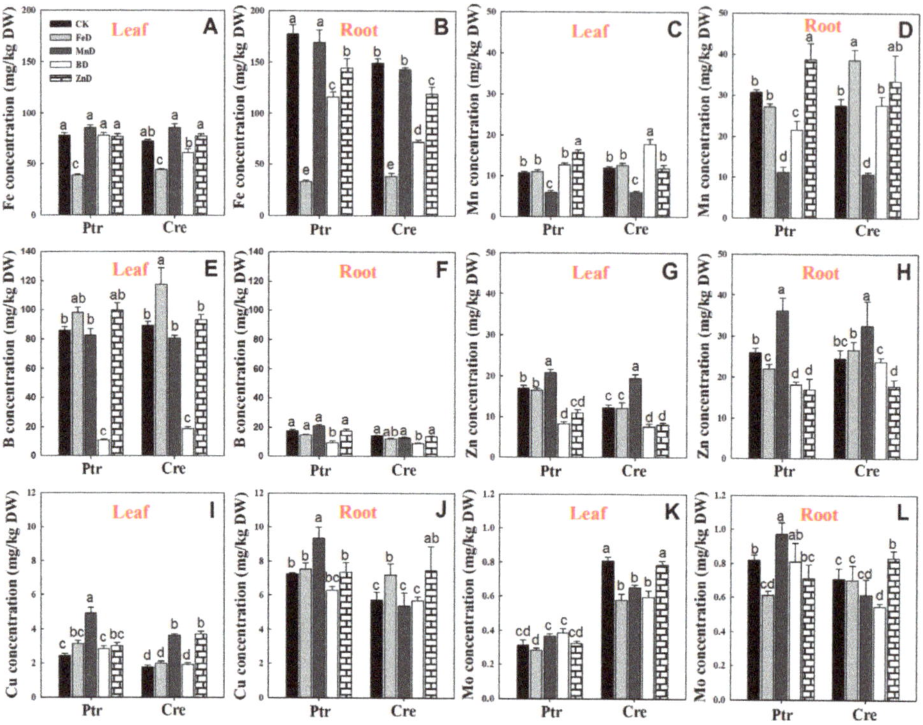

Figure 8. Effects of micronutrient deficiency on the micronutrient concentrations (mg/kg DW) in the leaves and roots of trifoliate orange and red tangerine seedlings. Fe concentration in leaf (**A**) and root (**B**); Mn concentration in leaf (**C**) and root (**D**); B concentration in leaf (**E**) and root (**F**); Zn concentration in leaf (**G**) and root (**H**); Cu concentration in leaf (**I**) and root (**J**); Mo concentration in leaf (**K**) and root (**L**). Trifoliate orange (Ptr) and red tangerine (Cre) seedlings were grown under different micronutrient deficiency conditions for 12 weeks. Data are presented as means ± SE of nine replicates (n = 9, one plant for each replicate). Different lowercase letters above the bars indicate significant differences ($p < 0.05$) between different growth conditions. CK: control, FeD: iron deficiency, MnD: manganese deficiency, BD: boron deficiency, and ZnD: zinc deficiency.

2.6. Dynamic Analysis of Lateral Roots Development

A short-term experiment was conducted to investigate the effects of micronutrient deficiency on LR formation and growth. As shown in Table 2, the Ptr first LR number was significantly lower under the FeD and MnD treatments than under the CK treatment, although not under the BD and ZnD treatments. In contrast, there was no significant decline in the number of first LRs in Cre, except for under FeD stress. The lengths of all first and second LRs were inhibited under micronutrient deficiency conditions, except for in ZnD-treated Ptr seedlings. Under FeD and MnD conditions, the number of second LRs decreased significantly in both rootstock species. Under ZnD conditions, the number of second LRs only declined in Cre. Interestingly, no significant changes occurred in the root number in all cases (two species and two kinds of LRs) under BD stress. Therefore, the root density of second LRs increased under BD stress. In contrast, this density was inhibited under MnD and ZnD conditions only in Cre seedlings.

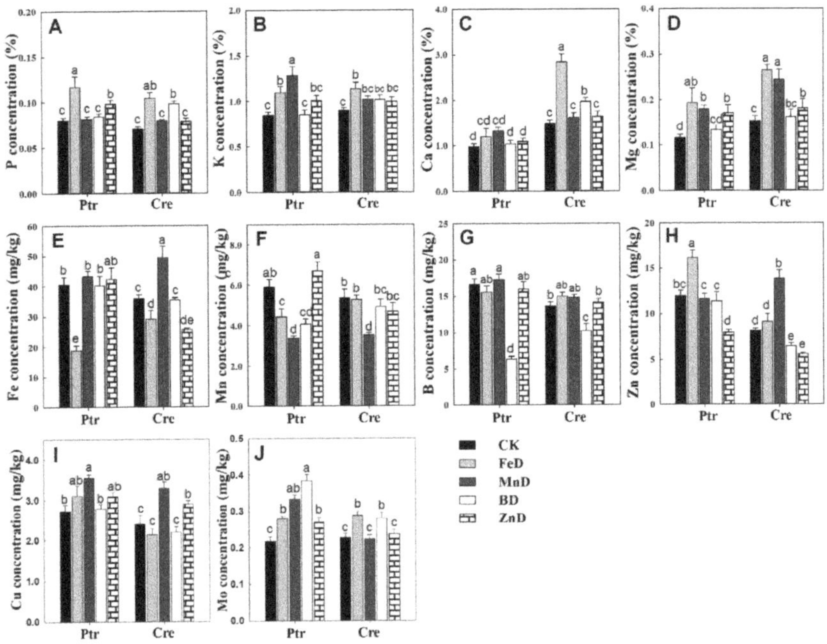

Figure 9. Effects of micronutrient deficiency on the mineral nutrient concentrations in the stems of trifoliate orange and red tangerine seedlings. (**A**) P concentration; (**B**) K concentration; (**C**) Ca concentration; (**D**) Mg concentration; (**E**) Fe concentration; (**F**) Mn concentration; (**G**) B concentration; (**H**) Zn concentration; (**I**) Cu concentration; (**J**) Mo concentration. Trifoliate orange (Ptr) and red tangerine (Cre) seedlings were grown under different micronutrient deficiency conditions for 12 weeks. Data are presented as means ± SE of nine replicates (n = 9, one plant for each replicate). Different lowercase letters above the bars indicate significant differences ($p < 0.05$) between different growth conditions. CK: control, FeD: iron deficiency, MnD: manganese deficiency, BD: boron deficiency, and ZnD: zinc deficiency.

Table 2. Effects of micronutrient deficiency on lateral root formation and growth in trifoliate orange and red tangerine seedlings.

Species	Treatment	Primary Lateral Roots		Secondary Lateral Roots		
		Number	Length (cm)	Number	Length (cm)	Density (Number/cm)
Ptr	CK	19.56 ± 1.17 a	8.23 ± 0.55 a	24.44 ± 2.30 a	2.90 ± 0.08 a	2.96 ± 0.21 b
	FeD	13.22 ± 1.11 c	4.85. ± 0.31 c	14.33 ± 0.74 c	1.92 ± 0.47 c	3.02 ± 0.32 b
	MnD	16.78 ± 1.29 b	7.07 ± 0.63 b	19.44 ± 2.85 b	2.48 ± 0.23 b	2.90 ± 0.38 b
	BD	20.67 ± 1.61 a	4.93 ± 0.34 d	23.89 ± 2.68 a	1.15 ± 0.13 d	4.93 ± 0.61 a
	ZnD	18.11 ± 2.42 ab	7.66 ± 0.60 ab	22.89 ± 4.74 ab	2.62 ± 0.76 ab	3.14 ± 0.48 b
Cre	CK	22.33 ± 1.61 a	7.67 ± 0.74 a	19.83 ± 1.74 a	2.66 ± 0.13 a	2.67 ± 0.26 b
	FeD	16.00 ± 1.18 b	6.18 ± 0.33 cd	15.83 ± 0.95 b	2.05 ± 0.14 b	2.57 ± 0.09 b
	MnD	18.17 ± 1.30 ab	6.54 ± 0.44 c	13.83 ± 1.08 c	1.78 ± 0.14 c	2.14 ± 0.17 c
	BD	21.17 ± 1.17 a	5.51 ± 0.57 d	20.67 ± 1.94 a	1.44 ± 0.17 d	3.96 ± 0.53 a
	ZnD	20.33 ± 2.09 a	6.88 ± 0.43 b	13.33 ± 2.54 c	1.79 ± 0.19 c	1.98 ± 0.20 c

Note: Trifoliate orange (Ptr) and red tangerine (Cre) seedlings were grown under different micronutrient deficiency conditions for 40 days. CK: control, FeD: iron deficiency, MnD: manganese deficiency, BD: boron deficiency, and ZnD: zinc deficiency. Different lowercase letters following the mean values indicate significant differences ($p < 0.05$) between different growth conditions.

Based on the above results, dynamic analyses of LR development were conducted in Ptr and Cre under micronutrient deficiency conditions by determining the different-order

LR growth rates (Figure 10). Under normal (CK) conditions, the growth rates of first and second LRs in both rootstock species displayed changes with treatment time. Under FeD conditions, the Ptr root elongation rate decreased significantly in first LRs, but it increased at first and then decreased in second LRs. For Cre, the rate remained unchanged after 10 days of FeD treatment in both LRs. Under MnD and ZnD conditions, the root growth rate did not change significantly in the first LR of Cre, but the root growth rate dropped significantly at the end of the treatments in the second LR of Cre and in both the first and second LRs of Ptr. For BD-treated Ptr, the root growth rate remained unchanged during the entire treatment period in the first LR, whereas the root growth rate first increased and then declined in the second LR. However, for BD-treated Cre, the root growth rate showed a downward trend in the first LR and a slowly increasing trend in the second LR during the treatment time (Figure 10).

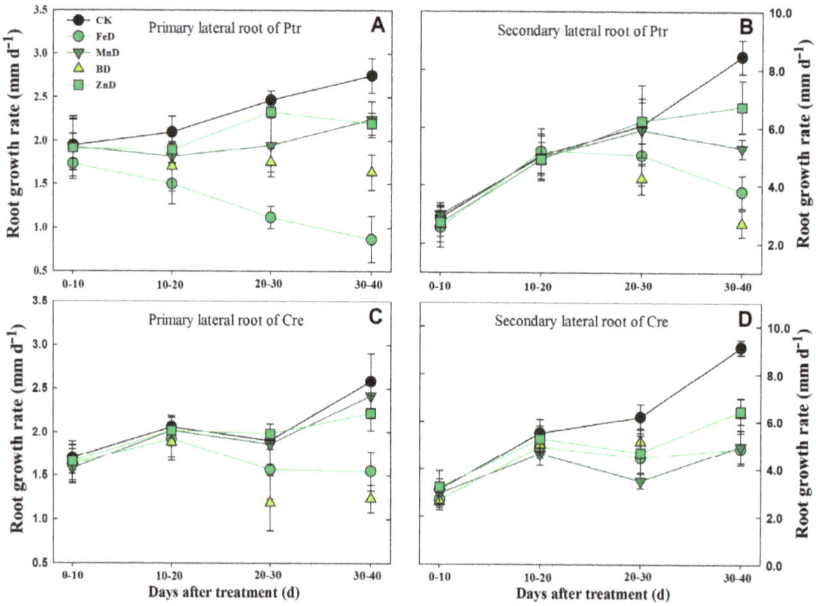

Figure 10. Dynamic analysis of the lateral root growth rates of trifoliate orange (Ptr) and red tangerine (Cre) seedlings under micronutrient deficiency conditions. (**A**) The primary lateral root growth rate of Ptr; (**B**) The secondary lateral root growth rate of Ptr; (**C**) The primary lateral root growth rate of Cre; (**D**) The secondary lateral root growth rate of Cre. Trifoliate orange and red tangerine seedlings were grown under different micronutrient deficiency conditions for 40 days. CK: control, FeD: iron deficiency, MnD: manganese deficiency, BD: boron deficiency, and ZnD: zinc deficiency. All results regarding the per-plant root growth rate data are the average value (±SD) from nine seedlings.

3. Discussion

3.1. Mineral Nutrient Distribution in Citrus Root

The present work analyzed the mineral nutrient concentrations in taproots and lateral roots to clarify the distribution of mineral nutrients in different parts of the citrus root system (Figures 1 and 2). The results suggested that there were obvious differences in the mineral nutrient distribution characteristics of root partitioning. Furthermore, the results revealed that the content of mineral nutrients in different parts of the roots varied and that the distribution of the same mineral nutrients in different parts of the roots was also different. This conclusion is similar to the findings of previous research on citrus leaves [25]. Investigating the spatial distribution characteristics of mineral nutrients in citrus roots can not only serve as a reference for effective nutrient management in citrus production but also shed light on the formation mechanism of root mineral nutrient deficiency symptoms, providing a theoretical basis for nutritional diagnosis in citrus.

The spatial distribution of mineral nutrients in different leaf parts is well known in higher plants, including citrus plants, based on research conducted using elemental imaging or ICP-MS [25–27]. However, very few studies have been performed on plant roots, and the spatial distribution of mineral nutrients in different root parts of citrus plants is unclear. As shown in Figure 1, there were significant differences in the spatial distribution of mineral nutrients in different root parts in both Ptr and Cre. The mechanism underlying these differences requires further study. There are two main sources of mineral nutrients in citrus roots: mineral nutrients absorbed by the roots themselves and mineral nutrients transported from elsewhere in the plant. It is well known that the younger the root system, the better its ability to absorb and preferentially use mineral nutrients. Previous research demonstrated that the P and K contents of roots were unaffected by P distribution patterns, whereas the Ca and Mg contents were markedly increased in older tissues when P was present in the ambient solution [28]. These data support the concept that internally translocated P and externally absorbed P have differing effects at a given root site because the former does not reach the site in the same quantity as the latter. In this study, except for the P concentrations in Ptr, the last two levels of LRs (second and third LRs) exhibited the highest macronutrient concentrations (Figure 1). In addition, the differential distribution of mineral nutrients may be influenced by variations in transpiration, differences in the root absorption of mineral elements, and the mobility of elements.

Interestingly, based on the results of PCA, B and P were very close in terms of their spatial distribution (Figure 2B). This result implies that P and B may play similar roles in root growth and development. The P nutrition status exerts very fine adjustment actions on plant root system growth. The soil P status induces or inhibits plant taproot or LR formation and development, thereby altering the root morphology and architecture [1,29–31]. Our previous studies showed that the P concentrations in the roots of seven types of citrus rootstock seedlings declined significantly under BD conditions [22]. This decrease in P associated with BD conditions may explain why the growth and development of citrus rootstock roots were significantly inhibited under BD. Nutrient interaction is an important factor affecting the efficient use of nutrients and yield formation of crops, and the results of this study suggest that the interaction relationship between B and P in citrus necessitates further study. In addition, there were no elements distributed in the negative area of PC1 (Figure 2B). This suggests that root mineral nutrients are mainly derived from uptake by roots rather than translocation from other parts of the plant. As shown in Figure 2B, Mo, Mn, Cu, and Fe contributed significantly to PC1, while B, Cu, Mo, and Zn contributed significantly to PC2 in both Ptr and Cre. Many previous studies have also demonstrated that micronutrients play important roles in the growth and development of plant roots. For example, under FeD conditions, the number of root tips and the length of the main root increased significantly, whereas the activity and the number of roots decreased significantly in young seedlings of Malus hupehensis [32]. In addition, compared with the Zn-rich conditions, low-Zn conditions reduced the taproot length per plant of 'Newhall' citrus by 19.30% and that of 'Ponkan' citrus by 20.07% [33]. Altogether, these results indicate that micronutrients play essential roles in citrus root growth and development.

3.2. Adaptive Growth of Root and Stem Under Micronutrient Deficiency Conditions

The results of the current study indicated that the growth and development of root and stem were significantly inhibited under micronutrient deficiency conditions (Figures 3–6 and Table 1). Previous studies have demonstrated that B deficiency treatment significantly decreased the stem length (seedling height) of 7 different citrus rootstock seedlings, including Ptr and Cre [22]. The stem morphologies were changed significantly after 100 days of iron deficiency, but there were no significant differences found in stem length and stem dry weight among Zhique (tolerant to iron deficiency), Xiangcheng (tolerant to iron deficiency), and trifoliate orange (sensitive to iron deficiency) [16]. Our results were consistent with those of previous studies (Table 1). However, stem growth and development did not show characteristics of adaptation to micronutrient deficiency stress. Compared with the stem, the growth and development of the root showed a strong adaptability to micronutrient deficiency stress. Under plant nutrient deficiency, the development of plant roots, root morphology, and architecture undergo a series of proactive changes to adapt to the stress of nutrient deficiency and maintain the normal growth of plants, and the same is true for micronutrient deficiency. To mitigate P scarcity, plants express an array of morphological, physiological, and metabolic changes known as the phosphate starvation response [30,34–36]. Previous studies have shown that short-term low P stress can promote the elongation of LRs and increase the number of LRs in tobacco. The increased LR length and number can alleviate the decreased P content in each part of the plant under P deficiency stress and may be useful in breeding tobacco varieties with stronger root systems [29]. The same conclusion was also found in a previous study on P deficiency in apples [34]. In the present study, a short-term experiment was conducted to investigate the effects of micronutrient deficiency on LR formation in Ptr seedlings. As shown in Table 2, there were no significant decreases in the numbers of first and second LRs in Ptr seedlings under BD and ZnD conditions. Dynamic analyses of LR development were conducted in Ptr and Cre under micronutrient deficiency conditions, and the results indicated that the growth rates of first and second LRs did not decrease significantly compared with the CK under short-term (10 days) B deficiency stress (Figure 10). However, our previous research demonstrated that the LR number and length in Ptr were significantly reduced under long-term (8 or 10 weeks) B deficiency stress [22]. In addition, plants can respond to nutrient deficiency stress by regulating phytohormone levels, but long-term nutrient deficiency stress can lead to root tip necrosis and, therefore, a decrease in the number of root tips, thus affecting the synthesis and transport of cytokines and auxin. Therefore, the root number, root length, and growth rate of LRs of citrus at all levels (first and second LRs) increased significantly under short-term micronutrient deficiency stress to adapt to the nutrient-limited environment, but this adaptability would be weakened under long-term stress. Overall, citrus alters its root morphology and biological traits as a nutrient acquisition strategy to maintain maximal micronutrient acquisition and growth.

Based on previous studies, field rootstock root management was closely related to micronutrient fertilization. But, the growers did not pay much attention to micronutrient fertilizer application in citrus field production. A recent study suggested that the number of root tips and the total root surface area were important parameters of root morphology in relation to the drought tolerance of plants. They could be the target traits for breeding and selection of rootstocks for drought tolerance. In addition, the trifoliate orange hybrid ZZ-022 demonstrated good potential as a superior rootstock due to its good root system and drought tolerance [37]. Selecting the right rootstock is the best way to obtain excellent root morphology, as different rootstocks have different types of root morphology. Ptr, which had the lowest canopy volume and the highest yield efficiency, presented the highest total soluble solids. The results indicated that Ptr would be the best rootstock for navel oranges in terms of fruit quality [38]. Similar studies have also been conducted on the effects of different rootstocks on 'Ehime 28' orange vegetative organs and the effects of approach-grafting rootstock combinations on the fruit quality of Lane Late navel orange [39,40]. The results of our current study provided a theoretical basis for the selection of rootstock in citrus production.

3.3. Micronutrient Deficiency Affects the Formation and Growth of LR

Several studies have demonstrated that micronutrients are involved in the growth and development of plant roots [1,3,33], with micronutrient deficiency or excess altering root morphology and having a pronounced effect on the formation and growth of LRs [8,41]. The LR formation or emergence mechanisms are well known under several macronutrient-limited conditions, such as P, K, and S deficiency [11], but are less well studied under micronutrient-limited conditions.

In citrus, B is one of the most extensively studied micronutrients and has the most significant effect on root growth and development, morphology, and architecture. Previous studies on citrus rootstock roots have revealed that the LR number increases significantly under B deficiency conditions [20,22]. For example, B deficiency significantly inhibited the growth of Poncirus trifoliata seedlings, significantly decreased the mean LR length and root number, and significantly enhanced the root diameter and LR primordial density after 30 days of treatment; furthermore, under 90 days of B deficiency treatment, the root number, length, and surface area were significantly reduced in fragrant citrus and trifoliate orange [20]. These results were consistent with the present findings. In the current study, a short-term experiment was conducted to investigate the effects of BD on LR formation in Ptr and Cre seedlings (Table 2). However, the formation mechanism of citrus LRs under B starvation conditions remains unclear. Therefore, dynamic analyses of LR development were conducted in Ptr and Cre under BD conditions by determining the different-order LR growth rates (Figure 10). The results further clarified the dynamic effects of BD on citrus LR growth.

For the deficiency of other micronutrients, few studies focusing solely on LR growth and development have been reported in citrus. In this work, FeD had the most significant effect on the growth and development of citrus roots among all micronutrient deficiencies in both Ptr and Cre seedlings (Figures 3–5 and 10). This may be due to the significantly higher concentration of Fe in plants than other micronutrients. In addition, the Fe concentration was significantly higher in citrus roots than in stems and leaves (Figure 8). The Fe concentration was remarkably higher in both the second LR and third LR than in other parts of the root of Ptr. This trend was also observed in Cre. The Fe concentration was significantly higher in the first LR than in the root collar for both rootstocks (Figure 1). Our previous work showed that, under normal conditions, the Fe concentrations in Ptr and Cre roots were 10.95 and 14.11 times that in the leaves, respectively [22]. Recent research has demonstrated that the LR number is significantly reduced under FeD conditions in trifoliate orange [16], and these results confirm that FeD can significantly affect the growth, development, and morphology of citrus roots.

As mentioned previously, extensive research has been conducted on citrus roots. However, the mechanism through which micronutrient deficiency affects citrus LR development remains unclear. LR formation is a multistep developmental process in which auxin and peptide hormones play essential roles [42]. Recent studies in Arabidopsis have revealed that the mitogen-activated protein kinase (MAPK) cascade MKK4/MKK5–MPK3/MPK6 functions in both a noncanonical auxin signaling pathway and the IDA peptide signaling pathway to regulate LR morphogenesis and emergence, respectively [43,44]. Therefore, investigating the effects of micronutrient deficiency on the growth, development, and morphology of citrus LRs from the perspective of auxin regulation is a promising avenue for future research.

4. Materials and Methods

4.1. Plant Materials and Treatment

Two citrus rootstocks, trifoliate orange (*Poncirus trifoliata* L. Raf.) and red tangerine (*Citrus reticulata* Blanco), were employed in the present study. The seeds of these two rootstocks were germinated, and seedling culture proceeded following a previously described method [45]. For the determination of the root system mineral nutrient distribution in different root parts (the root collar, taproot, and three levels of lateral roots), the seedlings were cultured in 15 L black plastic pots (with 30 cm height and 25 cm internal diameter) containing soil for 6 months. The soil characteristics were as follows: organic matter 19.91 ± 0.49 g/kg, total N 0.97 ± 0.12 g/kg, available P 21.14 ± 0.64 mg/kg, available K 115.25 ± 11.36 mg/kg, and pH 5.92 ± 0.52. For the micronutrient (Fe, Mn, B, and Zn) deficiency experiment, 4-week-old seedlings of similar taproot length were selected and then transferred to 5 L black pots containing nutrient-free medium composed of quartz sand–perlite (1:1, *v/v*) according to a published method [46,47]. Plants were cultured in a growth chamber with normal culture parameters according to the method of Zhou et al. [22]. Plants were irrigated twice a week with modified Hoagland's No. 2 nutrient solution containing 6 mM KNO_3, 4 mM $Ca(NO_3)_2$, 1 mM $NH_4H_2PO_4$, 2 mM $MgSO_4$, 9 μM $MnCl_2$, 0.8 μM $ZnSO_4$, 20 μM H_3BO_3, 0.3 μM $CuSO_4$, 0.01 μM H_2MoO_4, and 50 μM Fe-EDTA [45]. Irrigating with this control nutrient solution (CK), seedlings were precultured for 2–3 weeks until new white roots appeared. Then, seedlings were irrigated with 0 μM Fe-EDTA nutrient solution for the Fe deficiency (FeD) treatment, 0 μM $MnCl_2$ nutrient solution for the Mn deficiency (MnD) treatment, 0 μM H_3BO_3 nutrient solution for the B deficiency (BD) treatment, and 0 μM $ZnSO_4$ nutrient solution for the Zn deficiency (ZnD) treatment. The treatments began on 17 April 2023 and concluded on 4 July 2023 (12 weeks), when visible symptoms had appeared. For the dynamic analysis of LR development, a short-term (40-day) experiment was conducted to investigate the effects of micronutrient deficiency on the LR formation and growth rate in Ptr and Cre seedlings. In this experiment, the plant material culture and micronutrient deficiency treatments were the same as described previously.

4.2. Sampling and Measurement of Plant Growth Parameters

At the end of the experiment, nine plants per treatment were harvested randomly, rinsed in deionized water, and carefully blotted with tissue paper. Then, the materials were divided into leaf, stem, and root tissues. The fresh materials were placed into a forced air oven at 105 °C for 15 min and then heated at 75 °C until constant weights were reached to determine their stem and root dry weights (g). All dried samples were then ground into fine powder to determine the mineral nutrient concentrations in tissues. The stem length (cm) and taproot length (cm) were measured using a scaled ruler. Stem diameter (mm) was measured using a vernier caliper.

4.3. Root Morphology and Analysis

Nine seedlings were randomly sampled from each treatment group and rinsed with deionized water. For root morphology analysis, the root samples were scanned using an Epson digital scanner Expression 10000XL 1.0 (Epson Inc., Suwa City, Nagano Prefecture, Japan), and the image was analyzed using WinRhizo Pro (S) v. 2009c software (Regent Instruments Inc., Québec, QC, Canada). The root traits examined in this study included the total root length, root surface area, root volume, and root number. The root length and root surface area within a certain root diameter (0.0–0.5 mm, 0.5–1.0 mm, 1.0–1.5 mm, and >1.5 mm) were also analyzed using this software.

4.4. Dynamic Analysis of LR Development

A short-term (40-day) experiment was conducted to investigate the effects of micronutrient deficiency on the LR formation and growth rate in Ptr and Cre seedlings. After 40 days of experimental treatment, the 1st LR and 2nd LR were separated. The root number and length of the 1st and 2nd LRs were scanned with an Epson digital scanner, and the image was analyzed using WinRhizo Pro (S) v. 2009c software. The dynamic analysis of LR development in Ptr and Cre was conducted under micronutrient deficiency conditions by determining the growth rates of different LRs. The lengths of 1st and 2nd LRs were measured every 10 days until day 40, and the LR growth rates were analyzed.

4.5. Determination of Mineral Nutrients

The mineral nutrient concentrations of P, K, Ca, Mg, Fe, Mn, Zn, B, Cu, and Mo in different plant tissues were determined following the method described by Storey and Treeby [48]. In brief, 0.50 g of each sample was dry-ashed in a muffle furnace at 500 °C for 6 h, followed by dissolution in 0.1 N HCl. The mineral nutrient concentrations were then determined using ICP-MS 7900 (Agilent Technologies Inc., Santa Clara, CA, USA).

4.6. Experimental Design and Statistical Analysis

The experiment was set up in a completely randomized 5×2 factorial design with one control (CK) and four micronutrient deficiency solutions (FeD, MnD, BD, and ZnD) combined with two citrus rootstocks. The values are presented as means ± SE of nine seedlings. The data underwent analysis of variance (ANOVA) in SAS 8.1 software (SAS Institute Inc., Cary, NC, USA), and the differences were compared using Duncan's test with a significance level of $p < 0.05$. PCA was applied to analyze the dimensionality reduction of the ion group of the root based on the content correlation matrix of 10 elements in the root. The first two principal components were extracted to design the score diagram and the load diagram.

5. Conclusions

This study demonstrated that citrus root growth and development, morphology, and architecture were affected by micronutrients, as well as the morphology of the stem. In the present work, the distribution of mineral nutrients in different parts of the citrus root system suggested that there were obvious differences in the element distribution characteristics under root partitioning and that the element concentrations and ion composition in the second LR and third LR were very different compared with other root parts. The analysis of root morphology under micronutrient deficiency demonstrated that the root growth was significantly inhibited in both Ptr and Cre under FeD and BD conditions compared with the MnD and ZnD treatments, but Cre roots exhibited better performance than Ptr roots. The nutrient concentrations in Ptr and Cre seedlings under micronutrient deficiency revealed that single micronutrient deficiencies affected both their own concentrations and the concentrations of other mineral nutrients, whether in the roots or in stems and leaves. The dynamic analysis of LR development revealed that the growth rates of the first LR and second LR did not decrease significantly compared with CK under short-term micronutrient deficiency stress. To summarize, these results suggest that micronutrients play important roles in citrus root growth and development as well as RSA modulations. The present work on the spatial distribution characteristics and micronutrient deficiency of citrus roots provides a theoretical basis for effective micronutrient fertilization and the diagnosis of micronutrient deficiency symptoms in citrus.

Author Contributions: Conceptualization and supervision, G.Z. and J.Z.; methodology, Y.F.; validation, Y.L. and M.Y.; data curation, G.Z. and M.Y.; original draft preparation, G.Z.; writing—review and editing, G.Z. and J.Z. All authors have read and agreed to the published version of the manuscript.

Funding: This research was funded by the National Natural Science Foundation of China (Nos. 32160680 and 31960573).

Data Availability Statement: Data are contained within the article.

Conflicts of Interest: The authors declare no conflicts of interest.

References

1. López-Bucio, J.; Cruz-Ramírez, A.; Herrera-Estrella, L. The role of nutrient availability in regulating root architecture. *Curr. Opin. Plant Biol.* **2003**, *6*, 280–287. [CrossRef] [PubMed]
2. Meister, R.; Rajani, M.S.; Ruzicka, D.; Schachtman, D.P. Challenges of modifying root traits in crops for agriculture. *Trends Plant Sci.* **2014**, *19*, 779–788. [CrossRef]
3. White, P.J. Root traits benefitting crop production in environments with limited water and nutrient availability. *Ann. Bot.* **2019**, *124*, 883–890. [CrossRef] [PubMed]
4. Kumar, A.; Shahbaz, M.; Koirala, M.; Blagodatskaya, E.; Seidel, S.J.; Kuzyakov, Y.; Pausch, J. Root trait plasticity and plant nutrient acquisition in phosphorus limited soil. *J. Plant Nutr. Soil Sci.* **2019**, *182*, 945–952. [CrossRef]
5. Péret, B.; Clément, M.; Nussaume, L.; Desnos, T. Root developmental adaptation to phosphate starvation: Better safe than sorry. *Trends Plant Sci.* **2011**, *16*, 442–450. [CrossRef]
6. Williamson, L.C. Phosphate availability regulates root system architecture in Arabidopsis. *Plant Physiol.* **2001**, *126*, 875–882. [CrossRef]
7. Tyburski, J.; Dunajska, K.; Tretyn, A. A role for redox factors in shaping root architecture under phosphorus deficiency. *Plant Signal. Behav.* **2010**, *5*, 64–66. [CrossRef] [PubMed]
8. Tian, H.Y.; De Smet, I.; Ding, Z.J. Shaping a root system: Regulating lateral versus primary root growth. *Trends Plant Sci.* **2014**, *19*, 426–431. [CrossRef] [PubMed]
9. Zhang, H.; Barlow, P.W. Dual pathways for regulation of root branching by nitrate. *Proc. Natl. Acad. Sci. USA* **1999**, *96*, 6529–6534. [CrossRef] [PubMed]
10. Vidal, E.A.; Araus, V.; Lu, C.; Parry, G.; Green, P.J.; Coruzzi, G.M.; Gutiérrez, R.A. Nitrate-responsive miR393/AFB3 regulatory module controls root system architecture in *Arabidopsis thaliana*. *Proc. Natl. Acad. Sci. USA* **2010**, *107*, 4477–4482. [CrossRef]
11. Perez-Torres, C.A.; López-Bucio, J.; Cruz-Ramírez, A.; Ibarra-Laclette, E.; Dharmasiri, S.; Estelle, M.; Herrera-Estrella, L. Phosphate availability alters lateral root development in *Arabidopsis* by modulating auxin sensitivity via a mechanism involving the TIR1 auxin receptor. *Plant Cell* **2008**, *20*, 3258–3272. [CrossRef]
12. Reymond, M.; Svistoonoff, S.; Loudet, O.; Nussaume, L.; Desnos, T. Identification of QTL controlling root growth response to phosphate starvation in *Arabidopsis thaliana*. *Plant Cell Environ.* **2010**, *29*, 115–125. [CrossRef]
13. Krouk, G.; Lacombe, B.; Bielach, A.; Perrine-Walker, F.; Malinska, K.; Mounier, E.; Hoyerova, K.; Tillard, P.; Leon, S.; Ljung, K. Nitrate-regulated auxin transport by NRT1.1 defines a mechanism for nutrient sensing in plants. *Dev. Cell* **2010**, *18*, 927–937. [CrossRef] [PubMed]
14. Chen, L.S.; Han, S.; Qi, Y.P.; Yang, L.T. Boron stresses and tolerance in citrus. *Afr. J. Biotechnol.* **2012**, *11*, 5961–5969.
15. Fu, X.Z.; Peng, L.Z.; Xing, F.; Ling, L.L.; Chun, C.P.; Jing, C.L.; Cao, L. Zinc deficiency in citrus: Current studies and future perspectives. *J. Fruit Sci.* **2014**, *31*, 132–139.
16. Fu, L.N.; Zhu, Q.Q.; Sun, Y.Y.; Du, W.; Pan, Z.Y.; Peng, S.A. Physiological and transcriptional changes of three citrus rootstock seedlings under iron deficiency. *Front. Plant Sci.* **2017**, *8*, 01104. [CrossRef] [PubMed]
17. Forner-Giner, M.A.; Alcaide, A.; Primo-Millo, E.; Forner, J.B. Performance of 'Navelina' orange on 14 rootstocks in Northern Valencia (Spain). *Sci. Hortic.* **2003**, *98*, 223–232. [CrossRef]
18. Sharma, R.M.; Dubey, A.K.; Awasthi, O.P.; Kaur, C. Growth, yield, fruit quality and leaf nutrient status of grapefruit (*Citrus paradisi* Macf.): Variation from rootstocks. *Sci. Hortic.* **2016**, *210*, 41–48. [CrossRef]
19. Li, Q.H.; Liu, Y.Z.; Pan, Z.Y.; Xie, S.; Peng, S.A. Boron deficiency alters root growth and development and interacts with auxin metabolism by influencing the expression of auxin synthesis and transport genes. *Biotechnol. Biotechnol. Equip.* **2016**, *30*, 661–668. [CrossRef]
20. Mei, L.; Li, Q.H.; Wang, H.; Sheng, O.; Peng, S.A. Boron deficiency affects root vessel anatomy and mineral nutrient allocation of *Poncirus trifoliata* (L.) raf. *Acta Physiol. Plant.* **2016**, *38*, 86. [CrossRef]

21. Wu, X.W.; Riaz, M.; Yan, L.; Du, C.Q.; Liu, Y.L.; Jiang, C.C. Boron deficiency in trifoliate orange induces changes in pectin composition and architecture of components in root cell walls. *Front. Plant Sci.* **2017**, *8*, 1882. [CrossRef]
22. Zhou, G.F.; Peng, S.A.; Liu, Y.Z.; Wei, Q.J.; Han, J.; Islam, M.Z. The physiological and nutritional responses of seven different citrus rootstock seedlings to boron deficiency. *Trees* **2014**, *28*, 295–307. [CrossRef]
23. Marschner, H. *Mineral Nutrition of Higher Plants*, 2nd ed.; Academic Press: London, UK, 1995.
24. Chapman, N.; Miller, A.J.; Lindsey, K.; Whalley, W.R. Roots, water, and nutrient acquisition: Let's get physical. *Trends Plant Sci.* **2012**, *17*, 701–710. [CrossRef]
25. Liu, G.D.; Hu, P.; Zeng, Y.; Zeng, Z.J.; Guan, G.; Yao, F.X.; Zhou, G.F. Mineral nutrients distribution in leaf different parts in navel orange plant. *Acta Hortic. Sin.* **2019**, *46*, 47–56.
26. Callahan, D.L.; Hare, D.J.; Bishop, D.P.; Doble, P.A.; Roessner, U. Elemental imaging of leaves from the metal hyperaccumulating plant *Noccaea caerulescens* shows different spatial distribution of Ni, Zn and Cd. *RSC Adv.* **2016**, *6*, 2337. [CrossRef]
27. Vogel-Mikuš, K.; Simcic, J.; Pelicon, P.; Budnar, M.; Kump, P.; Necemer, M.; Mesjasz-Przybylowicz, J.; Przybylowicz, W.J.; Regvar, M. Comparison of essential and non-essential element distribution in leaves of the Cd/Zn hyperaccumulator *Thlaspi praecox* as revealed by micro-PIXE. *Plant Cell Environ.* **2008**, *31*, 1484–1496. [CrossRef]
28. McClure, G.W. Nutrient distribution in root zones. III. Further studies of the effects of phosphorus distribution on corn and wheat. *Can. J. Bot.* **1972**, *50*, 2275–2282. [CrossRef]
29. Zheng, X.; Zhang, Z.; Chen, X.; Jia, Z.; Su, Y. Responses of root morphology and architecture to phosphorus deficiency at seedling stage of tobacco (*nicotiana tabacum*) growth. *Aust. J. Crop Sci.* **2013**, *7*, 1967–1972.
30. Péret, B.; Desnos, T.; Jost, R.; Kanno, S.; Berkowitz, O.; Nussaume, L. Root architecture responses: In search of phosphate. *Plant Physiol.* **2014**, *166*, 1713–1723. [CrossRef]
31. Silva, A.D.; Bruno, I.P.; Franzini, V.I.; Marcante, N.C.; Benitiz, L.; Muraoka, T. Phosphorus uptake efficiency, root morphology and architecture in Brazilian wheat cultivars. *J. Radioanal. Nucl. Chem.* **2016**, *307*, 1055–1063. [CrossRef]
32. Qiao, H.T.; Yang, H.Q.; Sheng, E.B.; Jiang, Q.Q.; You, S.Z.; Zhang, L.; Ran, K.; Zhang, X.R. Effect of nitrogen deficient and iron deficient on root architecture of young seedlings of *Malus hupehensis* (Pamp) Rehd. *Acta Hortic. Sin.* **2009**, *36*, 321–326.
33. Chen, Y.Y.; Hu, C.Y.; Xiao, J.X. Effects of arbuscular mycorrhizal inoculation on the growth, zinc distribution and photosynthesis of two citrus cultivars grown in low-zinc soil. *Trees* **2014**, *28*, 1427–1436. [CrossRef]
34. Fan, W.G.; Yang, H.Q.; Han, X.J. Changes of root architecture and phosphorus uptake by roots of *Malus hupehensis* (Pamp) Rehd. under the condition of phosphorus deficiency. *Acta Hortic. Sin.* **2007**, *34*, 1341–1346.
35. Plaxton, W.C.; Tran, H.T. Metabolic adaptations of phosphate-starved plants. *Plant Physiol.* **2011**, *156*, 1006–1015. [CrossRef] [PubMed]
36. Fabiańska, I.; Gerlach, N.; Almario, J.; Bucher, M. Plant-mediated effects of soil phosphorus on the root-associated fungal microbiota in *Arabidopsis thaliana*. *New Phytol.* **2019**, *221*, 2123–2137. [CrossRef]
37. Luo, G.T.; Liu, X.N.; Zhang, M.M.; Yu, H.; Hu, Z.; Wang, F.S.; Zhu, S.P.; Zhao, X.C. Root morphology of citrus rootstocks and drought tolerance evaluation of their grafted plants. *J. Fruit Sci.* **2020**, *37*, 1314–1325.
38. Zhu, S.P.; Huang, T.J.; Yu, X.; Hong, Q.B.; Xiang, J.S.; Zeng, A.Z.; Gong, G.Z.; Zhao, X.C. The effects of rootstocks on performances of three late-ripening navel orange varieties. *J. Integr. Agric.* **2020**, *19*, 1802–1812. [CrossRef]
39. He, M.; Tian, Y.; Yu, Y.; Pu, J.N.; Zhang, C.Y.; He, R.J.; Zhang, Y.Q.; Lyu, Q.; Xie, R.J.; Ma, Y.Y.; et al. Effect of different root stocks on nutrient absorption and utilization of vegetative organs in 'Ehime 28'. *J. Southwest Univ.* **2022**, *44*, 80–87.
40. Song, F.; Qin, W.; Kuang, B.; Sun, H.; Chen, Q.X.; Jiang, Y.C.; Liu, J.H.; Wu, L.M. Effects of approach-grafting rootstock combinations on nutrition, growth and development, and fruit quality of Lane Late navel orange. *J. Huazhong Agric. Univ.* **2022**, *41*, 116–121.
41. Malamy, J.E. Intrinsic and environmental response pathways that regulate root system architecture. *Plant Cell Environ.* **2005**, *28*, 67–77. [CrossRef]
42. Jourquina, J.; Fukakic, H.; Beeckman, T. Peptide-receptor signaling controls lateral root development. *Plant Physiol.* **2020**, *182*, 1645–1656. [CrossRef]
43. Huang, R.; Zheng, R.; He, J.; Zhou, Z.; Xu, T. Noncanonical auxin signaling regulates cell division pattern during lateral root development. *Proc. Natl. Acad. Sci. USA* **2019**, *116*, 21285–21290. [CrossRef]
44. Zhu, Q.; Shao, Y.; Ge, S.; Zhang, M.; Zhang, T.; Hu, X.; Liu, Y.; Walker, J.; Zhang, S.; Xu, J. A MAPK cascade downstream of IDA–HAE/HSL2 ligand–receptor pair in lateral root emergence. *Nat. Plants* **2019**, *5*, 414–423. [CrossRef] [PubMed]
45. Zhou, G.F.; Wei, Q.J.; Li, B.X.; Zeng, X.L.; Liu, G.D. Establishment and optimization of a hydroponic system for root morphological and nutritional analysis of citrus. *Sci. Agric.* **2020**, *77*, e20180261. [CrossRef]
46. Papadakis, I.E.; Dimassi, N.; Bosabalidis, A.M.; Therios, I.N.; Patakas, A.; Giannakoula, A. Boron toxicity in 'Clementine' mandarin plants grafted on two rootstocks. *Plant Sci.* **2004**, *166*, 539–547. [CrossRef]

47. Sheng, O.; Song, S.W.; Peng, S.A.; Deng, X.X. The effects of low boron on growth, gas exchange, boron concentration and distribution of 'Newhall' navel orange (*Citrus sinensis* Osb.) plants grafted on two rootstocks. *Sci. Hortic.* **2009**, *121*, 278–283. [CrossRef]
48. Storey, R.; Treeby, M.T. Nutrient uptake into navel orange during fruit development. *J. Hortic. Sci. Biotechnol.* **2000**, *77*, 91–99. [CrossRef]

plants

MDPI

Article

Hyperspectral Imaging and Machine Learning for Huanglongbing Detection on Leaf-Symptoms

Ruihao Dong [1], Aya Shiraiwa [2], Katsuya Ichinose [3], Achara Pawasut [4], Kesaraporn Sreechun [4], Sumalee Mensin [4] and Takefumi Hayashi [1,*]

1 Faculty of Informatics, Kansai University, Osaka 569-1095, Japan; k975945@kansai-u.ac.jp
2 Electrical Engineering and Computer Science, Tottori University, Tottori 680-8552, Japan; shiraiwa@tottori-u.ac.jp
3 Independent Researcher, Tsukuba 300-1252, Japan
4 Royal Project Foundation, 910 Moo 3, T. Maehia, Muang, Chiang Mai 50200, Thailand; acrpwst@gmail.com (A.P.); ketcnx@gmail.com (K.S.); linly317@gmail.com (S.M.)
* Correspondence: haya@kansai-u.ac.jp

Abstract: Huanglongbing is one of the most destructive diseases of citrus worldwide. Infected trees die due to the absence of practical cures. Thus, the removal of HLB-infected trees is one of the principal HLB managements for the regulation of disease spread. Here, we propose a non-destructive HLB detection method based on hyperspectral leaf reflectance. In total, 72 hyperspectral leaf images were collected in an HLB-invaded citrus orchard in Thailand and each image was visually distinguished into either any HLB symptom appearance (symptomatic) or no symptoms (asymptomatic) on the leaf. Principal component analysis was applied on the hyperspectral data and revealed 16 key wavelengths at red-edge to near-infrared regions (715, 718, 721, 724, 727, 730, 733, 736, 930, 933, 936, 939, 942, 945, 957, and 997 nm) that were characteristically differentiated in the symptomatic group. Seven models learnt on the spectral data at these 16 wavelengths were examined for the potential to separate these two image groups: random forest, decision tree, support vector machine, k-nearest neighbor, gradient boosting, logistic regression, linear discriminant. F1-score was employed to select the best-fit model to distinguish the two categories: random forest achieved the best score of 99.8%, followed by decision tree and k-nearest neighbor. The reliability of the visual grouping was evaluated by nearest neighbor matching and permutation test. These three models separated the two image categories as precisely as PCR results, indicating their potential as alternative tool instead of PCR.

Keywords: citrus greening disease; *Citrus reticulata*; disease diagnosis; feature extraction; multivariate analysis

1. Introduction

Huanglongbing (HLB), or citrus greening disease, is caused by the bacterium *Candidatus* Liberibacter asiaticus, which is transmitted between plants by the vector insect *Diaphorina citri* in Asian countries [1]. HLB is recognized as the most destructive disease of citrus worldwide. As no effective measures have been established yet [1], the current HLB management consists of three major measures: use of pathogen-free materials and sanitation of fields to plant trees, vector control, and removal of infected trees. The conventional detection of HLB is based primarily on regular visual inspection of trees and DNA polymerase chain reaction (PCR) [2]. Visual inspection needs skills to determine HLB infection, while PCR requires specialized laboratory equipment with much cost [3]. Alternative diagnostic measures that are cost-effective, efficient, and highly accurate are urgent issues to develop. Optical image analyses could be a potential candidate for HLB detection [4].

Diagnosing plant diseases with RGB image analyses, particularly with convolutional neural networks (CNNs), has been enthusiastically explored in this decade [4]. These analyses rely on HLB symptoms appearing on trees: yellow blotchy mottling, thickening, and enlarged or corky veins of leaves [5], and wilting, stunted growth, or yellow shoots of the canopy [1]. Kalim et al. [6] achieved 87% precision in distinguishing HLB from other five diseases with combining machine learning and deep learning on 594 leaf images with these symptoms. Elaraby et al. [7] reached 94.3% accuracy in diagnosing six citrus diseases with an SGDM+AlexNet model that used 4745 leaf images, while Gómez-Flores et al. [8] used a VGG19 learning model in which they transferred 953 leaf image data distinguished into HLB-infected and not infected groups. This model correctly identified the HLB infection with 95% precision. Deep learning with numerous RGB images thus achieves high precision in detecting the HLB-infected trees. However, various background environmental noises are also taken in the images and the elimination of such noises negatively interferes with the image analyses [9]. Furthermore, the success in image analyzing depends on how appropriately the targeted samples are photographed, being determined by the experience on visual disease inspection [10].

Finer spectral data with finer division of wavelengths than RGB contribute to successfully excluding environmental noises and provide more effective image data for crop health, pest, and soil assessment in agriculture [4]. This technique utilizes the electromagnetic intensity of spectra that are determined by the chemical and physical properties in plants [11]. Two measures are adopted, depending on the differences in the number of wavelengths: multispectral analyses in which 4–16 wavelengths are used and hyperspectral analyses that the optical data are collected at 200–1000 wavelengths divided at every 1 to 7 nm (Figure 1). Owing to its finer partitioning of wavelengths to analyze, hyperspectral imaging usually gives more accuracy in separating different groups, in particular detecting asymptomatic infection [12]. For example, multispectral analyses with three spectral regions (520–600 nm, 630–690 nm, 760–900 nm) succeed in detecting HLB in an accuracy > 90% with four machine learning techniques on the image data that were obtained from aerial viewing of an orchard by a drone [13]. On the other hand, Deng et al. [14] analyzed hyperspectral data and extracted 13 specific wavelengths using six machine learning models, successfully distinguishing HLB infection with a precision of 96%, being more sensitive by 6% than the former.

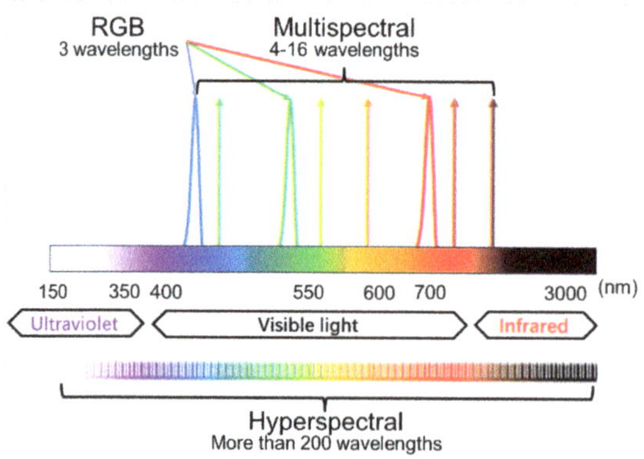

Figure 1. The schematic of RGB, multispectral, and hyperspectral images.

Although hyperspectral cameras may specify the disease infection in an early stage, including HLB, they are costly and impractical for large-scale use [15]. Confining the data in wavelengths that contribute to the differentiation in the reflectance properties in HLB-infected trees, the size-down in image data could be possible, and more handy equipment for HLB detection may be able to develop. Various authors have published a number of articles devoted to the determination of HLB using hyperspectral remote sensing and have attempted to extract some characteristic wavelengths (Table 1, [14,16–24]). In this paper, we propose statistical methods to efficiently confine useful wavelengths from hyperspectral data and machine learning models to identify HLB infection.

Table 1. Previous studies using hyperspectral remote sensing for HLB detection.

Publication Year	Equipment Wavelengths (nm)	Feature Wavelengths (nm)	Study Type
2011–2013 [16–18]	350–2500	537, 612, 638, 662, 688, 713, 763, 813, 998, 1066, 1120, 1148, 1296, 1445, 1472, 1546, 1597, 1622, 1746, 1898, 2121, 2172, 2348, 2471, 2493	field
2012 [19]	457–921	650–850	field and lab
2012 [20]	457–921	410–432, 440–509, 634–686, 734–927, 932, 951, 975, 980	field
2018 [21]	379–1023	493, 515, 665, 716, 739	lab
2019 [14]	400–1000	544, 718, 753, 760, 764, 930, 938, 943, 951, 969, 985, 998, 999	field
2020 [22]	450–950, 325–1075	468, 504, 512, 516, 528, 536, 632, 680, 688, 852	field
2024 [23]	400–1000	560, 678, 726, 750	lab
2025 [24]	325–1075	375–425, 650–750, 890–925	lab

2. Results

2.1. PCA for the Explanation of the Variance in the Data

The contributions of the first eight principal components (PCs) revealed by the PCA (principal component analysis) were 83.44%, 12.83%, 2.17%, 0.84%, 0.14%, 0.08%, 0.06%, and 0.04%, respectively (Figure 2a). Since the first four PCs explained over 99% of the variance in the data, we output their loading plot as shown in Figure 2b. The loadings represent the contribution of each wavelength to the distinction of image groupings, asymptomatic and symptomatic. Figure 2b shows relatively large loading variations in the 500–600 nm, near 700 nm, and 900–1000 nm regions. This suggests that the reflectance properties of the foliar hyperspectral data vary the most at these wavelengths, which may be critical for HLB detection classification. The loading curve for PC1 was smoother and usually represented the main differences in overall reflection or intensity. PC2–4 explained smaller proportions of the variance and showed more peaks and troughs, often representing minor but still important differences or possibly noise.

We then compared the distribution of asymptomatic and symptomatic data on PC1–4 two by two and output their scatter plots (Figure 3). Regardless of which two PCs are compared, there is a large overlap region, which indicates that PCA only is not yet able to separate the differences between the two categories well. Further wavelength optimization may be needed to exclude some redundant information. We analyzed the top eight wavelengths of PC1–4 and their loadings as shown in Table 2. Wavelengths closely regressed on PC1 in the range 715–736 nm, which corresponded to the red-edge region of citrus leaves. On PC2, the wavelengths at the 930–960 nm range were regressed, expanding over the near-infrared region. Wavelengths on PC3 included shorter wavelengths, 400, 710, and 933 nm, while on PC4 near infrared wavelengths 933–957 nm were related, mostly overlapping with PC2. Since PC1 and PC2 explained more than 96% of the variance and there was no overlap in the top eight wavelengths, a combination of these two PCs was chosen for further wavelength optimization. The PC1 score of the symptomatic group was negatively correlated against the PC2, while asymptotic groups were scattered near the origin in the coordinates with no apparent correlation.

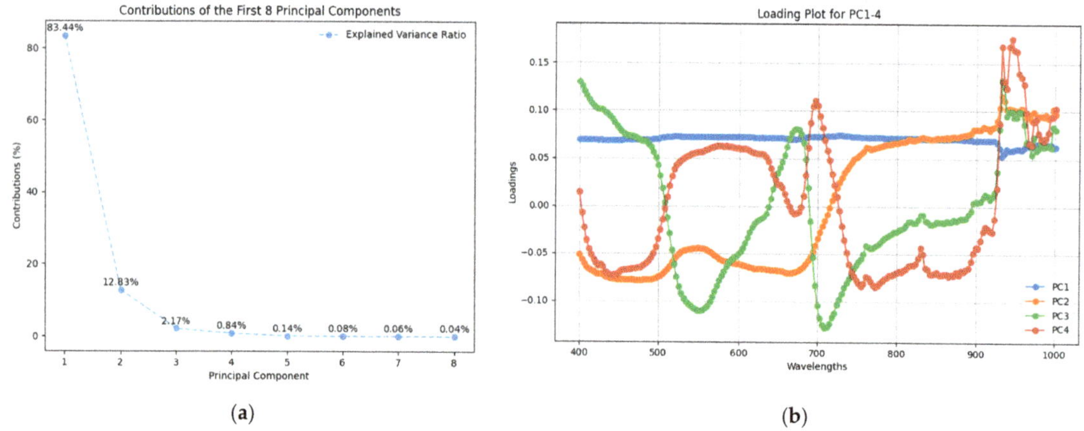

Figure 2. The contributions of the first eight principal components (**a**) and the loading plot for principal components 1–4 (**b**).

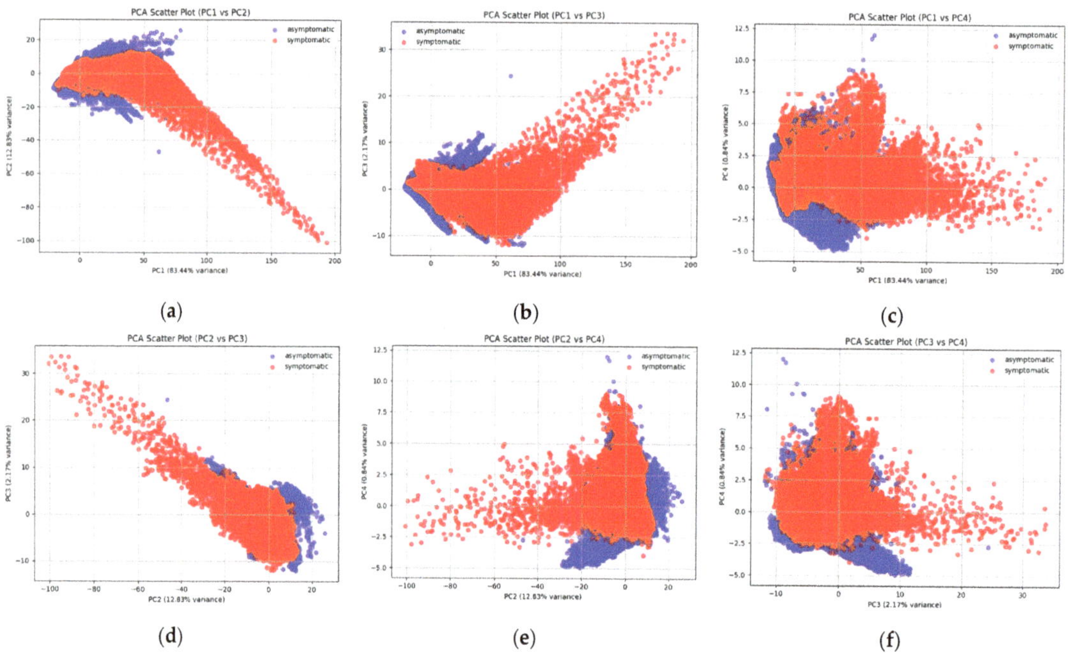

Figure 3. The scatter plots of different PCs between PC1 and PC4 (**a**–**f**).

Table 2. Top eight wavelengths of principal components 1–4 and their loadings.

PC1		PC2		PC3		PC4	
Wavelength	Loading	Wavelength	Loading	Wavelength	Loading	Wavelength	Loading
727	0.0741	933	0.1165	933	0.1324	945	0.1764
724	0.0741	936	0.1111	400	0.1303	942	0.1686
730	0.0740	930	0.1040	709	−0.1276	933	0.1683
721	0.0739	957	0.1032	712	−0.1271	948	0.1648
733	0.0739	942	0.1031	936	0.1260	951	0.1636
718	0.0738	939	0.1025	403	0.1254	954	0.1400
736	0.0737	945	0.1025	706	−0.1247	957	0.1359
715	0.0735	997	0.1025	715	−0.1232	936	0.1312

2.2. Hyperspectral Data Wavelength Optimization

To further refine HLB characteristic wavelengths, we applied the wavelength optimization method based on incremental feature selection. By sequentially increasing the number of selected wavelengths according to their loading scores, we obtained a total of 64 wavelength combinations. These combinations were then input into the seven models for training and the F1 scores were compared for three conditions: full wavelengths, feature extraction, and wavelength optimization.

Nonlinear models (random forest (RF), decision tree, k-nearest neighbor (KNN), gradient boosting, and support vector machine (SVM)) consistently achieved F1 scores above 95%, while linear models (linear discriminant analysis (LDA) and logistic regression) performed poorly, with F1 scores below 40% after data reduction (Table 3). Among the nonlinear models, only RF, decision tree, KNN, and gradient boosting maintained high F1 scores (>95%) after wavelength optimization. The RF model integrating nine wavelengths achieved the highest F1 score of 99.8%, and the gradient boosting model integrating twelve wavelengths achieved the lowest F1 score of 96.2%. We took these four models and discussed them further in terms of their reliability.

Table 3. F1 scores of seven classification models with symptomatic and asymptomatic datasets for full wavelengths, feature extraction, and wavelength optimization.

Classification Model		Full Wavelengths (204 Wavelengths)	Feature Extraction (16 Wavelengths)	Wavelength Optimization (No. of Wavelengths)
Nonlinear	RF	99.5%	99.6%	99.8% (9)
	Decision tree	99.1%	99.2%	99.3% (6)
	KNN	98.8%	96.3%	97.9% (4)
	Gradient boosting	97.7%	95.6%	96.2% (12)
	SVM	98.4%	89.2%	89.2% (16)
Linear	LDA	89.3%	38.8%	38.8% (15)
	Logistic regression	91.8%	38.3%	38.4% (14)

2.3. Reliability of Expert System

A PCR-based study relied on precise PCR results to build a hyperspectral dataset of foliar leaves, which was learned and tested using six different machine learning models and achieved 96% accuracy on a 13-wavelength combination. As the four nonlinear models mentioned above performed well after data reduction, we conducted nearest neighbor matching for comparing wavelengths extracted in wavelength optimization by them with this 13-wavelength combination reported in the PCR-based study (Table 4). The discrepancies between PCR-based and model-selected wavelengths were 2.78 nm for RF, 3.14 nm for decision tree, 3.5 nm for KNN, and 5.0 nm for gradient boosting (Figure 4). Owing to their discrepancies smaller than the hyperspectral camera's resolution of 7 (0 ± 3.5 nm), the RF, decision tree, and KNN models can be taken as effective alternatives to the PCR test for HLB detection. Therefore, these three models were selected for further analyses.

Table 4. Extracted wavelengths by RF, decision tree, KNN, and gradient boosting models and their wavelength discrepancies between the study by Deng et al. [14].

Classification Model	Extracted Wavelengths (nm)	Discrepancies
Deng et al. [14]	544, 718, 753, 760, 764, 930, 938, 943, 951, 969, 985, 998, 999	
RF	727, 930, 933, 936, 939, 942, 945, 957, 997	2.78
Decision tree	727, 930, 933, 936, 939, 942, 957	3.14
KNN	727, 930, 933, 936	3.5
Gradient boosting	721, 724, 727, 730, 733, 930, 933, 936, 939, 942, 945, 957	5.0

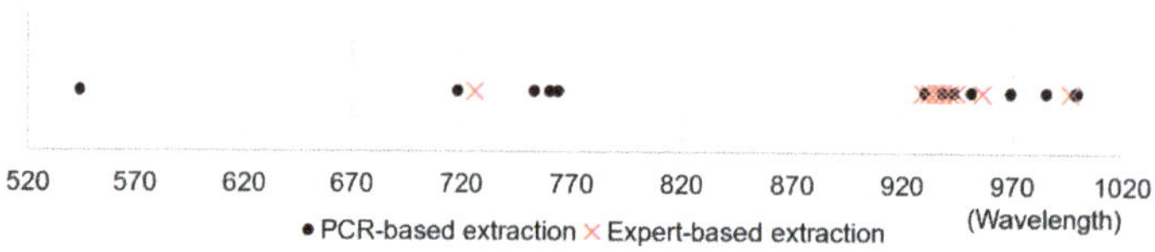

Figure 4. Comparison of key wavelengths identified by the PCR-based study and our expert-based RF model.

We analyzed the F1 scores, wavelengths used, and test time of these three models (Table 5). The decision tree with seven wavelengths completed the calculation in the shortest time, 7 ms, but the precision was comparable to that of the RF model. Although the KNN model required the longest time in the calculation, it utilized only four wavelengths with keeping the F1-score higher. These findings indicate that the RF, decision tree, and KNN models in which red to infrared reflectance data are integrated can recognize HLB infection with more than 97% precision. In particular, the RF and decision tree models almost perfectly separated image groups.

Table 5. The F1-scores, wavelengths used and test time achieved by RF, decision tree and KNN models.

Classification Model	F1-Score	Wavelengths Used	Test Time (ms)
RF	99.8%	9	869
Decision tree	99.3%	7	7
KNN	97.9%	4	899

The availability of models may be considered in three aspects: precision in HLB determination, quickness in calculation time, and the reduction of used wavelengths (Figure 5). If focusing only on the precision, the RF model will be selected. On the other hand, if calculation time is the priority, the decision tree model should be the first choice. If simpler equipment is preferred, the KNN model should be considered because it uses only four wavelengths, which may be easily integrated into the multispectral analyzing system. Since all three models have high-level precision while the decision tree model has an outstanding processing speed, we would recommend prioritizing this model.

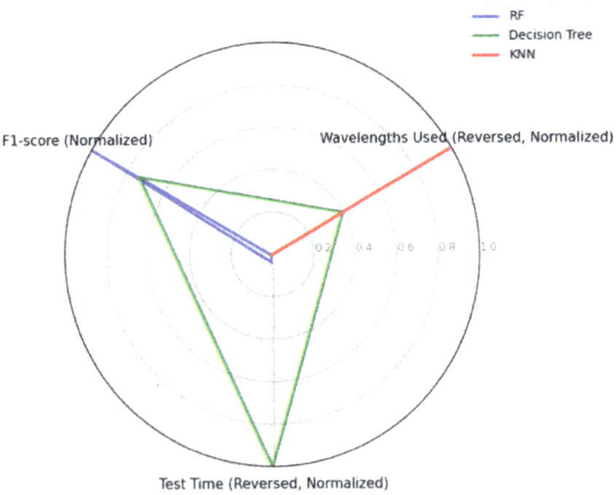

Figure 5. Normalized performance comparison of classification models with F1 score, reversed test time, and reversed wavelengths used.

Permutation tests were conducted on these three models selected above to validate the statistical significance in their HLB determination. We randomly shuffled the true labels of the test set 10,000 times and evaluated them using the same trained models. All three models were significant ($p < 0.01$), meaning that RF, decision tree, and KNN models all performed HLB detection significantly better than random performance.

3. Discussion

The RF, decision tree, and KNN models indicated that red-edge to infrared wavelengths at 727 to 960 nm significantly contributed to determine leaves' HLB symptoms from symptomatic leaves (Table 4). The 727 nm wavelength lies on the red-edge of visual light, which is typically closely related to the chlorophyll a content in plant leaves and their photosynthetic activities [25,26]. HLB infection disrupts chlorophyll metabolism, leading the optical characteristics at this wavelength to change and, accordingly, other waves around it also may be modified. Wavelengths at 930–960 nm fall within the near-infrared region and are associated with variations in leaf water content and cellular structure [27]. The results in this present study indicate that wavelengths differentiated in these bands may reflect the changes in either or both characters in HLB-infected leaves. The three models selected in our study thus could be integrated into the HLB-infection system as more cost-effective and efficient sensing technologies.

Meanwhile, there are some limitations in our study as well. We chose the study of Deng et al. [14] for comparison because their equipment, imaging conditions, and feature selection methods are similar to ours. This alignment can provide a more meaningful and fair comparison to highlight the reliability and applicability of our approach. However, accurate PCR tests should have been performed on the collected leaf samples to provide a more objective corroboration of the results of the expert inspection. Although our models successfully achieved almost perfect distinction between symptomatic and asymptomatic leaves with the integration of the two major components, in some cases, minor components often make important contributions to category-distinction. Ignorance of such minor components sometimes reduces the performance of models [28]. The existence of a partial overlap in the scatter plot of PC1 and PC2 (Figure 3a) suggests that leaf refraction in some HLB-infected leaves could not be differentiated from non-infected leaves based solely on the wavelengths explained by PC1. This insufficient separation in the HLB-infected leaves may be apparent in the early stages of infection, where few wavelengths lead noticeable changes in this stage. Such incomplete differentiation, if any, would result in the failing in the distinction of HLB infection by these three models. Therefore, the overlap might be attributed to how long the leaf has been infected by HLB: the shorter the time since the HLB infection, the less the wavelength traits, making early differentiation more challenging. Minor components other than PC1 and PC2 may contribute to distinguishing early infected leaves, which deserves further investigation. In addition, a mentioned nonlinear dimensionality reduction technique, t-SNE, has shown better results than PCA in feature extraction of some hyperspectral data [29,30]. We also plan to try to use this method to solve the overlapping problem mentioned above.

Our expert system is trained with images taken from multiple leaves and applied to the real-world recognition. Thus, multiple leaves can be determined into a presumed category at once by taking similar images. Compared to models that require taking pictures of one leaf at a time, the detection efficiency can be improved several times more precisely. Nonetheless, it is still inevitable to reach an incorrect determination or perform the determination less effectively. Recently, eye-tracking studies reveal experts assess HLB by scanning broader areas, suggesting the use of holistic cues not captured by current deep learning models [31]. Given the successful application of attention mechanisms in deep learning models in recent years [32], we propose that integrating expert diagnostic strategies can enhance model performance. By learning from the judgment experience of seasoned experts, deep learning models can establish key attention areas on the whole tree, improving learning efficiency and diagnostic accuracy. This approach has the potential to simplify the HLB diagnosis process and contribute to more effective disease management strategies, which is our future work.

4. Materials and Methods

Figure 6 illustrates our data processing flow, where foliar images were obtained with a hyperspectral camera (Specim IQ, SPECIM, Oulu, Finland) of 204 wavelengths partitioned at an interval of about 7 nm at 400–1000 nm. Based on expert visual diagnosis, hyperspectral pixel data were divided into two categories: any symptoms on leaves (symptomatic) or no symptom (asymptomatic). The pixel data were randomly split into training (80%) and test (20%) sets by label proportions. PCA was applied on every individual wavelength to extract specific wavelengths that were characteristic in foliar symptomatic appearances. PCA was applied to the training set only to avoid overfitting due to data leakage [33]. Then, an incremental feature selection method was conducted to explore the possible combinations of the PCA-determined key wavelengths, and the optimal wavelength combinations were determined on each of the seven machine learning models based on the F1-score metric. These models included five nonlinear models: RF, decision tree, SVM, KNN, and gradient boosting, and two linear models: logistic regression and LDA. Finally, nearest neighbor matching and permutation test methods were used to compare the HLB characteristic wavelengths extracted by the expert visual inspection and PCR-based detection to validate the reliability of our expert system.

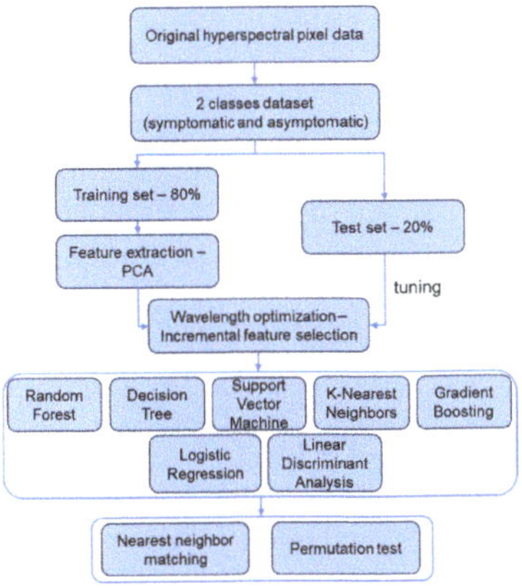

Figure 6. The flowchart of hyperspectral data analysis and modeling in this study.

4.1. Hyperspectral Image Data Collection

All images in this study were obtained in a citrus orchard, approximately 490 m^2 located in Kuet Chang Sub-district, Mae Taeng District, Chiang Mai, Thailand (Figure 7). In this orchard, 7-year-old trees of the local tangerine cultivar, Sai Num Phung (*Citrus reticulata*), were cultivated, spaced 4 m from each other on a row and 3 m between rows.

Figure 7. The orchard in Thailand for this study (19°14′02″ N 98°53′14″ E, 611 m AMSL). This image is available from Google Earth. The Thai word in the figure is the local place name.

Hyperspectral images were collected at 10:30 to 11:30 am on a sunny morning on 7 August 2024, with the hyperspectral camera fixed on a tripod. The distance between the front lens and imaged leaves was 40 cm and a neutral density filter (Ni060-40.5, MIDOPT, Palatine, IL, USA) was attached on the top of the objective lens to reduce light intensity. This filter reduced light intensity which would adversely affect image captures. A white reference panel (IQ White Reference, SPECIM, Oulu, Finland) was set beside plants for reflectance calibration every time the plants were imaged (Figure 8). Table 6 shows the devices and their specifications in this study. A total of 22 branch images were collected from 10 trees in this study.

Table 6. Specification of the hyperspectral camera used in this study.

Device	Specification	Value
Specim IQ	Resolution	512 × 512 pix
	Wavelength range (204)	397–1004 nm
	Dimension	207 × 91 × 74 mm
	Pixel size	17.58 μm × 17.58 μm

Table 6. *Cont.*

Device	Specification	Value
Calibration whiteboard	Reflectivity Size	100% 10 × 10 cm
Neutral density filter	Average Transmission	25%

(a) (b)

Figure 8. RGB images of branches in symptomatic (**a**) and asymptomatic (**b**) groups created from the composite of hyperspectral images.

4.2. *Hyperspectral Data Preprocessing*

A python-based preprocessing software developed by us was utilized to eliminate background noises in the images and extract area needed for the analyses using HSV (hue, saturation, value) thresholds and manual selection (Figure 9). This software extracted spectral data of 72 leaves in the 22 branch images. Expert visual assessment identified HLB symptoms in four out of the ten trees, resulting in 12 symptomatic and 60 asymptomatic leaves. All pixels from symptomatic leaves were labeled as "symptomatic" and those from asymptomatic leaves as "asymptomatic". We eventually extracted 63,056 "symptomatic" pixels and 294,886 "asymptomatic" pixels. Considering the sample imbalance, we used stratified sampling in dividing the dataset, 80% from each of "symptomatic" and "asymptomatic" as the training set and the remaining 20% as the test set.

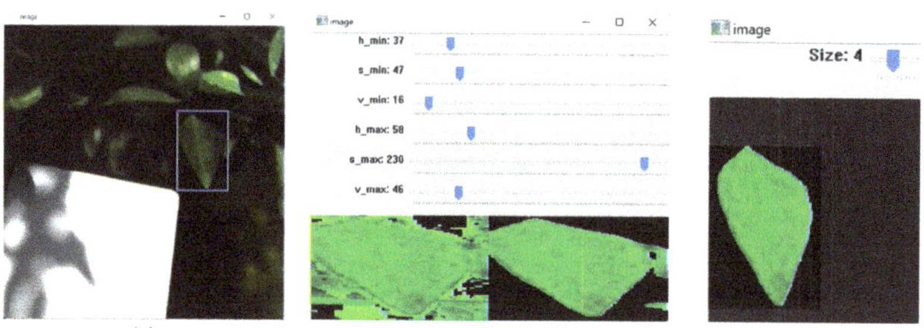

(a) (b) (c)

Figure 9. Schematic of our preprocessing software. First, the region of interest is manually selected (blue region) (**a**). Background noise is then largely removed by HSV thresholding (**b**), with manual adjustment for finer noise removal (**c**). The blue boxes can be used to control the threshold in (**b**) and the size of the pixels when manually removing in (**c**). Finally, hyperspectral data of the selected region are extracted and saved.

4.3. Hyperspectral Data Analysis and Modeling

4.3.1. Feature Extraction—PCA

PCA is a statistical technique used to simplify complex datasets while retaining as much important information as possible, widely used in fields like image processing, genetics, and finance to reduce data complexity and uncover meaningful patterns [34]. In this study, it was used to focus on wavelengths that could have differentiated optical characteristics of leaf reflectance due to the HLB symptoms. The PCs were created and calculated, and the proportion of variance explained by each component was used as a criterion for selecting key features. Often, just a few of these new components capture most of the information in the dataset. By focusing on these few components, the dataset can be simplified while preserving its essence. Before performing PCA, all spectral data were normalized to ensure uniform scaling across wavelengths.

4.3.2. Wavelength Optimization—Incremental Feature Selection

Since 16 spectral are the maximum wavelengths available for multispectral cameras, the extraction of wavelengths in the hyperspectral data was limited to or fewer than 16 wavelengths. However, 16 wavelengths rise to $2^{16} - 1 = 65{,}535$ combinations, except for the null combination, which impose heavy computational burden. In this study, this problem was avoided by adopting an incremental feature selection approach to generate targeted wavelength combinations (Table 7). This approach aligns with the data-driven insights provided by PCA, focusing on the most informative wavelengths from PCs rather than all available features [35]. Specifically, this method systematically explores combinations by gradually increasing the number of wavelengths from each component, ensuring a balanced representation of features while maintaining a logical structure.

Table 7. An example of an incremental feature selection approach focusing on the first two components, the primary and secondary principal components, PC1 and PC2, respectively.

PC1	PC2	Wavelength Selection	Counts
A_1	B_1	$A_1B_1, A_1B_1B_2, \ldots, A_1B_1B_2B_3\ldots B_n$	n
A_2	B_2	$A_1A_2B_1, A_1A_2B_1B_2, \ldots, A_1A_2B_1B_2B_3\ldots B_n$	n
A_3	B_3	$A_1A_2A_3B_1, A_1A_2A_3B_1B_2, \ldots, A_1A_2A_3B_1B_2B_3\ldots B_n$	n
……		……	n
A_n	B_n	$A_1A_2A_3\ldots A_nB_1, A_1A_2A_3\ldots A_nB_1B_2, \ldots, A_1A_2A_3\ldots A_nB_1B_2B_3\ldots B_n$	n

4.3.3. Machine Learning Models

Seven machine learning models were assessed for how precisely they distinguished images in two groups, symptomatic or asymptomatic. Models examined in this study were five nonlinear models (RF [36], decision tree [37], KNN [38], gradient boosting [39], and SVM [40]) and two linear models (logistic regression [41] and LDA [42]). These methods are well-established in the current machine learning and were chosen due to their classification capabilities [14].

4.4. Evaluation of Models for Leaf Image Separation

4.4.1. F1 Score

F1 score was used to evaluate the precision of machine learning models in separating image data into the groups. This is a powerful tool for imbalanced datasets, as it adopts the harmonic mean of precision and recall, and effectively regulates both false positives and false negatives to provide a comprehensive performance measure [43]. The F1 score is calculated through the following three equations:

$$\text{Precision } (\%) = \frac{\text{TP}}{\text{TP} + \text{FP}} \times 100 \tag{1}$$

$$\text{Recall } (\%) = \frac{\text{TP}}{\text{TP} + \text{FN}} \times 100 \tag{2}$$

$$\text{F1} - \text{score } (\%) = \frac{2 \times \text{Precision} \times \text{Recall}}{\text{Precision} + \text{Recall}} \times 100 \tag{3}$$

where TPs (True Positives) means correctly identified positive cases; TNs (True Negatives) means correctly identified negative cases; FPs (False Positives) means negative cases incorrectly identified as positive; FNs (False Negatives) means positive cases incorrectly identified as negative [44].

4.4.2. Nearest Neighbor Matching

Nearest neighbor matching is a statistical technique commonly used in observational studies to pair each unit in a treatment group with the closest unit in a control group based on a predefined set of covariates or characteristics [45]. Given two wavelength sets, say A = $\{a_1, a_2, \ldots a_n\}$ and B = $\{b_1, b_2, \ldots, b_m\}$, the matching calculates the mean of the differences between each wavelength in A and its nearest wavelength in B as a measure of similarity:

$$\text{Distance} = \frac{1}{n}\sum_{i=1}^{n}\left(\min\left|a_i - b_j\right|\right) \text{ for each } a_i \in \text{A}, b_j \in \text{B} \tag{4}$$

To validate the effectiveness of our expert's visual inspection, we used nearest neighbor matching to compare the HLB characteristic wavelengths extracted by our models with those reported in a PCR-based study. That study utilized hyperspectral imaging based on PCR testing to analyze the optimal wavelength combinations, and 13 HLB characteristic wavelengths (544, 718, 753, 760, 764, 930, 938, 943, 951, 969, 985, 998, and 999 nm) achieved the best performance of an accuracy of 96% [14]. Using nearest neighbor matching, we calculated the distance errors between the wavelengths extracted by each model and those from the PCR-based study.

4.4.3. Permutation Test

The permutation test is a robust, non-parametric statistical method which assesses whether the performance of a machine learning model is significantly better than random performance [46]. The procedure involves shuffling target labels to break the relationship between the input feature and the target variable, followed by re-evaluating the model on the permuted dataset multiple times (at least 1000 times is required for reliability) [46]. In this study, we randomly shuffled the true labels of the test set and evaluated the shuffled labels using the same trained models. This process was repeated 10,000 times. The frequencies of F1 scores greater than or equal to the scores achieved with the true labels were counted and the model's statistical significance in the HLB determination was examined at $p < 0.01$ [46].

5. Conclusions

The following conclusions can be drawn from the study:

- Our RF, decision tree, and KNN models are as reliable as PCR in identifying HLB.
- Nonlinear models outperform linear models for HLB spectral data.
- Using PCA for nonlinear models is effective for HLB feature extraction.

- Decision tree model provides high accuracy with faster prediction, suitable for real-time applications.
- KNN model shows promising potential for multispectral imaging applications.
- The red-edge and near-infrared regions may be critical for HLB detection.

Future work will focus on applying these findings to multispectral cameras and analyzing expert visual patterns to develop deep learning models for whole-tree HLB diagnosis.

Author Contributions: Conceptualization, R.D., K.I. and T.H.; methodology, R.D. and K.I.; software, R.D.; validation, R.D., K.I. and A.P.; formal analysis, R.D., A.S. and K.I.; investigation, R.D., K.I., A.P., K.S., S.M. and T.H.; resources, R.D., K.S., S.M. and T.H.; data curation, R.D. and A.S.; writing—original draft preparation, R.D.; writing—review and editing, K.I. and T.H.; visualization, R.D. and A.S.; supervision, A.S. and K.I.; project administration, A.P. and T.H.; funding acquisition, A.S. All authors have read and agreed to the published version of the manuscript.

Funding: This research was funded by JSPS KAKENHI in Japan, Grant Number JP22K12075.

Data Availability Statement: The data presented in this study are available on request from the corresponding author. Software: PyCharm Community Edition 2024.2.1 for hyperspectral data analysis and machine learning modeling; Anaconda3 for managing library files.

Acknowledgments: We would like to express our gratitude to the Highland Research and Development Institute (Thailand) for the guidance and field applied for our data collection.

Conflicts of Interest: The authors declare no conflicts of interest.

Abbreviations

CNN (convolutional neural network); HLB (Huanglongbing); KNN (k-nearest neighbors); LDA (linear discriminant analysis); PC (principal component); PCA (principal component analysis); PCR (polymerase chain reaction); RF (random forest); SVM (support vector machine).

References

1. Tipu, M.M.H.; Masud, M.M.; Jahan, R.; Baroi, A.; Hoque, A.K.M.A. Identification of citrus greening based on visual symptoms: A grower's diagnostic toolkit. *Heliyon* **2021**, *7*, e08387. [CrossRef] [PubMed]
2. Deng, X.; Lan, Y.; Hong, T.; Chen, J. Citrus greening detection using visible spectrum imaging and C-SVC. *Compt. Electron. Agric.* **2016**, *130*, 177–183. [CrossRef]
3. Cap, H.Q.; Suwa, K.; Fujita, E.; Kagiwada, S.; Uga, H.; Iyatomi, H. A deep learning approach for on-site plant leaf detection. In Proceedings of the 2018 IEEE 14th International Colloquium on Signal Processing & Its Applications, Penang, Malaysia, 9–10 March 2018; pp. 18–122.
4. Yang, D.; Wang, F.; Hu, Y.; Lan, Y.; Deng, X. Citrus huanglongbing detection based on multi-modal feature fusion learning. *Front. Plant Sci.* **2021**, *12*, 809506. [CrossRef] [PubMed]
5. Gottwald, T.R. Current epidemiological understanding of citrus Huanglongbing. *Annu. Rev. Phytopathol.* **2010**, *48*, 119–139. [CrossRef]
6. Kalim, H.; Chug, A.; Singh, A.P. Citrus leaf disease detection using hybrid CNN-RF model. In Proceedings of the 4th International Conference on Artificial Intelligence and Speech Technology (AIST), Delhi, India, 9–10 December 2022; pp. 1–4.
7. Elaraby, A.; Hamdy, W.; Alanazi, S. Classification of citrus diseases using optimization deep learning approach. *Comput. Intell. Neurosci.* **2022**, *2022*, 9153207. [CrossRef] [PubMed]
8. Gómez-Flores, W.; Garza-Saldaña, J.J.; Varela-Fuentes, S.E. A huanglongbing detection method for orange trees based on deep neural networks and transfer learning. *IEEE Access* **2022**, *10*, 116686–116696. [CrossRef]
9. Liu, X.; Min, W.; Mei, S.; Wang, L.; Jiang, S. Plant disease recognition: A large-scale benchmark dataset and a visual region and loss reweighting approach. *IEEE Transact. Image Process.* **2021**, *30*, 2003–2015. [CrossRef]
10. Xia, F.; Xie, X.; Wang, Z.; Jin, S.; Yan, K.; Ji, Z. A novel computational framework for precision diagnosis and subtype discovery of plant with lesion. *Front. Plant Sci.* **2022**, *12*, 789630. [CrossRef] [PubMed]
11. Li, L.; Zhang, Q.; Huang, D. A Review of Imaging Techniques for Plant Phenotyping. *Sensors* **2014**, *14*, 20078–20111. [CrossRef] [PubMed]

12. Wang, K.; Dongmei, G.; Zhang, Y.; Deng, L.; Xie, R.; Lv, Q.; Yi, S.; Zheng, Y.; Ma, Y.; He, S. Detection of Huanglongbing (citrus greening) based on hyperspectral image analysis and PCR. *Front. Agric. Sci. Eng.* **2019**, *6*, 172–180. [CrossRef]
13. Lan, Y.; Huang, Z.; Deng, X.; Zhu, Z.; Huang, H.; Zheng, Z.; Lian, B.; Zeng, G.; Tong, T. Comparison of machine learning methods for citrus greening detection on UAV multispectral images. *Compt. Electron. Agric.* **2020**, *171*, 105234. [CrossRef]
14. Deng, X.; Huang, Z.; Zheng, Z.; Lan, Y.; Dai, F. Field detection and classification of citrus Huanglongbing based on hyperspectral reflectance. *Compt. Electron. Agric.* **2019**, *167*, 105006. [CrossRef]
15. Podlesnykh, I.; Kovalev, M.; Platonov, P. Towards the future of ubiquitous hyperspectral imaging: Innovations in sensor configurations and cost reduction for widespread applicability. *Technologies* **2024**, *12*, 221. [CrossRef]
16. Sankaran, S.; Mishra, A.; Maja, J.M.; Ehsani, R. Visible-near infrared spectroscopy for detection of Huanglongbing (HLB) Using a VIS-NIR Spectroscopy Technique. *Comput. Electron. Agric.* **2011**, *77*, 127–134. [CrossRef]
17. Sankaran, S.; Ehsani, R. Visible-near infrared spectroscopy based citrus greening detection: Evaluation of spectral feature extraction techniques. *Crop Prot.* **2011**, *30*, 1508–1513. [CrossRef]
18. Sankaran, S.; Maja, J.; Buchanon, S.; Ehsani, R. Huanglongbing (Citrus Greening) Detection Using Visible, Near Infrared and Thermal Imaging Techniques. *Sensors* **2013**, *13*, 2117–2130. [CrossRef] [PubMed]
19. Li, X.; Lee, W.S.; Li, M.; Ehsani, R.; Mishra, A.R.; Yang, C.; Mangan, R.L. Spectral difference analysis and airborne imaging classification for citrus greening infected trees. Comput. *Electron. Agric.* **2012**, *83*, 32–46. [CrossRef]
20. Kumar, A.; Lee, W.S.; Ehsani, R.J.; Albrigo, L.G.; Yang, C.; Mangane, R.L. Citrus greening disease detection using aerial hyperspectral and multispectral imaging techniques. *J. Appl. Remote Sens.* **2012**, *6*, 063542.
21. Weng, H.; Lu, J.; Cen, H.; He, M.; Zeng, Y.; Hua, S.; Li, H.; Meng, Y.; Fang, H.; He, Y. Hyperspectral reflectance imaging combined with carbohydrate metabolism analysis for diagnosis of citrus Huanglongbing in different seasons and cultivars. *Sens. Actuators B Chem.* **2018**, *275*, 50–60. [CrossRef]
22. Deng, X.; Zhu, Z.; Yang, J.; Zheng, Z.; Huang, Z.; Yin, X.; Wei, S.; Lan, Y. Detection of Citrus Huanglongbing Based on Multi-Input Neural Network Model of UAV HRS. *Remote Sens.* **2020**, *12*, 2678. [CrossRef]
23. Menezes, J.; Dharmalingam, R.; Shivakumara, P. HLB Disease Detection in Omani Lime Trees Using Hyperspectral Imaging Based Techniques. In *Recent Trends in Image Processing and Pattern Recognition*; Springer Nature: Cham, Switzerland, 2024; pp. 67–81.
24. Yan, K.; Song, X.; Yang, J.; Xiao, J.; Xu, X.; Guo, J.; Zhu, H.; Lan, Y.; Zhang, Y. Citrus huanglongbing detection: A hyperspectral data-driven model integrating feature band selection with machine learning algorithms. *Crop Prot.* **2025**, *188*, 107008. [CrossRef]
25. Carter, G.A.; Knapp, A.K. Leaf optical properties in higher plants: Linking spectral characteristics to stress and chlorophyll concentration. *Am. J. Bot.* **2001**, *88*, 677–684. [CrossRef] [PubMed]
26. Ustin, S.L.; Jacquemoud, S. How the optical properties of leaves modify the absorption and scattering of energy and enhance leaf functionality. In *Remote Sensing of Plant Biodiversity*; Cavender-Bares, J., Gamon, J.A., Townsend, P.A., Eds.; Springer International Publishing: Cham, Switzerland, 2020; pp. 349–384.
27. Peñuelas, J.; Inoue, Y. Reflectance indices indicative of changes in water and pigment contents of peanut and wheat leaves. *Photosynthetica* **1999**, *36*, 355–360. [CrossRef]
28. Zheng, J.; Rakovski, C. On the Application of Principal Component Analysis to Classification Problems. *Data Sci. J.* **2021**, *20*, 26. [CrossRef]
29. Silva, R.; Melo-Pinto, P. t-SNE: A study on reducing the dimensionality of hyperspectral data for the regression problem of estimating oenological parameters. *Artif. Intell. Agric.* **2023**, *7*, 58–68. [CrossRef]
30. Kang, Z.; Fan, R.; Zhan, C.; Wu, Y.; Lin, Y.; Li, K.; Qing, R.; Xu, L. The Rapid Non-Destructive Differentiation of Different Varieties of Rice by Fluorescence Hyperspectral Technology Combined with Machine Learning. *Molecules* **2024**, *29*, 682. [CrossRef]
31. Kobayash, O.; Dong, R.; Shiraiwa, A. Eye-tracking data analysis for improving diagnosis performance in citrus greening disease. *IEICE Tech. Rep.* **2024**, *124*, 35–38.
32. Dong, R.; Shiraiwa, A.; Pawasut, A.; Sreechun, K.; Hayashi, T. Diagnosis of Citrus Greening Using Artificial Intelligence: A Faster Region-Based Convolutional Neural Network Approach with Convolution Block Attention Module-Integrated VGGNet and ResNet Models. *Plants* **2024**, *13*, 1631. [CrossRef]
33. Rosenblatt, M.; Tejavibulya, L.; Jiang, R.; Noble, S.; Scheinost, D. Data leakage inflates prediction performance in connectome-based machine learning models. *Nat. Comm.* **2024**, *15*, 1829. [CrossRef]
34. Jolliffe, I. Principal component analysis. In *International Encyclopedia of Statistical Science*; Lovric, M., Ed.; Springer: Berlin/Heidelberg, Germany, 2011; pp. 1094–1096.
35. Liu, H.; Setiono, R. Incremental feature selection. *Appl. Intell.* **1998**, *9*, 217–230. [CrossRef]
36. Breiman, L. Random Forests. *Mach. Learn.* **2001**, *45*, 5–32. [CrossRef]
37. Breiman, L.; Friedman, J.; Olshen, R.A.; Stone, C.J. *Classification and Regression Trees*, 1st ed.; Chapman and Hall/CRC: Boca Raton, FL, USA, 1984. [CrossRef]
38. Cover, T.; Hart, P. Nearest neighbor pattern classification. *IEEE Trans. Inf. Theory* **1967**, *13*, 21–27. [CrossRef]
39. Jerome, H. Friedman. Greedy function approximation: A gradient boosting machine. *Ann. Stat.* **2001**, *29*, 1189–1232. [CrossRef]

40. Cortes, C.; Vapnik, V. Support-vector networks. *Mach. Learn.* **1995**, *20*, 273–297. [CrossRef]
41. Kleinbaum, D.G. *Logistic Regression: A Self-Learning Text*; Springer: Berlin/Heidelberg, Germany, 1994.
42. Liu, Z.P. Linear Discriminant Analysis. In *Encyclopedia of Systems Biology*; Dubitzky, W., Wolkenhauer, O., Cho, K.H., Yokota, H., Eds.; Springer: New York, NY, USA, 2013; pp. 1132–1133. [CrossRef]
43. Powers, D. Evaluation: From precision, recall and F-measure to ROC, informedness, markedness & correlation. *J. Mach. Learn. Technol.* **2011**, *2*, 37–63.
44. Fawcett, T. An introduction to ROC analysis. *Pattern Recogn. Lett.* **2006**, *27*, 861–874. [CrossRef]
45. Clark, P.J.; Evans, F.C. Distance to Nearest Neighbor as a Measure of Spatial Relationships in Populations. *Ecology* **1954**, *35*, 445–453. [CrossRef]
46. Ojala, M.; Garriga, G.C. Permutation tests for studying classifier performance. In Proceedings of the 2009 Ninth IEEE International Conference on Data Mining, Miami, FL, USA, 6–9 December 2009; pp. 908–913.

plants

MDPI

Article

Effects of Different Postharvest Treatments on Fruit Quality, Sucrose Metabolism, and Antioxidant Capacity of 'Newhall' Navel Oranges During Storage

Bo Xiong *,†, Linlv Han †, Yinghong Ou, Wenjia Wu, Jialu Wang, Junfei Yao, Yisong Li, Siyu Chen, Taimei Deng, Hongzhen Chen, Chenming Wang, Qingqing Ma, Yujing Fan, Yixuan Li and Zhihui Wang

College of Horticulture, Sichuan Agricultural University, Chengdu 611130, China; 18989192178@163.com (L.H.); 18284187563@163.com (Y.O.); 202302925@stu.sicau.edu.cn (W.W.); wangjial0321@163.com (J.W.); 19845921559@163.com (J.Y.); liyisong0629@163.com (Y.L.); csy18190735953@163.com (S.C.); dtm123692023@163.com (T.D.); 15775554829@163.com (H.C.); wangchenming0104@163.com (C.W.); 122005185923@163.com (Q.M.); fan0712150024@163.com (Y.F.); liyixuan112025@163.com (Y.L.); wangzhihui318@sicau.edu.cn (Z.W.)

* Correspondence: xiongbo1221@sicau.edu.cn

† These authors contributed equally to this work.

Abstract: During the post-harvest storage of citrus, the flavor of fruit gradually fade. In this study, we investigated the effects of different treatments—control check (CK), heat treatment (HT), salicylic acid treatment (SA), and 1-methylcyclopropene treatment (1-MCP)—on the quality of 'Newhall' navel oranges, particularly focusing on sucrose metabolism and related gene expression during storage. Combining the experimental data, we compared the three different treatments with CK. The results showed that the oranges subjected to HT had a significantly higher flavonoid content (26.40 μg) and total phenolic content (19.42 μg) than those used for the CK at the late storage stage, and was also the most effective in slowing the decline in sugar, titratable acid and other indexes, followed by SA, with 1-MCP performing poorly. Quantitative results showed that the three treatments contributed to the increase in sucrose content by elevating the expression of the *SPS1* and *SPS2* genes involved in sucrose synthesis compared to the CK. However, no clear pattern was observed between the genes involved in sucrose catabolism (*SUS1* and *SUS3*) and sucrose content. These results provided a rationale for the selection of post-harvest treatments to extend the storage life and maintain the quality of 'Newhall' navel oranges, with broader implications for the citrus industry.

Keywords: 'Newhall' navel orange; postharvest storage; antioxidant properties; sugar metabolism; citrus preservation

1. Introduction

The 'Newhall' navel orange, a citrus plant belonging to the *Rutaceae* family, originates from the United States and is currently cultivated in regions such as Sichuan, Chongqing, Jiangxi, and other places in China. Despite its notable resistance to storage and transportation, the fruit's quality still deteriorates to varying degrees during post-harvest handling and distribution, thereby impacting its taste [1]. Sugar content, a critical determinant of fruit flavor, frequently becomes the focal point in research concerning postharvest storage.

Soluble sugars in citrus fruits mainly consist of sucrose, fructose and glucose, with their contents and proportions varying according to the citrus variety [2]. Sucrose is one of the most abundant soluble sugars in citrus, and serves as the main form of sugar transport in different organs [3], playing a crucial role in citrus sucrose metabolism. Sucrose phosphate synthase (SPS) and sucrose synthase (SS) are key enzymes regulating sucrose fluxes in sink tissues for sucrose accumulation in citrus [4]. A genome-wide characterization of citrus sucrose synthase revealed that SS catabolic activity was significantly correlated with sucrose content, suggesting that SS influences sucrose levels and thus sugar accumulation in the fruit [5]. SPS catalyzes the conversion of fructose-6-phosphate and UDP-Glc to sucrose-6-phosphate, which is subsequently hydrolyzed to sucrose by sucrose–phosphatase. Consequently, SPS synchronizes sucrose metabolism and accumulation [6].

In an ever-changing global marketplace with increasing competition, the citrus industry is continuously seeking new varieties of fruit to extend the marketing period, diversify and differentiate the product and, importantly, provide the fruit with a higher nutritional value and additional health-related properties [7]. The application of GRAS salts, phytopharmaceuticals and yeasts represents a promising integrated approach for managing postharvest diseases in citrus, addressing consumer concerns about the toxicity of residuals in edible fruits [8]. Mechanistic, observational and intervention studies have demonstrated that the bioactive compounds present in fruits, vegetables and whole grains prevent oxidative damage to cells by detoxifying free radicals [9], thereby minimizing the incidence of many diseases associated with metabolic syndrome, including neurodegenerative diseases, cardiovascular disease (CVD), type 2 diabetes and cancer [10]. The antioxidant properties of foods have received increasing attention as an important parameter for evaluating fruit quality. Flavonoids and total phenols play a direct role in scavenging reactive oxygen species, inhibiting lipid oxidation in vitro, enhancing the activity of antioxidant enzymes in the body and reducing peroxide formation [11]. It is important to note that the antioxidant activity of plant extracts can be determined using various methods, such as ABTS [2,2′-Azinobis-(3-ethyl-benzthiazoline-6-sulphonate)], DPPH (2,2-diphenyl-1-picrylhydrazyl) and FRAP (Ferric Reducing Antioxidant Power) [12].

Common postharvest fruit storage methods can be categorized into chemical, physical and biological preservation. There are currently three chemical fungicides commonly used for postharvest citrus preservation: benzimidazole, imidazole and bisguanidine salts [13]. Peach fruits are susceptible to cold damage during low-temperature storage. However, treatments with hot air or hot water have been shown to preserve fruit quality, reduce reactive oxygen species (ROS) and enhance antioxidant activity [14]. *Bacillus pelliculatus* HY19 can inhibit Penicillium infestations in citrus peel by releasing volatile organic compounds, thereby reducing the incidence of Penicillium and Green Mold, and decreasing the decay rate of citrus fruits [15]. For environmental and health reasons, scientists are now focusing on the development of new, environmentally friendly postharvest treatments. Yellow peel leaves show promise as a natural preservative for postharvest citrus treatments [16].

Previous studies have demonstrated that treatment with physical treatments (e.g., low temperature storage, heat treatment, etc.) or non-toxic chemical reagents (e.g., salicylic acid, 1-Methylcyclopropene (1-MCP)) can effectively preserve the freshness of citrus fruits. The decay rate and water loss rate of navel oranges stored at 5 °C were significantly lower than those stored at 26 °C [17]. Citrus fruits treated with salicylic acid (SA) maintained a better fruit firmness, total soluble solids and vitamin C (Vc) content during storage [18]. The pre-harvest application of SA alone or in combination with ascorbic acid (ASA) was effective in preventing pomegranate frost damage and maintaining pomegranate quality. It was determined that a combination of SA 250 ppm and ASA 250 ppm was the most effective for protecting plants from environmental stresses such as heavy metals, high temperatures and salinity [19]. Date palm fruits exhibited higher organoleptic attributes after 3 min and 5 min of treatment, which improved ripening and preserved nutritional quality postharvest [20]. After storage, 1-MCP-treated citrus fruits showed significantly higher levels of total phenol, antioxidant activity, flavonoids, proteins and free amino acids compared to untreated citrus fruits [21].

Research on citrus antioxidants has garnered significant attention. However, limited studies have focused on 'Newhall' navel orange varieties. Additionally, there is a lack of in-depth analyses on the effects of different treatments on sucrose metabolism during storage, particularly from the perspectives of enzyme activities related to sucrose metabolism and gene expression. In this study, we investigated the impact of various postharvest treatments on citrus fruit quality by evaluating antioxidant properties such as flavonoid content, physiological indexes (fructose, glucose, sucrose), and molecular parameters including the activities of enzymes associated with sucrose synthesis and gene expression. The aim was to identify a treatment that can effectively mitigate the decline in citrus fruit quality postharvest.

2. Materials and Methods

2.1. Plant Materials and Reagent Material

Newhall navel oranges were harvested from a commercial orchard in Leibo County, Sichuan Province, China, ensuring uniformity in size and color and the absence of visible damage. Fruits were transported to the laboratory for analysis at 0, 6, 18 and 30 days postharvest. Upon arrival, pulp samples were immediately snap frozen in liquid nitrogen, homogenized, and stored at −80 °C for subsequent analysis. Each analysis included three replicates, with each replicate consisting of six fruits. Salicylic acid was analytically pure, while methanol, isopropanol and acetonitrile were chromatographically pure. The RNAprep Pure Plant Total RNA Extraction Kit for polysaccharides and polyphenols was purchased from TIANGEN Biochemical Technology Co., Ltd. (Beijing, China), and the M5 Sprint qPCR RT Kit was obtained from JUHEMEI Biotechnology Co., Ltd. (Beijing, China).

2.2. Experimental Treatment

Control check (CK): no treatment. Heat treatment (HT): navel oranges were soaked in hot water at 50 °C for 5 min and then dried [22]. Salicylic acid treatment (SA): a 0.3 g/L salicylic acid solution was prepared, and the fruits were soaked in this solution for 20 min before drying [23]. 1-MCP treatment (1-MCP): a commercial 1-MCP solution at a concentration of 0.2 g/L was prepared, and the fruits were immersed in the solution for 3 min, followed by drying. After the treatments, the first sampling was conducted immediately and designated as Day 0. For each sampling, six to nine fruits were selected. The pulp was separated from the skin, chopped, and stored in an ultra-low-temperature freezer at −80 °C. The remaining fruits were uniformly placed in a refrigerator set at 5 ± 1 °C with a relative humidity of 90–95%. Samples were taken and processed on the 0, 6th, 18th, and 30th storage days.

2.3. Measurement of Flavonoids and Total Phenols

The flavonoid content was determined using the aluminum chloride colorimetric method. Specifically, 75 μL of 95% ethanol, 10 μL of 10% aluminum chloride, 10 μL of 1.0 M potassium acetate and 140 μL of distilled water were added to the extract. After 15 min of incubation, the absorbance was measured at 415 nm. A standard curve was prepared using rutin, and the results were expressed as μg/g FW (fresh weight) [24].

The total phenol content was quantified by the Folin–Ciocalteu phenol method. Specifically, 12.5 μL aliquots of water-soluble extracts were mixed with 250 μL of 2% sodium carbonate solution in a 96-well microtiter plate and incubated for 5 min at room temperature. Subsequently, 12.5 μL of 50% Folin–Ciocalteu phenol reagent was added, and the mixture was allowed to stand at room temperature for 30 min. The absorbance of the reaction mixture was then measured at 650 nm using a plate reader. Calibration was performed using a gallic acid aqueous solution (100–1000 μg/mL) (Figure 1) [25].

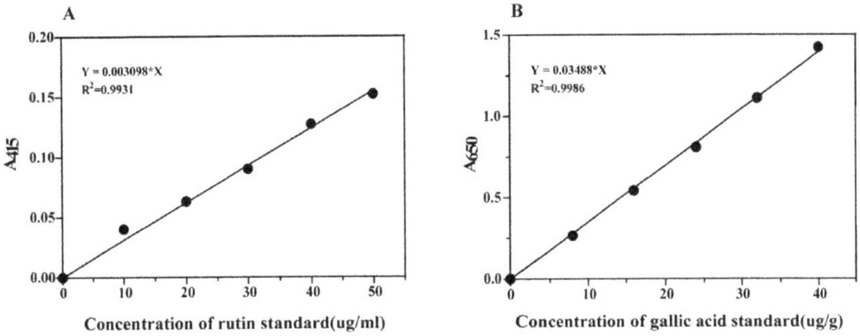

Figure 1. Standard curve range and R^2 for rutin (**A**); standard curve range and R^2 for gallic acid (**B**).

2.4. Measurement of DPPH, ABTS and FRAP Radical Scavenging

The experiments were primarily conducted using total antioxidant capacity assay kits (DPPH, ABTS, FRAP) provided by P Yeast Bioengineering Co., Ltd. (Wuhan, China).

2.5. Measurement of Fruit Weight Loss

Separately, 6 fruits were randomly selected from each treatment group and weighed using an electronic balance. Formula (1) was then used:

$$Weightlessness\ (\%) = \frac{(Fruit\ quality\ before\ storage\ -\ Fruit\ quality\ before\ storage)}{Fruit\ quality\ before\ storage} \times 100\% \quad (1)$$

2.6. Measurement of Total Soluble Solid, Titratable Acidity, and Total Sugar

'Newhall' navel orange fruits were selected at each development stage to determine total soluble solids (TSSs) and titratable acidity (TA). At each developmental stage, more than 18 fruits were used for quality assessment, with three replicates established. Total soluble solids content (%) and titratable acidity (%) were measured using a digital acidometer (model Pocket PAL-BXIACID1, Atago, Tokyo, Japan) according to the manufacturer's instructions.

Total sugars were quantified using the sulfuric acid–anthrone method [26] using Formula (2):

$$Total\ sugar\ content = \frac{(Sucrose\ content\ in\ sdandard\ curve \times volume \times dilution)}{Volume\ of\ sample\ extract\ added \times Sample\ quality \times 10^6} \times 100\% \quad (2)$$

2.7. Determination of Glucose, Fructose, and Sucrose Contents

The contents of glucose, fructose and sucrose were determined according to the method described by Liu et al. [27]. Two grams of pulp were homogenized with 10 mL of double-distilled water (ddH_2O) and incubated at 80 °C for 15 min. The mixture was then centrifuged and filtered. The resulting supernatant was analyzed using an Agilent 1260 high performance liquid chromatography system (Agilent Technologies, Santa Clara, CA, USA) equipped with a refractive index detector and an Innova NH_2 column (4.6 mm × 250 mm, 5 μm) from Aeger Technology (Shanghai, China). The mobile phase consisted of acetonitrile–water (80:20, v/v) at a flow rate of 1 mL/min.

2.8. Determination of Enzyme Activities Related to Sucrose Metabolism

The activities of sucrose metabolism-related enzymes were determined with reference to the method of Zhang et al. [28]. All operations were conducted at 0~4 °C, and the absorbance values were measured immediately after adding the reaction substrate.

2.9. qRT-PCR Analysis

RNA was extracted from the pulp at different times of the year using the RNAprep Pure Total RNA Extraction Kit for Polysaccharide–Polyphenol Plants (TIANGEN). The extracted RNA was reverse transcribed into cDNA using the M5 Sprint qPCR RT kit (JUHEMEI) according to the instructions.

The RT-qPCR gene-specific primers are listed in Table 1, with Actin serving as the endogenous control gene. The cDNA was used as a template for amplification using the 2× M5 HiPer Dual SYBR Green Real-Time PCR Super mix kit (JUHEMEI). The amplification reaction system is detailed in Table 2, and the total volume was 12.5 μL. RT-qPCR was performed on a CFX96 Real-Time PCR Detection System (Bio-Rad, Shanghai, China).

Table 1. Sequence of primers for real-time fluorescence quantification.

Target Gene	Forward Primer (5′-3′)	Reverse Primer (5′-3′)
SPS1	TGTAACAGTGGCAGTGAT	GTGTGAGTGGTAATAGAAGTC
SPS2	GTTACAACACAAGACACAAT	ATCACCTCAGACCACATT
SUS1	GGCCTTTGGCTTGACTGTTG	GGGATCTGCCTTGCACTTCT
SUS3	ATTCCGAGCATCAGAGAGCG	TCGTCGATCAGTACATGCGG

Table 2. Real-time fluorescence quantitative PCR reaction system.

Ingredient	Volume (μL)
cDNA	1.00
PCR Forward Primer (10 μM)	0.50
PCR Reverse Primer (10 μM)	0.50
2× M5SYBR	6.25
dd H_2O	4.25
Total	12.50

Program: pre-denaturation at 95 °C for 2 min, denaturation at 95 °C for 5 s, annealing at 57 °C for 40 s, cycling 39 times. The melt curve method was conducted at 95 °C for 5 s, 65 °C for 1 min, followed by continuous collection at 97 °C, and cooling at 40 °C for 10 s. Each sample was analyzed in triplicate, and the relative gene expression levels were calculated using the $2^{-\Delta\Delta Ct}$ method.

2.10. Data Analysis

GraphPad Prism (version 9) was used for all data analyses, including construction and analysis of variance (ANOVA). Means and standard deviations were calculated, and the final results were presented as the mean ± SD. Post hoc multiple comparisons were conducted using the LSD test. The significance level was indicated by the p-value; $p < 0.05$ denoted a significant difference, and $p < 0.01$ denoted a highly significant difference between the compared groups. Data visualization was performed using Origin (version 2017), with significance levels marked in the figures.

3. Results

3.1. Changes in Flavonoids and Total Phenolic Content of Fruit Pulp During Storage

It was observed that the flavonoid content peaked on the 30th storage day and generally exhibited a decreasing and then increasing trend (Figure 2A). At day 30 postharvest, the flavonoid contents in the heat treatment (HT) and salicylic acid (SA) groups were significantly higher than that in the control check(CK) group ($p < 0.01$), indicating that these treatments effectively inhibited the decline in flavonoid content during the later stages of storage. Additionally, the flavonoid content in the 1-methylcyclopropene (1-MCP) group was significantly higher than that in the other three groups on the 18th storage day ($p < 0.01$), likely due to the superior effect of 1-MCP in inhibiting the decrease in flavonoid content during the middle storage stage.

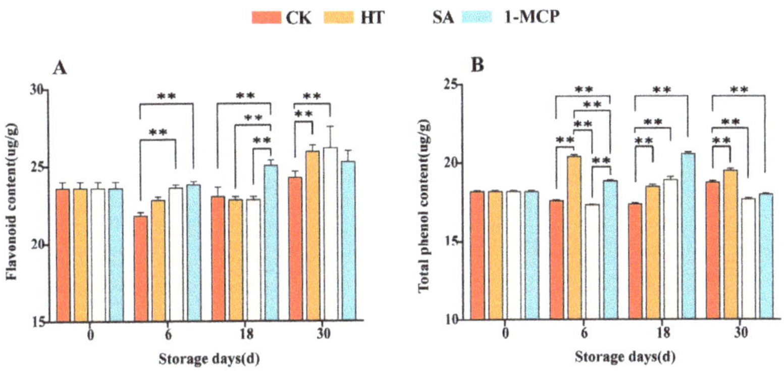

Figure 2. Changes in flavonoid (**A**) and total phenolic contents (**B**) of fruit pulp during storage. "**" means the result of significance analysis is highly significant ($p < 0.01$).

Regarding total phenol content, all three treatments (HT, SA and 1-MCP) were significantly different from the CK group in the later stage ($p < 0.01$) (Figure 2B). This suggested that all three treatments effectively inhibited the decline in total phenol content, with the heat treatment performing best on 6 storage days and 1-MCP on the 18th storage day.

3.2. Changes in DPPH Free Radicals Scavenging, ABTS Free Radicals Scavenging and FRAP Reduction Capacity in Fruit Pulp During Storage

DPPH radical scavenging, ABTS radical scavenging and FRAP total antioxidant capacity exhibited an increasing trend in the early postharvest period, followed by a gradual decrease (Figure 3). This suggested that the antioxidant performance of these samples peaked in the early postharvest period. The initial increase in antioxidant capacity was likely due to the gradual release or activation of antioxidant components, while the subsequent decline might be attributed to the depletion of these components over time. The results showed that on day 30, the ABTS radical scavenging rate of CK was 9.16%, which was significantly ($0.01 < p < 0.05$) different from HT (7.67%) and highly significant ($p < 0.01$) compared to SA and 1-MCP, suggesting that HT delayed the decline in ABTS radical scavenging rate. On day 30, the FRAP radical scavenging rate of SA was 18.05%, which was much higher than that of HT and 1-MCP, and could effectively inhibit the decrease in FRAP radical scavenging capacity.

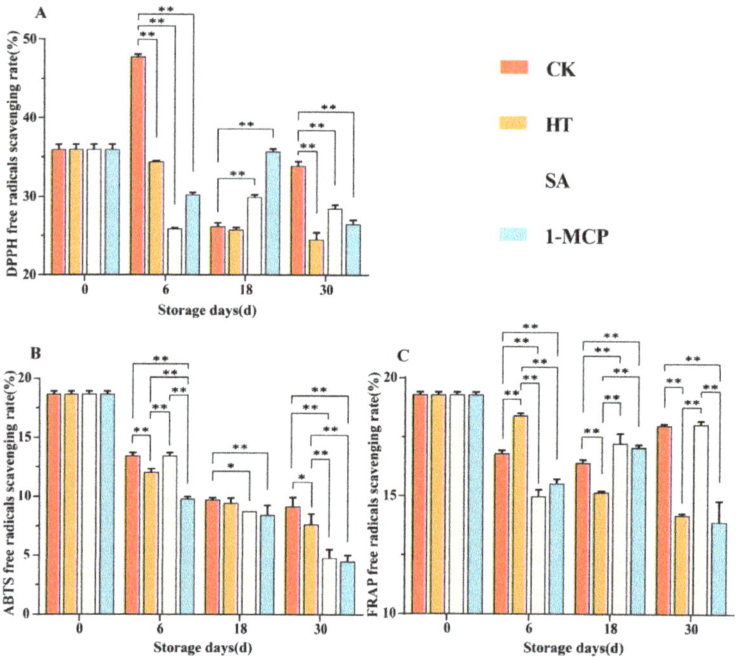

Figure 3. Changes in DPPH free radicals scavenging (**A**), ABTS free radicals scavenging (**B**) and FRAP free radicals scavenging (**C**) in fruit pulp during storage. "**" means the result of significance analysis is highly significant ($p < 0.01$), "*" means the result of significance analysis is significant ($0.01 < p < 0.05$).

3.3. Effect of Different Treatments on the Rate of Fruit Weight Loss During Storage

The rate of weight loss increased with the extension of storage time (Figure 4). On the 30th day, the weight loss rates were 3.48% for the heat treatment (HT) group, 3.10% for the salicylic acid (SA) group and 4.56% for the 1-methylcyclopropene (1-MCP) group, all of which were lower than that for the control (CK) group (5.16%). The rate in the HT group was significant ($0.01 < p < 0.05$) lower than that of the CK group, while that of the SA group was significantly ($p < 0.01$) lower than that of the CK group. There was a non-significant difference between the 1-MCP and CK groups ($p > 0.05$). These findings indicate that the heat treatment (HT) and salicylic acid (SA) treatment were more effective in retarding fruit quality decline during storage, whereas 1-MCP performed less effectively in reducing weight loss.

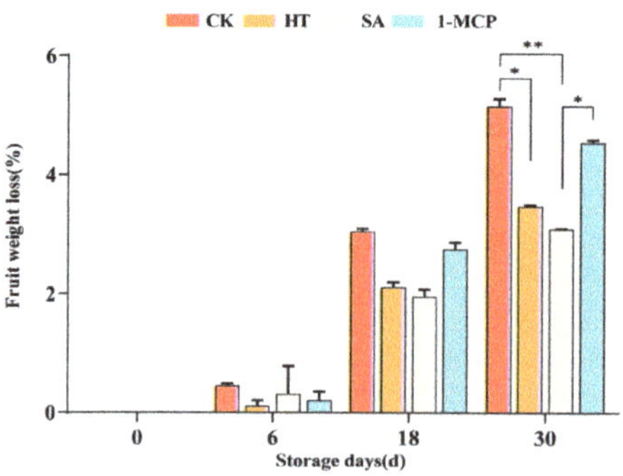

Figure 4. Effect of different treatments on the rate of fruit weight loss during storage. "**" means the result of significance analysis is highly significant ($p < 0.01$), "*" means the result of significance analysis is significant ($0.01 < p < 0.05$).

3.4. Changes in Titratable Acid and Total Soluble Solids Content of Fruit Pulp During Storage

In the significance analysis, no significant differences were observed in the titratable acid contents among the different treatments (Figure 5A). However, our overall observations indicated that the SA treatment better stabilized the levels of titratable acids, with the CK group showing 0.42 g/100 mL of titratable acid at storage day 0, while the SA treatment maintained a level of 0.37 g/100 mL even on the 30th storage day, demonstrating minimal variation between the two groups (Figure 5B). An analysis of total soluble solids content revealed significant differences among all treatment groups. Specifically, the HT group exhibited a significantly higher total soluble solids content compared to the other three groups across all four storage periods ($p < 0.01$), indicating that the HT positively influenced fruit quality by promoting an increase in the total soluble solids content.

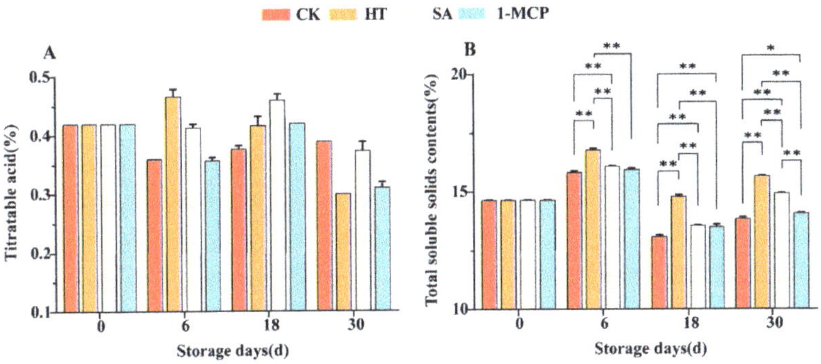

Figure 5. Changes in the titratable acid (**A**) and total soluble solids contents (**B**) of fruit pulp during storage. "**" means the result of significance analysis is highly significant ($p < 0.01$), "*" means the result of significance analysis is significant ($0.01 < p < 0.05$).

3.5. Changes in Total Sugar and Sugar Components During Storage

The total sugar content reached its peak on the sixth storage day and then gradually decreased (Figure 6A). The total sugar content in the heat treatment (HT) group was significantly higher than that in the control (CK) group at all time points except for the sixth storage day, when the difference was not significant ($p > 0.05$). During the later stages of storage, the salicylic acid (SA) treatment significantly inhibited the decrease in total sugar content, indicating that both heat treatment and salicylic acid treatment primarily delayed the decline in total sugar content during the later stages of storage. On the 30th storage day, the total sugar content in the 1-methylcyclopropene (1-MCP) treatment group was significantly lower than that in the CK group ($p < 0.01$). Overall, heat treatment and salicylic acid treatment to some extent delayed the decrease in total sugar content, with heat treatment being the most effective.

The fructose and glucose contents generally showed a trend of first increasing and then decreasing (Figure 6B). On the sixth storage day, the fructose content in the HT group was significantly higher than that in the CK group ($0.01 < p < 0.05$). However, by the 30th day of storage, only the SA group showed a significantly higher fructose content compared to the CK group ($0.01 < p < 0.05$), while the differences between the other treatment groups and the CK group were not significant ($p > 0.05$), with the results for some treatments being even lower than for the CK group, indicating that these treatments were less effective in promoting fructose accumulation. For glucose, the results for the HT group were extremely significantly higher than those for the CK group on the 6th and 18th storage days ($p < 0.01$), but the difference was not significant on the 30th day ($p > 0.05$) (Figure 6C). These results indicate that these three treatments mainly promoted glucose accumulation during the middle stage of storage but were less effective in the later stages.

The trend of the sucrose content in the CK group was similar to that of the total sugar content. On the 18th storage day, the sucrose content in the CK group was significantly lower than that in the SA group ($0.01 < p < 0.05$), and on the 30th day, it was significantly lower than that in the HT group ($p < 0.01$). In contrast, the sucrose content in the 1-MCP group did not differ significantly from that in the CK group ($p > 0.05$), indicating that heat treatment and salicylic acid treatment promoted sucrose accumulation during storage (Figure 6D).

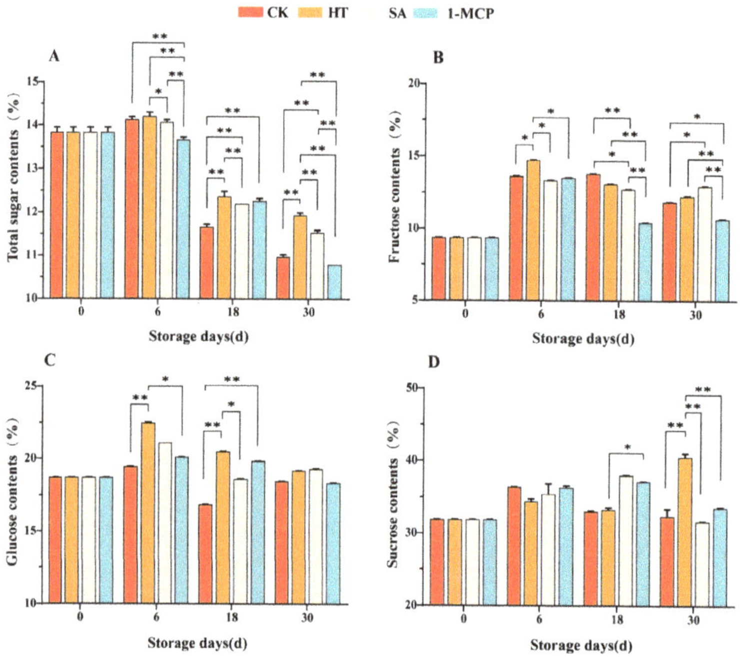

Figure 6. Changes in total sugar (**A**), fructose (**B**), glucose (**C**) and sucrose (**D**) contents of fruit pulp during storage. "**" means the result of significance analysis is highly significant ($p < 0.01$), "*" means the result of significance analysis is significant ($0.01 < p < 0.05$).

3.6. Changes in the Activities of Enzymes Related to Sucrose Metabolism

During the postharvest storage period, except for the SPS activity in the control (CK) group, which decreased from 25.73 U/g to 22.01 U/g, the activities of sucrose phosphate synthase (SPS) and sucrose synthase (SS) in the synthesis direction showed a general upward trend with increasing storage duration (Figure 7A,B). This indicates that the activities of enzymes related to sucrose synthesis increased after a period of storage, and that the three treatments enhanced these enzyme activities to some extent. The SS activity in the catabolic direction activity was relatively low, but in the later stages, the activity in the treatment groups was significantly higher than that in CK ($p < 0.01$), indicating that the three treatments increased the activity of SS-catabolism direction enzymes (Figure 7C). On the sixth storage day, when the total sugar content reached its peak, the activities of all three enzymes were elevated, suggesting that enzymes involved in sucrose synthesis might play a major role during this period. On the 30th storage day, the activities of the enzymes related to sucrose synthesis had increased compared to those on day 18, while the activities of enzymes related to sucrose decomposition, except for SA, had decreased compared to those on the 18th storage day. The SPS activity in the treatment groups was significantly higher than that in the CK group at the same time point ($p < 0.01$), which indicated that all three treatments better promoted SS activity in the catabolic direction. The SPS activity of the HT group was significantly higher than that of the SA and 1-MCP ($p < 0.05$) groups, except on the sixth storage day, indicating that heat treatment was the most effective among the three treatments. Overall, the different treatments increased the activities of enzymes related to sucrose metabolism, with a more pronounced effect on the synthesis aspect. This suggested that the treatments could maintain fruit flavor by enhancing the activities of enzymes involved in sucrose synthesis.

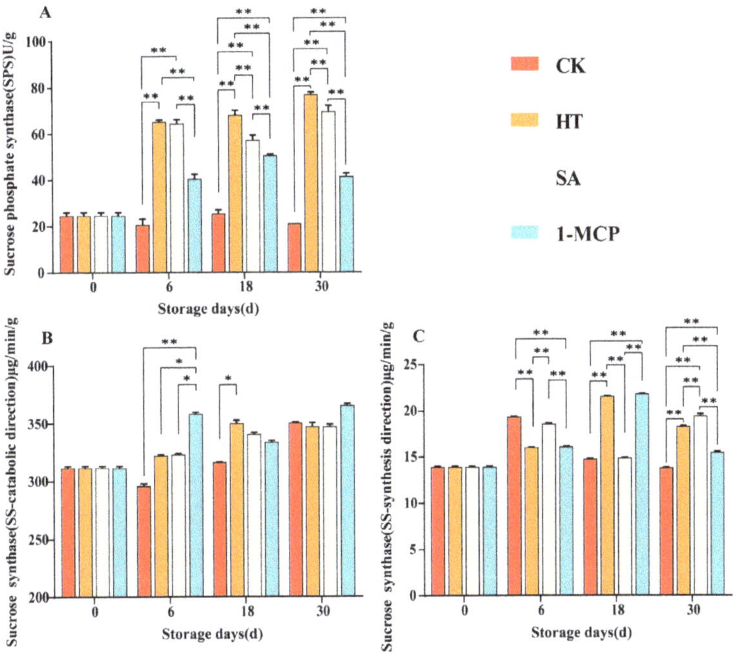

Figure 7. Changes in SPS (**A**), SS synthesis direction (**B**) and SS catabolism direction (**C**) enzyme activities in fruit pulp during storage. "**" means the result of significance analysis is highly significant ($p < 0.01$), "*" means the result of significance analysis is significant ($0.01 < p < 0.05$).

3.7. Expression of Genes Related to Rate-Limiting Enzymes of Sucrose Metabolism

The expression peaks of *SPS2* and *SUS3* were relatively high, while those of *SPS1* and *SUS1* were relatively low. This suggested that *SPS2* and *SUS3* might play more important role during the storage process (Figure 8). The expression levels of *SPS1* and *SPS2* exhibited an increasing and then decreasing trend. In the early stage of storage, the expression of *SPS1* and *SPS2* was higher in the treatment groups compared to the control (CK), while in the late stage, their expression levels decreased. It was speculated that exogenous substances might be involved in gene induction through alternative pathways. In the different treatments, the expression of *SUS1* was higher than that in the CK group, while in the late stage, the expression in was lower in all treatments except SA. The expression of *SUS3* also showed an increasing and then decreasing trend. Except for SA, other treatments partially suppressed the expression of this gene in the early stage of storage, which is consistent with the trend of enzyme activity in the SS catabolic direction. However, in the late stage, *SUS3* expression was almost negligible, suggesting that *SUS3* primarily regulates SS catabolic activities. Notably, the SA treatment significantly inhibited the reduction in *SUS1* and *SUS3* expression.

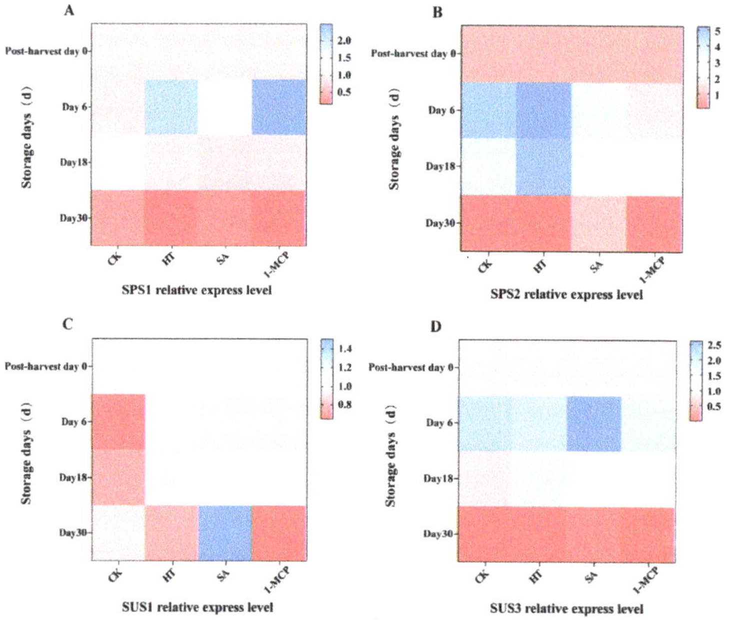

Figure 8. Expression of genes involved in sucrose synthesis and catabolism in Newhall navel orange. (**A**) Expression levels of *SPS1* in navel orange pulp. (**B**) Expression level of *SPS2* in navel orange pulp. (**C**) Expression level of *SUS1* in navel orange pulp. (**D**) Expression level of *SUS3* in navel orange pulp.

4. Discussion

The antioxidant properties of fruits have become a focal point for assessing the quality of fruits and their potential health benefits. In this study, we investigated flavonoids and total phenols as the main antioxidants in 'Newhall' navel oranges. Complex changes in the content of these compounds occurred during the different postharvest treatments. Flavonoid content exhibited a trend of initially decreasing and then increasing, with heat treatment (HT) and salicylic acid treatment (SA) showing an advantage in inhibiting the decline in the later stages. This may be due to the activation of certain defense mechanisms in the fruit by these treatments. Heat stress can trigger the synthesis of stress-responsive proteins such as the heat shock proteins (HSPs) involved in flavonoid biosynthesis pathways, as reported in studies on heat-treated fruits [29]. In addition, to counteract the negative effects of heat stress on cells, plants may also stimulate antioxidant enzymes, such as superoxide dismutase (SOD), peroxidase (POD) and catalase (CAT), as well as non-enzymatic antioxidants, such as ascorbic acid and carotenoids [30,31]. However, SA can regulate many aspects of plants at the gene level, thereby increasing their tolerance to abiotic stresses. SA has been reported to induce several genes responsible for encoding chaperones, heat shock proteins (HSPs), antioxidants and secondary metabolites [32].

The three treatments significantly inhibited the decrease in total phenolic content in the different storage periods. Compared with CK, the effect of HT was excellent on day 6 and 30, while SA and 1-MCP performed better on day 18. Which may be related to their different modes of action. Heat treatment likely increased the activity of enzymes involved in phenolic compound synthesis through heat-activated metabolic pathways. In contrast, 1-MCP may enhance antioxidant capacity by interfering with the ethylene signaling pathway [33].

Citrus fruits are high in antioxidants [34], and by quantifying their free radical scavenging ability, we can predict the beneficial effects of citrus on human health. The scavenging capacity for DPPH, ABTS and FRAP radicals decreased over time, with HT showing a slower decrease in scavenging capacity during the late storage period. This might be attributed to the induction of heat shock proteins during heat treatment, which protect cellular components from oxidative damage and maintain the integrity of the antioxidant enzyme system [35].

The weight lost from the fruits increased with storage time. Heat and salicylic acid treatments were more effective in delaying weight loss and maintaining fruit quality. This might be attributed to the following mechanism: heat treatment altered the cell membrane fluidity and integrity, thereby reducing water loss. The salicylic acid treatment enhanced the water potential and increased the plant water content by increasing osmotic pressure [36].

Total soluble solids (TSSs) and titratable acidity (TA) are important parameters affecting fruit quality. Maintaining an appropriate sugar–acid ratio while suppressing the decline in TA content is crucial [37]. Heat treatment significantly increased the total soluble solids content compared to the CK, consistent with the findings of Zhou et al. [38]. Although no significant differences were observed among the treatments in terms of titratable acid content, the SA treatment was sufficient to better stabilize the titratable acid content, which is consistent with findings on the effects of SA on the postharvest storage of table grapes, presumably because SA slows the rate of TA decline by inhibiting respiration rates and ethylene biosynthesis.

Sugar metabolism is a key determinant of flavor and quality during fruit storage. In the 'Newhall' navel oranges, the total sugar content initially increased and then decreased, reaching a peak on the sixth storage day. The changes in the soluble sugar content were different under different treatments. HT and SA were more effective in inhibiting the decline of total sugar content during the later storage period. The different treatments exhibited distinct effects on fructose and glucose. Notably, HT had a significant impact on sucrose content (40.51%), which was significantly higher compared to the other three treatments. This finding is similar to previous findings on the regulation of sugar concentration by heat stress in tobacco [39], the significant changes in sugar concentration observed after HT confirm the very complex metabolic response of plants to these abiotic stresses.

Overall, the different treatments increased the activities of enzymes related to sucrose metabolism, with effects on the direction of synthesis mainly in the pre-storage period and on the direction of catabolism throughout the storage period. This suggested that the treatments could maintain fruit flavor by enhancing the activities of enzymes involved in sucrose synthesis. The SPS activity of HT was 68.12 U/g on day 18 and 77.03 U/g on day 30, which was significantly higher than the remaining three treatments, indicating that heat treatment was the most effective among the three treatments. The results for sucrase activity and sucrose content were consistent; heat treatment increased sucrose content by enhancing the activity of enzymes related to sucrose synthesis.

SPS1 and *SPS2* are genes involved to the regulation of sucrose phosphate synthase SPS, which plays a key role in sucrose synthesis, while *SUS1* and *SUS3* are genes related to the regulation of sucrose synthase SS, which is primarily responsible for sucrose breakdown. The expression levels of *SPS1* and *SPS2* exhibited an increasing and then decreasing trend, indicating that these genes mainly function during the early stages of sucrose synthesis. In contrast, the expression of *SUS3* was almost negligible in the late stage, with *SUS1* being the predominant gene at this time. Sucrose content is known to increase during the ripening stage. While the expression of all invertase isozymes decreased significantly during this period, SPS expression remained higher. Therefore, the accumulation of sucrose at maturity can be attributed to increased SPS activity and reduced invertase activity (vesicle, cytoplasmic and cell wall isoforms). An analysis of SPS1 and SPS2 gene expression showed that the expression of SPS1 in 1-MCP on the sixth day of storage was much higher than that of CK during the same period, and the expression of SPS2 in HT was much higher than that of SA and 1-MCP during the same period. Heat stress has been shown to enhance sucrose content, enzyme expression, proline levels and sugar recovery [40]. Similarly, salicylic acid (SA) treatment has been demonstrated to significantly modulate the transcript abundance of genes related to sucrose biosynthesis and degradation in peach fruit [41]. Norwegian apples treated with 1-methylcyclopropene (1-MCP) exhibited lower sucrose hydrolysis and delayed ripening compared to the control check (CK) group [42]. However, in the present study, 1-MCP treatment was not as effective for the storage of 'Newhall' navel oranges. It is hypothesized that this may be due to differences in plant material. 'Newhall' navel oranges were used in this experiment, and their growing environment may have determined the effectiveness of the different treatments.

5. Conclusions

In this study, we comprehensively investigated the effects of various postharvest treatments, including control check (CK), heat treatment (HT), salicylic acid treatment (SA) and 1-methylcyclopropene treatment (1-MCP) on the quality of 'Newhall' navel oranges during storage (Figure 9). HT is often used in postharvest storage management because of its environmental and easy-to-use advantages. HT was the most effective among the three treatments in enhancing the fruit's antioxidant properties, reducing fruit weight loss, increasing total soluble solids and promoting sucrose metabolism, and it has high application value. SA also showed significant benefits by stabilizing antioxidant capacity, titratable acid and total sugar content, but performed poorly in retarding the decline in total phenolic content and sucrose content. Although the 1-MCP performed the worst in some aspects, it still had a positive effect on antioxidant capacity and sucrose metabolism during storage. These results provide a rationale for selecting postharvest treatments to extend the storage life and maintain the quality of Newhall navel oranges, with broader implications for the citrus industry. However, each of the three treatments has its own strengths in terms of its antioxidant properties, sucrose metabolism and the final quality presented, and no one treatment showed a crushing advantage. Future research could explore combinations of these treatments or develop new approaches to further optimize postharvest fruit quality management to maintain fruit quality and postharvest freshness in citrus fruits.

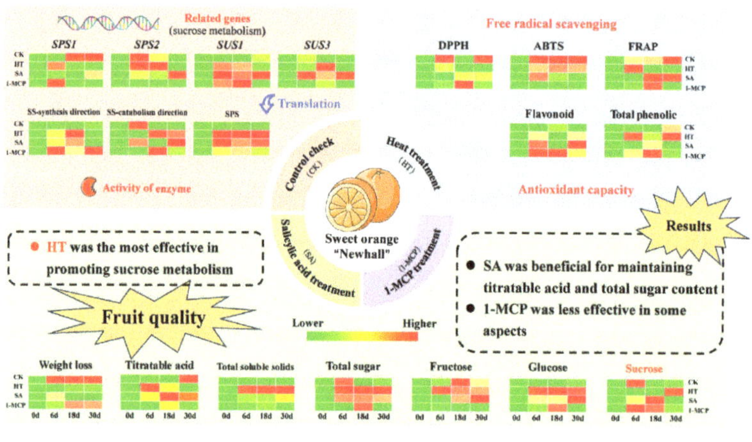

Figure 9. Effects of different postharvest treatments on fruit quality, sucrose metabolism and antioxidant capacity of 'Newhall' navel oranges during storage. Different colors represent different values of $\frac{X-X_{min}}{X_{max}-X_{min}}$ (X represents the measured value of each index, X_{min} represents the minimum value of the corresponding index, and X_{max} is the maximum value); higher values are redder and lower are greener.

Author Contributions: Conceptualization, B.X., L.H., Y.O., T.D. and Y.L. (Yixuan Li); Methodology, H.C.; Software, L.H., Y.O. and Y.L. (Yisong Li); Validation, W.W. and Y.L. (Yisong Li); Formal analysis, W.W. and C.W.; Investigation, S.C. and Q.M.; Resources, J.W. and J.Y.; Data curation, L.H. and Y.O.; Writing—original draft, L.H.; Writing—review & editing, L.H.; Visualization, L.H. and Y.F.; Supervision, B.X. and Z.W.; Project administration, B.X. and Z.W.; Funding acquisition, Z.W. All authors have read and agreed to the published version of the manuscript.

Funding: This study was funded by the National Key R&D Program of China (2023YFD2300601-05, 2024YFD2300803, 2021YFD1600802-02).

Data Availability Statement: Data are contained within the article.

Conflicts of Interest: The authors declare no conflicts of interest.

References

1. Zhang, J.; Liang, Y.; He, L.; Kaliaperumal, K.; Tan, H.; Jiang, Y.; Zhong, B.; Zhang, J. Effects of storage time and temperature on the chemical composition and organoleptic quality of Gannan navel orange (*Citrus sinensis* Osbeck cv. Newhall). *J. Food Meas. Charact.* **2021**, *16*, 935–944. [CrossRef]
2. Lowell, C.A.; Tomlinson, P.T.; Koch, K.E. Sucrose-Metabolizing Enzymes in Transport Tissues and Adjacent Sink Structures in Developing Citrus Fruit 1. *Plant Physiol.* **1989**, *90*, 1394–1402. [CrossRef] [PubMed]
3. Legua, P.; Modica, G.; Porras, I.; Conesa, A.; Continella, A. Bioactive compounds, antioxidant activity and fruit quality evaluation of eleven blood orange cultivars. *J. Sci. Food Agric.* **2022**, *102*, 2960–2971. [CrossRef] [PubMed]
4. Lu, W.; Hao, W.; Liu, K.; Liu, J.; Yin, C.; Su, Y.; Hang, Z.; Peng, B.; Liu, H.; Xiong, B.; et al. Analysis of sugar components and identification of SPS genes in citrus fruit development. *Front. Plant Sci.* **2024**, *15*, 1372809. [CrossRef]
5. Islam, M.Z.; Hu, X.-M.; Jin, L.-F.; Liu, Y.-Z.; Peng, S.-A. Genome-Wide Identification and Expression Profile Analysis of Citrus Sucrose Synthase Genes: Investigation of Possible Roles in the Regulation of Sugar Accumulation. *PLoS ONE* **2014**, *9*, e113623. [CrossRef]
6. Hussain, S.B.; Guo, L.-X.; Shi, C.-Y.; Khan, M.A.; Bai, Y.-X.; Du, W.; Liu, Y.-Z. Assessment of sugar and sugar accumulation-related gene expression profiles reveal new insight into the formation of low sugar accumulation trait in a sweet orange (*Citrus sinensis*) bud mutant. *Mol. Biol. Rep.* **2020**, *47*, 2781–2791. [CrossRef]
7. Zacarías-García, J.; Rodrigo, M.J.; Rambla, J.L.; Granell, A.; Zacarías, L. Physiological responses, nutritional quality and aroma volatiles of the red-fleshed kirkwood navel and ruby valencia oranges during postharvest cold storage. *Postharvest Biol. Technol.* **2023**, *199*, 112303. [CrossRef]
8. Aravinthkumar, A.; Raj, H.; Kumar, P.; Sharma, P.L.; Verma, S.; Sheela, J.; Parwan, S.; Shankar, S.V.; Ananthakrishnan, S.; Chauhan, A.; et al. Comparative efficacy of GRAS chemicals, botanicals and yeast in controlling green mould and fruit nutritional quality enhancement in Kinnow mandarin (*Citrus nobilis* Lour x *Citrus deliciosa* Tenora). *Sci. Hortic.* **2025**, *339*, 113869. [CrossRef]
9. Saini, R.K.; Ranjit, A.; Sharma, K.; Prasad, P.; Shang, X.; Gowda, K.G.M.; Keum, Y.-S. Bioactive Compounds of Citrus Fruits: A Review of Composition and Health Benefits of Carotenoids, Flavonoids, Limonoids, and Terpenes. *Antioxidants* **2022**, *11*, 239. [CrossRef]
10. Lapuente, M.; Estruch, R.; Shahbaz, M.; Casas, R. Relation of Fruits and Vegetables with Major Cardiometabolic Risk Factors, Markers of Oxidation, and Inflammation. *Nutrients* **2019**, *11*, 2381. [CrossRef]
11. Nakao, K.; Murata, K.; Itoh, K.; Hanamoto, Y.; Masuda, M.; Moriyama, K. Anti-hyperuricemia effects of extracts of immature Citrus unshiu fruit. *J. Trad. Med.* **2011**, *28*, 10–15. [CrossRef]
12. Sarvarian, M.; Jafarpour, A.; Awuchi, C.G.; Adeleye, A.O.; Okpala, C.O.R. Changes in Physicochemical, Free Radical Activity, Total Phenolic and Sensory Properties of Orange (*Citrus sinensis* L.) Juice Fortified with Different Oleaster (*Elaeagnus angustifolia* L.) Extracts. *Molecules* **2022**, *27*, 1530. [CrossRef] [PubMed]
13. Gündüz, G.T.; Juneja, V.K.; Pazir, F. Application of ultraviolet-C light on oranges for the inactivation of postharvest wound pathogens. *Food Control* **2015**, *57*, 9–13. [CrossRef]
14. Huan, C.; Han, S.; Jiang, L.; An, X.; Yu, M.; Xu, Y.; Ma, R.; Yu, Z. Postharvest hot air and hot water treatments affect the antioxidant system in peach fruit during refrigerated storage. *Postharvest Biol. Technol.* **2017**, *126*, 1–14. [CrossRef]
15. Li, S.; Hu, J.; Ning, S.; Li, W.; Jiang, R.; Huang, J.; Li, Y. Bacillus velezensis HY19 as a sustainable preservative in post-harvest citrus (*Citrus reticulata Blanco* L.) fruit management. *Food Control* **2024**, *155*, 110068. [CrossRef]
16. Yang, C.; Lin, Z.; Luo, Z.; Wang, Z.; Liu, P.; Xu, R.; Zhu, F.; Cheng, Y. The volatile compound (E)-2-hexenal in wampee (*Clausena lansium*) represses the development of *Penicillium italicum* and enhances the disease resistance of postharvest citrus fruit. *Postharvest Biol. Technol.* **2025**, *219*, 113241. [CrossRef]

17. Sun, Y.; Li, Y.; Xu, Y.; Sang, Y.; Mei, S.; Xu, C.; Yu, X.; Pan, T.; Cheng, C.; Zhang, J.; et al. The Effects of Storage Temperature, Light Illumination, and Low-Temperature Plasma on Fruit Rot and Change in Quality of Postharvest Gannan Navel Oranges. *Foods* **2022**, *11*, 3707. [CrossRef]
18. Ennab, H.A.; El-Shemy, M.A.; Alam-Eldein, S.M. Salicylic Acid and Putrescine to Reduce Post-Harvest Storage Problems and Maintain Quality of Murcott Mandarin Fruit. *Agronomy* **2020**, *10*, 115. [CrossRef]
19. Maghoumi, M.; Amodio, M.L.; Cisneros-Zevallos, L.; Colelli, G. Prevention of Chilling Injury in Pomegranates Revisited: Pre- and Post-Harvest Factors, Mode of Actions, and Technologies Involved. *Foods* **2023**, *12*, 1462. [CrossRef]
20. Li, J.; Hussain, I.; Azam, M.; Khan, M.A.; Akram, M.T.; Naveed, K.; Asif, M.; Anjum, N.; Zeng, J.; Zhang, J.; et al. Hot Water Treatment Improves Date Drying and Maintains Phytochemicals and Fruit Quality Characteristics of Date Palm (*Phoenix dactylifera*). *Foods* **2023**, *12*, 2405. [CrossRef]
21. Baswal, A.K.; Dhaliwal, H.S.; Singh, Z.; Mahajan, B.V.C. Post-harvest Application of Methyl Jasmonate, 1-Methylcyclopropene and Salicylic Acid Elevates Health-promoting Compounds in Cold-stored 'Kinnow' Mandarin (*Citrus nobilis* Lour x *C. deliciosa* Tenora) Fruit. *Int. J. Fruit Sci.* **2021**, *21*, 147–157. [CrossRef]
22. Kaewsuksaeng, S.; Tatmala, N.; Srilaong, V.; Pongprasert, N. Postharvest heat treatment delays chlorophyll degradation and maintains quality in Thai lime (*Citrus aurantifolia* Swingle cv. Paan) fruit. *Postharvest Biol. Technol.* **2015**, *100*, 1–7. [CrossRef]
23. Huang, Q.; Huang, L.; Chen, J.; Zhang, Y.; Kai, W.; Chen, C. Maintenance of postharvest storability and overall quality of 'Jinshayou' pummelo fruit by salicylic acid treatment. *Front. Plant Sci.* **2023**, *13*, 1086375. [CrossRef] [PubMed]
24. Wang, T.; Zheng, Z.; Deng, L.; Li, W.; Yuan, Y.; Zhang, M.; Sun, G.; He, S.; Wang, J.; Wang, Z.; et al. Effect of Natural Variation and Rootstock on Fruit Quality and Volatile Organic Compounds of 'Kiyomi tangor' (*Citrus reticulata* Blanco) Citrus. *Int. J. Mol. Sci.* **2023**, *24*, 16810. [CrossRef]
25. Molan, A.L.; Flanagan, J.; Wei, W.; Moughan, P.J. Selenium-containing green tea has higher antioxidant and prebiotic activities than regular green tea. *Food Chem.* **2009**, *114*, 829–835. [CrossRef]
26. Guo, N.; Yan, Y.; Li, Q.; Yang, Y. Pyruvic acid improves cold-storage quality of plum fruit by stimulating cyanide-resistant respiration and regulating sugar metabolism. *Sci. Hortic.* **2025**, *340*, 113926. [CrossRef]
27. Liu, J.; Yue, R.; Si, M.; Wu, M.; Cong, L.; Zhai, R.; Yang, C.; Wang, Z.; Ma, F.; Xu, L. Effects of Exogenous Application of Melatonin on Quality and Sugar Metabolism in 'Zaosu' Pear Fruit. *J. Plant Growth Regul.* **2019**, *38*, 1161–1169. [CrossRef]
28. Zhang, P.; Shao, X.; Wei, Y.; Xu, F.; Wang, H. At-harvest fruit maturity affects sucrose metabolism during cold storage and is related to chilling injury in peach. *J. Food Sci. Technol.* **2020**, *57*, 2000–2009. [CrossRef]
29. Cui, M.; Liang, Z.; Liu, Y.; Sun, Q.; Wu, D.; Luo, L.; Hao, Y. Flavonoid profile of *Anoectochilus roxburghii* (Wall.) Lindl. Under short-term heat stress revealed by integrated metabolome, transcriptome, and biochemical analyses. *Plant Physiol. Biochem.* **2023**, *201*, 107896. [CrossRef]
30. Almeida, J.; Perez-Fons, L.; Fraser, P.D. A transcriptomic, metabolomic and cellular approach to the physiological adaptation of tomato fruit to high temperature. *Plant Cell Environ.* **2020**, *44*, 2211–2229. [CrossRef]
31. Wang, L.; Ma, K.-B.; Lu, Z.-G.; Ren, S.-X.; Jiang, H.-R.; Cui, J.-W.; Chen, G.; Teng, N.-J.; Lam, H.-M.; Jin, B. Differential physiological, transcriptomic and metabolomic responses of Arabidopsis leaves under prolonged warming and heat shock. *BMC Plant Biol.* **2020**, *20*, 86. [CrossRef] [PubMed]
32. Khan, M.I.R.; Fatma, M.; Per, T.S.; Anjum, N.A.; Khan, N.A. Salicylic acid-induced abiotic stress tolerance and underlying mechanisms in plants. *Front. Plant Sci.* **2015**, *6*, 462. [CrossRef] [PubMed]
33. Balaguera-López, H.E.; Espinal-Ruiz, M.; Rodríguez-Nieto, J.M.; Herrera-Arévalo, A.; Zacarías, L. 1-Methylcyclopropene inhibits ethylene perception and biosynthesis: A theoretical and experimental study on cape gooseberry (*Physalis peruviana* L.) fruits. *Postharvest Biol. Technol.* **2021**, *174*, 111467. [CrossRef]
34. Mare, R.; Pujia, R.; Maurotti, S.; Greco, S.; Cardamone, A.; Coppoletta, A.R.; Bonacci, S.; Procopio, A.; Pujia, A. Assessment of Mediterranean Citrus Peel Flavonoids and Their Antioxidant Capacity Using an Innovative UV-Vis Spectrophotometric Approach. *Plants* **2023**, *12*, 4046. [CrossRef]
35. Huang, Y.-C.; Liu, C.-C.; Li, Y.-J.; Liao, C.-M.; Vivek, S.; Chuo, G.-L.; Tseng, C.-Y.; Wu, Z.-Q.; Shimada, T.; Suetsugu, N.; et al. Multifaceted roles of Arabidopsis heat shock factor binding protein in plant growth, development, and heat shock response. *Environ. Exp. Bot.* **2024**, *226*, 105878. [CrossRef]
36. Kaya, C.; Akin, S.; Sarioğlu, A.; Ashraf, M.; Alyemeni, M.N.; Ahmad, P. Enhancement of soybean tolerance to water stress through regulation of nitrogen and antioxidant defence mechanisms mediated by the synergistic role of salicylic acid and thiourea. *Plant Physiol. Biochem.* **2024**, *207*, 108320. [CrossRef]
37. Li, Z.; Jin, R.; Yang, Z.; Wang, X.; You, G.; Guo, J.; Zhang, Y.; Liu, F.; Pan, S. Comparative study on physicochemical, nutritional and enzymatic properties of two Satsuma mandarin (*Citrus unshiu* Marc.) varieties from different regions. *J. Food Compos. Anal.* **2021**, *95*, 103614. [CrossRef]
38. Zhou, H.-j.; Ye, Z.-w.; Su, M.-s.; Du, J.-h.; Li, X.-w. Effect of Heat Treatment on Protein Content and the Quality of 'Hujingmilu' Peach [*Prunus persica* (L.) Batsch]. *HortScience Horts* **2015**, *50*, 1531–1536. [CrossRef]

39. Ancillotti, C.; Bogani, P.; Biricolti, S.; Calistri, E.; Checchini, L.; Ciofi, L.; Gonnelli, C.; Del Bubba, M. Changes in polyphenol and sugar concentrations in wild type and genetically modified Nicotiana langsdorffii Weinmann in response to water and heat stress. *Plant Physiol. Biochem.* **2015**, *97*, 52–61. [CrossRef]
40. Mehdi, F.; Liu, X.; Riaz, Z.; Javed, U.; Aman, A.; Galani, S. Expression of sucrose metabolizing enzymes in different sugarcane varieties under progressive heat stress. *Front. Plant Sci.* **2023**, *14*, 1269521. [CrossRef]
41. Zhao, Y.; Song, C.; Qi, S.; Lin, Q.; Duan, Y. Jasmonic acid and salicylic acid induce the accumulation of sucrose and increase resistance to chilling injury in peach fruit. *J. Sci. Food Agric.* **2021**, *101*, 4250–4255. [CrossRef] [PubMed]
42. Falagán, N.; Terry, L.A. 1-Methylcyclopropene maintains postharvest quality in Norwegian apple fruit. *Food Sci. Technol. Int.* **2020**, *26*, 420–429. [CrossRef] [PubMed]

plants

MDPI

Article

Variety Effect on Peelability and Mechanisms of Action of Late-Ripening Citrus Fruits

Ya Yuan [1,†], Ziyi Huang [1,†], Yihong Wang [1,†], Lijun Deng [1], Tie Wang [1], Defa Cao [1], Ling Liao [1], Bo Xiong [1], Meiyan Tu [2], Zhihui Wang [1,*] and Jun Wang [1,*]

1 College of Horticulture, Sichuan Agricultural University, Chengdu 611130, China; yuanya12321@163.com (Y.Y.); huangziyiyx2024@163.com (Z.H.); 13502145088@163.com (Y.W.); denglijun0919@163.com (L.D.); wangtie106@163.com (T.W.); 19108095703@163.com (D.C.); liao19910331@163.com (L.L.); xiongbo1221@sicau.edu.cn (B.X.)

2 Institute of Horticulture, Sichuan Academy of Agricultural Sciences Key Laboratory of Horticultural Crop Biology and Germplasm Creation in Southwest China, Ministry of Agriculture and Rural Affairs, Chengdu 610066, China; tumeiyan@scsaas.cn

* Correspondence: wangzhihui318@sicau.edu.cn (Z.W.); cnwangjun01@163.com (J.W.)

† These authors contributed equally to this work.

Abstract: Peelability, a crucial commercial trait for fresh-eating citrus, has received limited research attention regarding its underlying mechanisms. This study investigated three late-maturing citrus cultivars, namely 'Qingjian' (QJ), 'Mingrijian' (MRJ), and 'Chunjian' (CJ), analyzing their peelability development using texture analysis and exploring the physiological and biochemical factors influencing peeling difficulty. The results showed that peelability improved with fruit maturation, reaching its peak at full ripeness, with the following order of peeling difficulty: QJ (hardest) > MRJ (intermediate) > CJ (easiest). At full maturity, QJ (the most difficult to peel) exhibited more regularly shaped peel cells with fewer intercellular spaces, lower intracellular organic matter accumulation, and higher levels of cell wall polysaccharides, calcium (Ca), and abscisic acid (ABA). These characteristics may be linked to the lower relative expression of soluble sugar (TS)-related genes (*CCR4A*, *SPP1*) and the titratable acid (TA)-related gene (*CsCit1*), as well as the higher relative expression of ABA biosynthesis genes (*NCED1*, *NCED2*). Correlation analyses demonstrated that citrus peel firmness and adhesion strength are significantly associated with multiple growth and developmental characteristics, including fruit morphometric parameters, peel cellular architecture, intracellular organic compound content, cell wall polysaccharide levels and related degradative enzyme activities, calcium concentrations, and endogenous phytohormone profiles. These findings provide valuable insights for studying peelability mechanisms and improving fruit quality in citrus breeding.

Keywords: late-ripening citrus; peel properties; peelability

1. Introduction

Citrus (*Citrus reticulata Blanco*), as the world's most important fresh fruit category, holds significant economic and ecological value [1]. In recent decades, global citrus breeding programs have undergone notable structural changes, with quality improvement increasingly focusing on seedlessness, easy peelability, intense flavor, and rich aroma [2]. Peelability refers to the ease of separating the rind from the segment membranes. Market research in citrus production and sales reveals that, given similar quality traits, consumers strongly prefer easy-peeling cultivars due to convenience and safety considerations [3,4].

Previous studies have established relatively mature subjective evaluation systems for citrus peelability based on empirical assessments and manual peeling difficulty. For instance, Goldenberg [5] classified 46 citrus cultivars into five grades (1, extremely difficult, to 5, extremely easy), while Simons [6] categorized eight citrus cultivars into two levels: low-grade (rind tightly adhering to flesh, difficult to peel) and high-grade (loose connection between rind and flesh, easy to peel). With recent advancements in texture analyzers and other biomechanical instruments, peelability evaluation has progressively shifted toward objective, scientific, and standardized methodologies. Notably, Yu [7] successfully quantified peeling difficulty by decomposing manual peeling into two measurable parameters using texture analyzers and digital force gauges: rind adhesion strength and penetration resistance (rind hardness).

The citrus peel comprises three histologically distinct layers from exterior to interior: the exocarp (flavedo), mesocarp (albedo), and endocarp [8–10]. The penetration resistance of citrus peel primarily originates from the oil gland layer (essential oil-containing structures), exhibiting significant correlation with parenchyma textural modifications, whereas peel adhesion strength is predominantly governed by the spongy layer (albedo tissue), whose histoarchitectural characteristics directly mediate rind-segment membrane binding dynamics [11,12]. The cell, as the fundamental structural unit of fruit tissues, directly determines fruit mechanical properties and texture through its morphology, architecture, spatial arrangement, and integrity [13]. Current scientific consensus holds that fruits with uniform polyhedral cells in compact arrangements exhibit firm texture, high hardness, and crispness, whereas those with irregular cell sizes and large intercellular spaces develop soft, coarse textures [14,15].

The accumulation and compositional changes of intracellular organic solutes can reduce fruit osmotic potential, regulate osmotic homeostasis, influence cell membrane stability, and ultimately modify fruit textural properties [16,17]. Notably, starch serves dual physiological roles: beyond providing essential energy for plant growth and development [18], it functions as a crucial organic osmolyte that maintains cellular turgor pressure, provides mechanical support, and retards firmness loss during fruit ripening [19,20]. The maturation process involves starch hydrolysis into soluble sugars, concomitant with dynamic changes in sugar and organic acid accumulation. These metabolic shifts not only determine flavor profiles but also significantly impact cellular osmotic regulation [21]. However, the complex interactions among starch, sugars, and acids as intracellular organic components remain poorly understood. While current research primarily focuses on sugars and acids as flavor determinants in citrus, their combined effects with starch on peelability have not been systematically investigated.

The plant cell wall plays essential roles in maintaining cellular morphology and size, preserving intercellular adhesion and rigidity, and withstanding turgor pressure [22–24]. Structurally composed of the middle lamella, primary wall, and secondary wall, its main constituents include pectins, cellulose (CL), and hemicellulose (HCL) [25–27]. Current scientific consensus holds that elevated activities of cell wall polysaccharide-degrading enzymes—particularly cellulase (Cx), pectin methylesterase (PME), and pectin lyase (PL)—induce textural modifications in citrus fruits by altering the CL-HCL matrix architecture and degrading pectinaceous compounds in the middle lamella, thereby directly influencing peelability [28,29]. Recent studies have partially elucidated the impact of cell wall polysaccharides on citrus peelability.

Mineral elements and phytohormones critically regulate cellular architecture and fruit textural modifications. Calcium (Ca), as a fundamental structural component of cell walls, primarily exists in the form of calcium pectate within the middle lamella and cell walls. This configuration enhances intercellular adhesion while improving cellular toughness and elasticity, thereby stabilizing tissue structure [30]. Endogenous phytohormones—including ethylene (ETH), abscisic acid (ABA), gibberellins (GA), and auxins (IAA)—play pivotal regulatory roles in citrus fruit textural dynamics and ripening processes. However, the mechanistic contributions of calcium and these phytohormones to peelability development remain underexplored.

The late-maturing citrus cultivars 'Qingjian' (QJ), 'Mingrijian' (MRJ), and 'Chunjian' (CJ) have become predominant commercial cultivars in China due to their distinctive flavor profiles and favorable market timing. Our preliminary investigations revealed significant variations in manual peelability among these closely related cultivars. This study quantitatively characterized peelability development by measuring pericarp hardness and adhesive strength using a texture analyzer, establishing peeling difficulty classifications at maturity. Through comprehensive analyses of fruit growth characteristics, pericarp cell morphology, intracellular organic compounds, cell wall polysaccharides and degrading enzymes, mineral elements, phytohormones, and differential gene expression across pericarp tissues, we systematically elucidated the physiological and molecular mechanisms underlying peelability variation in late-maturing citrus, providing a foundation for further research on peelability formation.

2. Results

2.1. Comparison of Fruit External Quality

Changes in the external fruit quality of three late-maturing citrus cultivars (QJ, MRJ, and CJ) during different growth stages are shown in Figure 1 and Table 1. During 120–180 days after full bloom (DAF), all the cultivars showed rapid fruit expansion, with progressive increases in longitudinal diameter, transverse diameter, and pulp diameter, while maintaining relatively stable peel thickness. Concurrently, the peel color transitioned from dark green to orange–yellow, accompanied by the progressive intensification of pulp coloration. At full maturity (270 d), all the dimensional parameters reached their maximum values, with QJ fruits demonstrating a significantly greater longitudinal diameter, transverse diameter, and pulp diameter compared to both the MRJ and CJ cultivars.

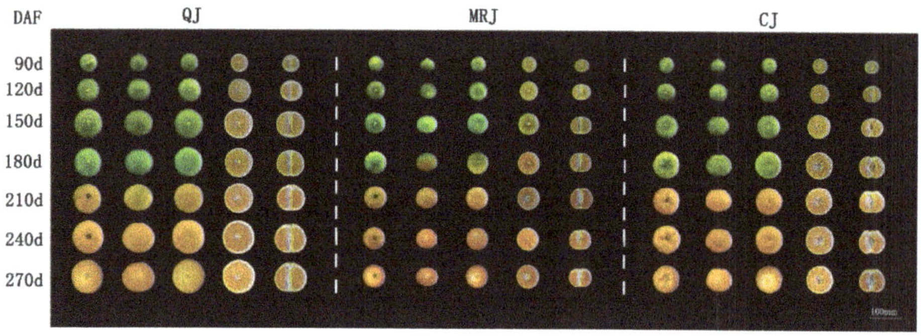

Figure 1. Changes in fruit phenotype of three late-maturing citrus cultivars at different growth and development stages.

Table 1. Changes in the external quality of three late-maturing citrus cultivars at different growth and development stages.

Cultivar	Days After Full Bloom (d)	VerticalDiameter (mm)	HorizontalDiameter (mm)	Fruit ShapeIndex	PulpDiameter (mm)	Total PeelThickness (mm)
QJ	90	53.92 ± 0.98 a	53.66 ± 1.08 a	1.00 ± 0.00 a	40.93 ± 1.17 a	6.37 ± 0.07 a
	120	63.19 ± 0.65 a	68.59 ± 1.05 a	0.92 ± 0.02 a	57.06 ± 0.67 a	5.76 ± 0.31 a
	150	72.80 ± 0.45 a	85.54 ± 1.07 a	0.85 ± 0.01 a	73.44 ± 1.03 a	6.05 ± 0.05 a
	180	74.15 ± 1.00 a	83.80 ± 0.46 a	0.88 ± 0.01 a	71.11 ± 0.73 a	6.34 ± 0.29 a
	210	74.94 ± 1.46 a	85.06 ± 1.37 a	0.88 ± 0.01 a	72.13 ± 1.94 a	6.47 ± 0.36 a
	240	83.61 ± 0.94 a	95.15 ± 1.24 a	0.88 ± 0.01 a	82.18 ± 0.68 a	6.49 ± 0.37 a
	270	84.22 ± 0.30 a	96.93 ± 0.79 a	0.87 ± 0.01 a	84.18 ± 0.84 a	6.37 ± 0.06 a
MRJ	90	40.59 ± 1.50 b	45.44 ± 1.57 b	0.89 ± 0.01 b	39.87 ± 1.20 ab	2.79 ± 0.18 c
	120	47.89 ± 0.13 c	55.60 ± 0.30 b	0.86 ± 0.00 b	51.67 ± 0.25 b	1.96 ± 0.03 c
	150	51.53 ± 1.25 c	61.90 ± 2.05 c	0.83 ± 0.02 a	58.08 ± 1.82 c	1.91 ± 0.13 c
	180	55.61 ± 0.48 c	64.36 ± 0.95 c	0.86 ± 0.02 a	60.31 ± 0.85 b	2.03 ± 0.06 c
	210	58.93 ± 1.28 c	66.23 ± 1.00 c	0.89 ± 0.01 a	61.57 ± 0.93 b	2.33 ± 0.03 c
	240	56.73 ± 2.41 c	64.93 ± 1.91 c	0.87 ± 0.01 a	59.75 ± 2.17 c	2.59 ± 0.16 c
	270	58.89 ± 0.60 c	66.88 ± 0.20 c	0.88 ± 0.01 a	61.07 ± 0.14 c	2.90 ± 0.06 c
CJ	90	41.95 ± 0.78 b	45.98 ± 0.70 b	0.91 ± 0.00 b	37.86 ± 0.97 b	4.06 ± 0.14 b
	120	51.32 ± 1.33 b	56.92 ± 1.09 b	0.90 ± 0.01 a	49.87 ± 1.04 c	3.53 ± 0.08 b
	150	55.23 ± 0.88 b	68.29 ± 0.43 b	0.81 ± 0.01 b	62.35 ± 0.41 b	2.97 ± 0.13 b
	180	62.33 ± 1.67 b	78.03 ± 0.97 b	0.80 ± 0.01 b	71.21 ± 1.02 a	3.41 ± 0.17 b
	210	62.78 ± 0.54 b	77.78 ± 1.05 b	0.81 ± 0.01 b	70.02 ± 1.32 a	3.88 ± 0.25 b
	240	64.60 ± 1.87 b	80.37 ± 1.27 b	0.80 ± 0.01 b	72.51 ± 0.98 b	3.93 ± 0.20 b
	270	64.46 ± 1.11 b	78.42 ± 0.70 b	0.82 ± 0.02 b	70.50 ± 0.63 b	3.96 ± 0.04 b

Note: Data are presented as mean ± standard deviation (n = 3). Different lowercase letters following similar data of different varieties in the same period indicate significant differences according to Duncan's test ($p < 0.05$).

2.2. Peelability Comparison of Citrus Fruits

Changes in the peel firmness and peel adhesion of three late-maturing citrus cultivars (QJ, MRJ, and CJ) during different growth stages are shown in Figure 2. A texture analysis demonstrated that both peel firmness and adhesion reached their maximum levels during the early fruit development stage, followed by a progressive decline to their lowest values at full maturity. The most substantial reductions in peel firmness and adhesion occurred during 150–210 and 120–150 d, respectively (peel adhesion measurements commenced at 120 d due to the technical challenge of separating peel from pulp at 90 d). At full maturity, CJ exhibited significantly lower peel firmness compared to both QJ and MRJ, while no significant difference was observed between QJ and MRJ. In terms of peel adhesion, MRJ displayed intermediate values, being significantly lower than QJ but significantly higher than CJ.

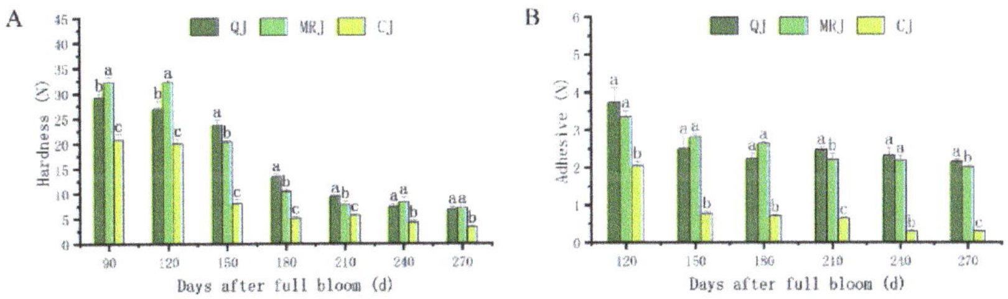

Figure 2. Changes in peel hardness and peel adhesion force of three late-maturing citrus cultivars at different growth and development stages. (**A**) Hardness. (**B**) Adhesive. The parameter values shown in each figure are expressed as mean ± standard deviation (n = 3). Different lowercase letters above the data of different varieties in the same period indicate statistically significant differences as determined by Duncan's test ($p < 0.05$).

2.3. *Comparison of Peel Characteristics and Cellular Morphology*

Developmental changes in the peel characteristics and cellular morphology of three late-maturing citrus cultivars (QJ, MRJ, and CJ) are illustrated in Figure 3, revealing distinct temporal patterns in the peel cellular architecture. Throughout the growth period, all the cultivars exhibited the progressive enlargement of oil glands, parenchyma cells, and spongy layer cells, though with marked structural differences: QJ maintained regular cell morphology with tightly packed spongy layer organization, showing only minimal intercellular spaces except for limited conspicuous gaps at full maturity, whereas MRJ and CJ displayed progressive cellular disorganization from 150 to 270 d, characterized by irregular cell shapes, increasingly loose arrangements, and expanded, unevenly distributed intercellular spaces that were macroscopically visible as prominent gaps in equatorial cross-sections. At full maturity, quantitative comparisons showed QJ with a significantly larger oil gland perimeter and area ($p < 0.05$) compared to both MRJ and CJ.

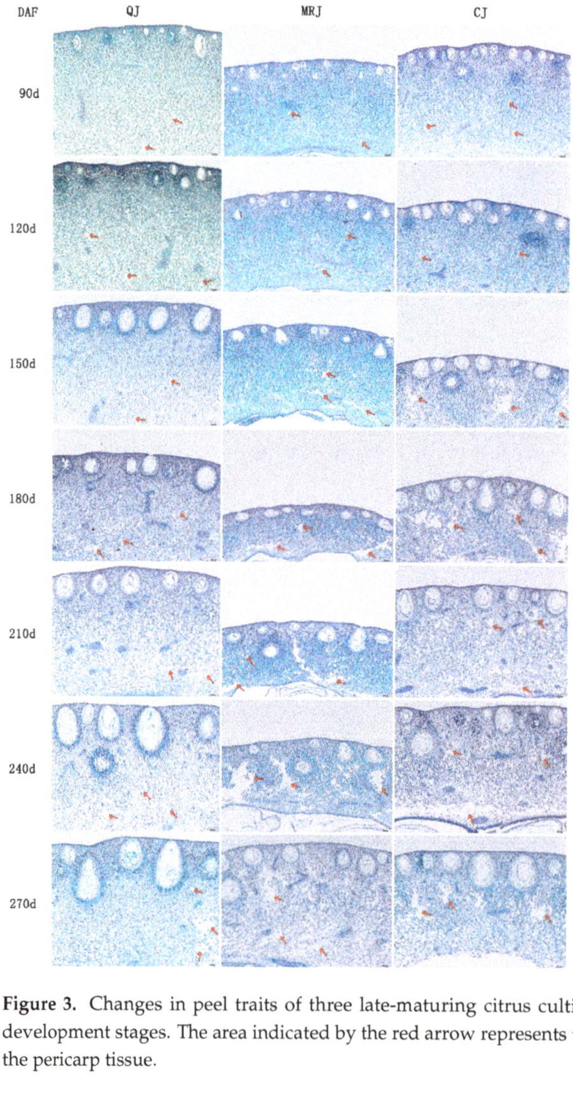

Figure 3. Changes in peel traits of three late-maturing citrus cultivars at different growth and development stages. The area indicated by the red arrow represents the intercellular spaces within the pericarp tissue.

2.4. Comparison of Intracellular Organic Matter and Component Contents

The dynamic changes in intracellular organic components within the peel of three late-maturing citrus cultivars (QJ, MRJ, and CJ) during different developmental stages are presented in Figures 4 and 5. Throughout fruit development, all the cultivars exhibited similar trends in component accumulation: both the oil gland layer and spongy layer showed an initial increase followed by stabilization in their total soluble solids (TS) and soluble tannins (ST) contents. The titratable acidity (TA) content in the oil gland layer initially decreased and then increased, while in the spongy layer it exhibited an overall gradual decline throughout the developmental stages. The critical phase of 150–210 d represented a pivotal metabolic transition period, during which the rapid accumulation of TS and ST occurred concurrently with a sharp reduction in the TA content, specifically within the oil gland layer. At full maturity, sucrose and citric acid emerged as the dominant components of TS and TA pools in both tissue layers across all the cultivars. A comparative analysis revealed that QJ consistently maintained significantly lower ($p < 0.05$) intracellular accumulation levels of the TS, ST, and TA components relative to both the MRJ and CJ cultivars, suggesting fundamental differences in carbon partitioning and organic acid metabolism during fruit ripening.

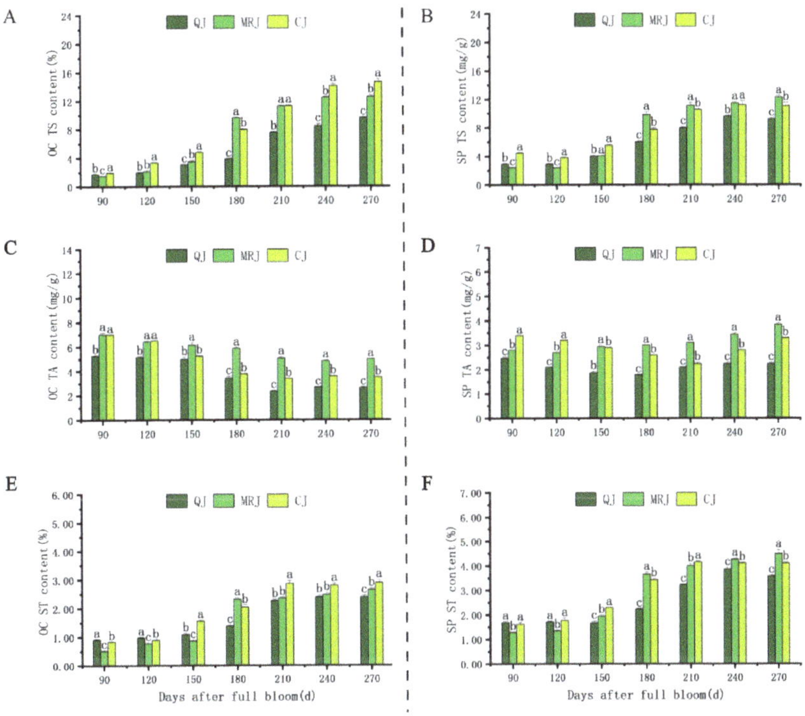

Figure 4. Changes in intracellular organic matter content in the peel of three late-maturing citrus cultivars at different growth and development stages. (**A**) Oil gland layer total sugar content. (**B**) Spongy layer total sugar content. (**C**) Oil gland layer total acid content. (**D**) Spongy layer total acid content. (**E**) Oil gland layer starch content. (**F**) Spongy layer total starch content. Different lowercase letters above the data of different varieties in the same period indicate statistically significant differences as determined by Duncan's test ($p < 0.05$).

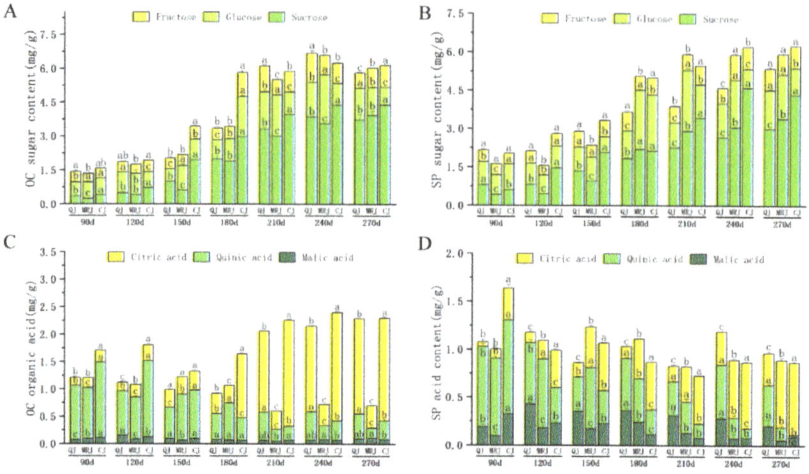

Figure 5. Changes in sugar and acid component content in the peel of three late-maturing citrus cultivars at different growth and development stages. (**A**) Oil gland layer sugar content. (**B**) Spongy layer sugar content. (**C**) Oil gland layer acid content. (**D**) Spongy layer acid content. Different lowercase letters above the data of different varieties in the same period indicate statistically significant differences as determined by Duncan's test ($p < 0.05$).

2.5. Comparison of Cell Wall Polysaccharide Content and Related Degrading Enzyme Activities

Changes in cell wall polysaccharide content and associated degrading enzyme activities in the peel of QJ, MRJ, and CJ during different developmental stages are shown in Figure 6. Throughout fruit development, all three cultivars exhibited consistent trends: The CL and HCL contents gradually decreased in both the oil gland layer and the spongy layer, while cellulase (Cx) and polygalacturonase (PG) activities progressively increased. From 150 to 270 d, the water-soluble pectin (WSP) content and the ion-bound pectin (ISP) content showed continuous accumulation, contrasting with the gradual decline in the chelator-soluble pectin (CSP) content. At full maturity, CJ displayed significantly lower ($p < 0.05$) CL, HCL, and CSP contents in both tissue layers compared to QJ and MRJ, while maintaining higher ISP accumulation. Notably, QJ exhibited significantly reduced ($p < 0.05$) pectate lyase (PL) and pectin methylesterase (PME) activities relative to both MRJ and CJ. Conversely, CJ demonstrated significantly enhanced ($p < 0.05$) Cx and PG activities in the oil gland layer compared to the other cultivars.

2.6. Comparison of Ca Content

The dynamic changes in Ca content within the peel of QJ, MRJ, and CJ during different developmental stages are presented in Figure 7. All three cultivars exhibited a consistent pattern of initial accumulation followed by gradual reduction in the Ca content across both the oil gland layer and the spongy layer, with the transition point occurring at approximately 180 d. This phase marked the beginning of significant Ca remobilization from peel tissues. At full maturity, the comparative analysis revealed that CJ maintained significantly lower ($p < 0.05$) Ca concentrations in both tissue layers relative to QJ and MRJ. The differential Ca retention capacity among the cultivars may contribute to observed variations in peel structural integrity and postharvest quality, as calcium plays crucial roles in cell wall stabilization and membrane function.

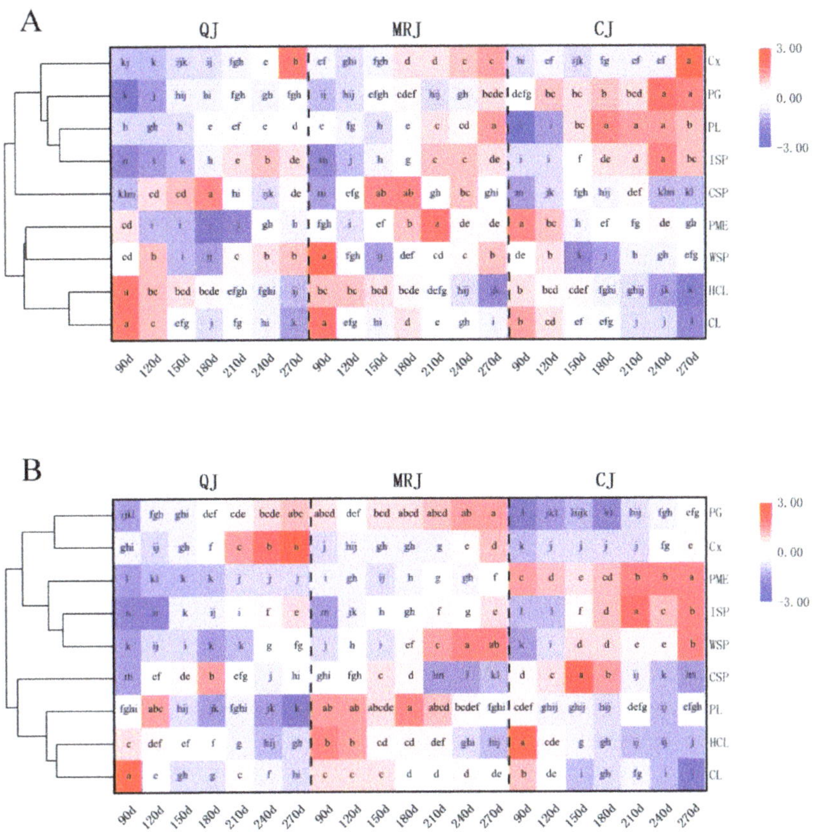

Figure 6. Changes in cell wall polysaccharide content and related degrading enzyme activities in the peel of three late-maturing citrus cultivars at different growth and development stages. (**A**) Oil gland layer. (**B**) Spongy layer. The data means were analyzed and visualized using the Heat Map with Dendrogram tool in Origin software. Different lowercase letters indicate statistically significant differences ($p < 0.05$) according to Duncan's multiple range test.

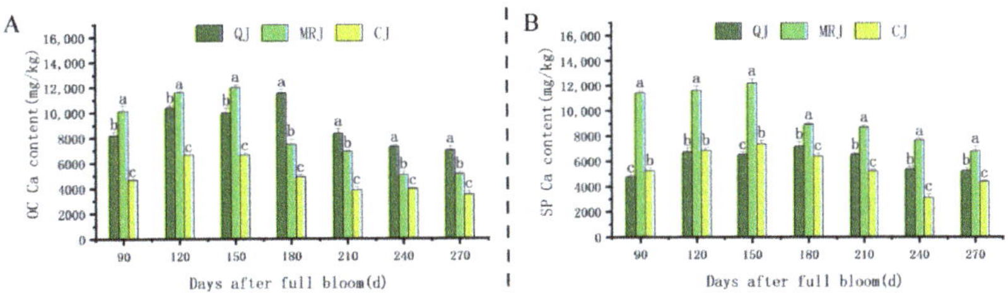

Figure 7. Changes in calcium content in the peel of three late-maturing citrus cultivars during different developmental stages. (**A**) Oil gland layer. (**B**) Spongy layer. Different lowercase letters above the data of different varieties in the same period indicate statistically significant differences as determined by Duncan's test ($p < 0.05$).

2.7. Comparison of Phytohormone Contents in Three Late-Maturing Citrus Cultivars

The dynamic changes in phytohormone contents within the peel of QJ, MRJ, and CJ at full maturity are presented in Figure 8. The ABA and ETH concentrations were highest across all the cultivars in both layers, followed by jasmonic acid (JA), salicylic acid (SA), and indole-3-acetic acid (IAA), with GA showing the lowest levels. Specifically, QJ's oil gland layer contained significantly higher ABA, JA, and SA but lower GA compared to MRJ and CJ ($p < 0.05$). In the spongy layer, QJ exhibited significantly greater ABA, ETH, and JA contents than the other cultivars ($p < 0.05$). However, the IAA levels showed no significant differences in either layer among the three late-maturing citrus cultivars ($p > 0.05$).

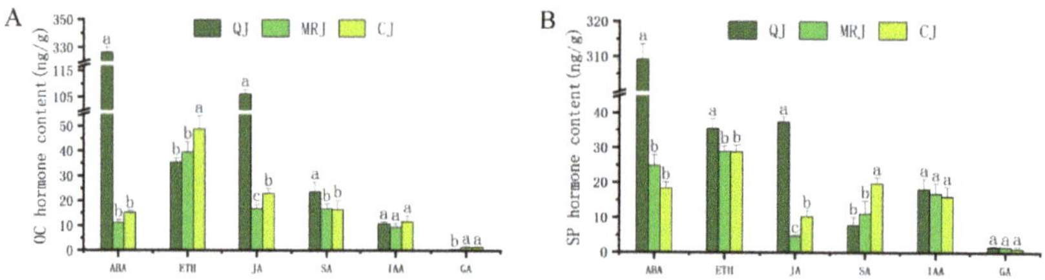

Figure 8. Changes in mesoelement contents in peel tissues of three late-maturing citrus cultivars at full maturity. (**A**) Oil gland layer. (**B**) Spongy layer. Different lowercase letters above the data of different varieties in the same period indicate statistically significant differences as determined by Duncan's test ($p < 0.05$).

2.8. Correlation Analysis of Peelability Determinants

The correlation analysis was conducted on key physiological indicators potentially influencing peelability formation in QJ, MRJ, and CJ cultivars throughout their developmental stages, based on current experimental data and previous studies on citrus peelability and fruit texture, as shown in Figure 9. The results demonstrated that peel hardness showed significant positive correlations with the TA, CL, HCL, and Ca contents in the oil gland layer ($p < 0.05$), while exhibiting significant negative correlations with the TS, ST, and ISP contents and the activities of Cx, PG, and PL ($p < 0.05$). No other indicators showed significant correlations with peel hardness. Similarly, peel adhesion was significantly positively correlated with the CL, HCL, and Ca contents in the spongy layer ($p < 0.05$), but negatively correlated with the TS, ST, WSP, and ISP contents and PME activity ($p < 0.05$), with no significant correlations observed for the other measured parameters.

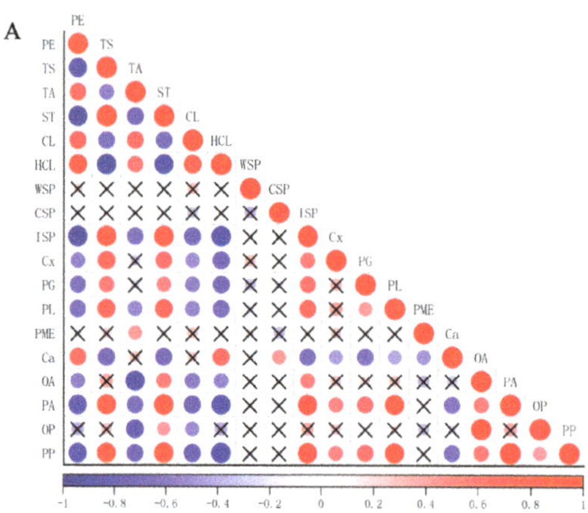

Figure 9. *Cont.*

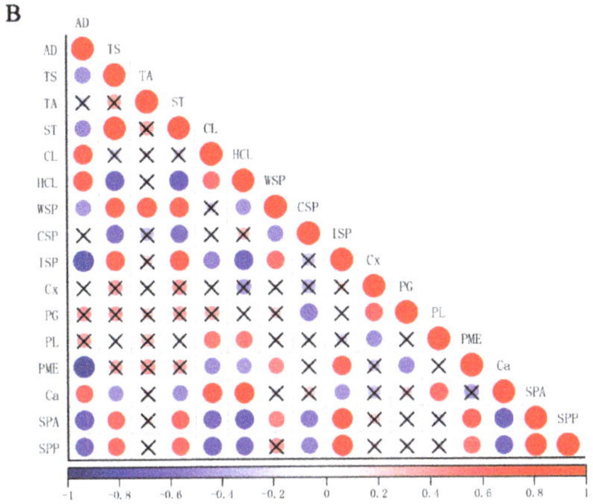

Figure 9. Correlation analysis of peel hardness and peel adhesion force with selected parameters at different growth and development stages. (**A**) Oil gland layer. (**B**) Spongy layer. PE: peel hardness; AD: peel adhesion.

2.9. Comparative Analysis of Gene Expression Patterns in Late-Maturing Citrus Cultivars

The gene expression profiles of TS, TA, and phytohormone biosynthesis-related genes were analyzed in the oil gland layers and the spongy layers of QJ, MRJ, and CJ cultivars at 90, 180, and 270 d using quantitative PCR in Figure 10. The results revealed distinct spatiotemporal expression patterns among the cultivars. At full maturity, QJ showed significantly higher expression levels of the TS synthesis gene *ALMT9* and the ABA biosynthesis genes *NCED1* and *NCED2* in the oil gland layers, while exhibiting lower expression of the GA biosynthesis genes *GA30X7* and *GA30X9*. Conversely, in the spongy layers, QJ demonstrated reduced expression of the ABA biosynthesis genes and the TA synthesis gene *CCR4A* compared to the other cultivars. These differential gene expression patterns provide molecular evidence for the observed physiological variations in peel characteristics during fruit maturation.

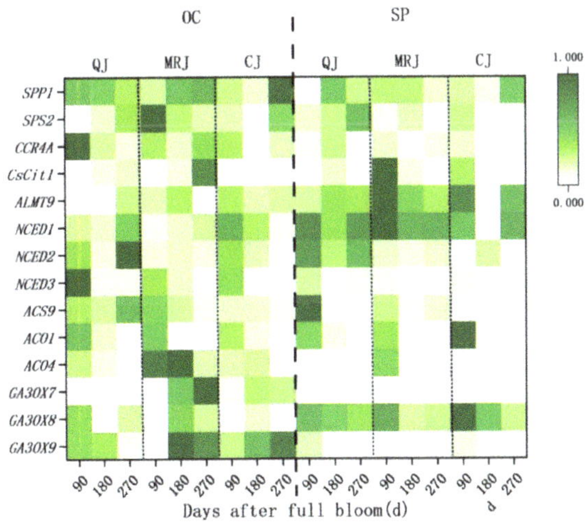

Figure 10. Changes in gene expression levels in the peel of three late-maturing citrus cultivars at different growth and development stages. OC: oil gland layer. SP: spongy layer. The mean values of relative expression data were normalized and analyzed using Origin software.

3. Discussion

3.1. Influence of Developmental Stages and Cellular Morphology on Peelability Formation in Late-Maturing Citrus Fruits

The maturation of citrus fruits is accompanied by continuous changes in fruit size, color, flavor, and peelability [31]. This experimental study found that the peel hardness and peel adhesion of QJ, MRJ, and CJ gradually decreased with fruit development and ripening, reaching their lowest values at full maturity, which represented the easiest peeling period for these three late-maturing citrus cultivars during the observation phase. These findings are consistent with previous research [32], with varietal differences showing QJ as the most difficult to peel, MRJ as the intermediate, and CJ as the easiest to peel.

The geometric characteristics of fruits, including their cellular arrangement, porosity, and tissue structure, significantly influence textural changes [33,34], as demonstrated by studies showing that apple fruit firmness correlates with parenchyma cell size, shape, intercellular strength, and bonding force in flesh tissue [35]. Our experimental results reveal that the peel cell perimeter and area progressively increase during fruit development and maturation, with QJ exhibiting larger oil glands but smaller, more uniform parenchyma and spongy layer cells with a regular morphology and more compact arrangement at full maturity compared to MRJ and CJ. These findings align with previous research demonstrating that exogenous calcium application promotes tighter cellular arrangement and reduced intercellular spaces in 'Bing tang' sweet orange, thereby enhancing fruit firmness and reducing cracking incidence [36]. The results collectively indicate that variations in peel characteristics among late-maturing citrus cultivars, combined with maturation-induced modifications in cellular morphology and arrangement, contribute to reduced peel hardness and adhesion, ultimately developing easy-peeling properties, while the distinct cellular morphology and arrangement patterns at full maturity account for the peelability differences observed among these late-maturing citrus cultivars.

3.2. Influence of Intracellular Organic Matter Accumulation on Peelability in Late-Maturing Citrus

The expansion of cell volume requires coordination between cell expansion rates and solute transport/accumulation. Increased solute accumulation and decreased solute potential lead to reduced turgor pressure, enhanced membrane permeability, and cell wall relaxation [37]. Soluble sugars, organic acids, and starch not only contribute to citrus fruit flavor but also serve as key solutes influencing fruit texture, with sucrose, glucose, and fructose being the primary osmotic regulators [38].

Our study revealed that during the observation period, TS and ST accumulation in both the oil gland and spongy layers of QJ, MRJ, and CJ increased as peelability improved, whereas TA accumulation in the spongy layer decreased with reduced peeling difficulty. These findings align with the upregulation of TS synthesis genes (*CsCit1* and *ALMT9*) and the downregulation of TA synthesis genes (*SPP1* and *CCR4A*) in QJ, which exhibited the most challenging peelability. During fruit maturation, the principal components of TS in the oil gland layer and spongy layer of QJ, MRJ, and CJ gradually shifted from glucose to sucrose, while the dominant constituent of TA transitioned from quinic acid to citric acid. At full maturity, sucrose and citric acid were the predominant TS and TA components in both tissue layers across all the cultivars, with QJ showing significantly lower TS, TA, and ST contents compared to MRJ and CJ, consistent with previous findings by Danilo et al. [39–41]. The results demonstrate that the accumulation of TS, TA, and ST cellular organic matter, combined with changes in the major components of TS and TA, may collectively reduce the osmotic potential and affect the membrane permeability. These changes decrease the cellular stability in the peels of QJ, MRJ, and CJ citrus varieties, leading to structural alterations that gradually develop easy-peeling characteristics. Moreover, differences in intracellular organic matter accumulation during full ripening cause variations in fruit peelability.

3.3. Influence of Cell Wall Polysaccharide Content and Related Degrading Enzyme Activities on Peelability in Late-Maturing Citrus

The plant cell wall is a complex structure primarily composed of cellulose, hemicellulose, and pectin polysaccharides, along with a small number of structural proteins, which collectively provide crucial mechanical support to plant cells [42,43]. Modifications in cell wall architecture and the depolymerization of its structural components are intimately linked to fruit softening, with this process being predominantly driven by the enzymatic degradation of polysaccharides, mediated by various cell wall hydrolases [44]. Throughout fruit development, the observed reduction in peel hardness and adhesion was associated with progressive decreases in CL and HCL content, coupled with elevated activities of cell wall-degrading enzymes, including Cx, PG, PME, and PL: findings that align with the established literature [17]. At full maturity, the comparative analysis revealed that CJ contained significantly lower CL and HCL levels in both the oil gland layer and the spongy layer compared to MRJ and QJ. These results demonstrate that the maturation-associated activation of Cx, PG, and other cell wall-modifying enzymes leads to progressive polysaccharide decomposition, resulting in cell wall loosening that reduces peel hardness and adhesion while enhancing peelability. Furthermore, the cultivar-specific variations in cell wall polysaccharide content and enzymatic degradation profiles ultimately determine the distinct peelability characteristics observed among these late-maturing citrus cultivars at commercial maturity.

3.4. Influence of Peel Ca Content on Peelability in Late-Maturing Citrus

Calcium, an essential plant nutrient and intracellular mineral element, plays a crucial role in fruit texture formation by participating in osmoregulation, influencing cell metabolism and division [17,36], while also reducing cell wall-degrading enzyme activity to maintain cell wall structure and function [45]. This study found that the calcium content in both the oil gland layer and the spongy layer of QJ, MRJ, and CJ initially increased and then decreased as peelability improved, with the correlation analysis revealing significant positive relationships between the calcium content and peel hardness, peel adhesion, and the CL and HCL contents, while showing significant negative correlations with TS and TA. At full maturity, CJ exhibited a significantly lower calcium content in both tissue layers compared to QJ and MRJ, indicating that reduced calcium levels influence the accumulation of cellular organic matter and cell wall polysaccharides, thereby affecting the cellular membrane and wall homeostasis and modifying the cell size and arrangement patterns, ultimately leading to decreased peel hardness and adhesion and facilitating peelability development, with the observed varietal differences in the calcium content accounting for peelability variations among the cultivars, consistent with calcium's established role as a signaling molecule that crosslinks homogalacturonan to enhance pectin stability, inhibit polygalacturonase activity, and delay cell wall degradation [46].

3.5. Influence of Phytohormones on Peelability in Late-Maturing Citrus

Fruit texture modification is coordinately regulated by multiple phytohormones. Previous studies have demonstrated that ETH and ABA can modulate fruit firmness in strawberries [47] and blueberries [48] by regulating pectin and other cell wall polysaccharides, while phytohormone levels may also influence cell wall stability through the modulation of cell wall-degrading enzyme activities [49,50]. Our results showed that at full maturity, QJ exhibited significantly higher ABA and JA contents in both the oil gland layer and the spongy layer compared to MRJ (moderately difficult to peel) and CJ. Conversely, CJ displayed a significantly higher ETH content in its oil glands and a significantly higher SA content in its spongy layer than both QJ and MRJ. These hormonal differences were accompanied by the elevated expression of key biosynthesis genes (*GA30X8*, *ACS9*, *NCED1*, *NCED2*, and *NCED3*), consistent with previous findings [51,52]. These results indicate that the differential accumulation of ABA, ETH, SA, and JA at full maturity likely affects the final determination of peel hardness and adhesion, thereby contributing to the observed peelability differences among the QJ, MRJ, and CJ cultivars.

4. Materials and Methods

4.1. Test Site and Materials

The experiment was carried out in the Institute of Fruit Trees, College of Horticulture, Sichuan Agricultural University (30.88°′ N, 104.25°′ E) from June 2023 to October 2024. The fruit samples of QJ, MRJ, and CJ were collected from 8-year-old adult fruit trees on orange rootstocks in the same citrus orchard in Sunjia Village, Xie'an Town, Renshou County, Sichuan Province. The area has a subtropical monsoon humid climate with an altitude of 450–880 m and an average annual temperature of 17.4 °C. The average annual rainfall is 945.9 mm, and the average sunshine duration is 1196.6 h. The orchard employed sprinkler irrigation systems, typically operating from 9:00 PM to 6:00 AM, with the irrigation frequency being adjusted according to the prevailing weather conditions. Sampling began at 90 days after full bloom, and then 36 fruits of 7 developmental stages were selected every 30 days. The peels were stored at −80 °C for the experimental analysis.

4.2. Determination of Mechanical Indexes of Peel

Peel hardness was measured on the equatorial surface of the citrus fruit using a texture analyzer (Food Technology Corporation, Dulles Town Center, VA, USA) and a cylindrical probe TA/5 (diameter 5 mm). The texture condition setting conditions were as follows: the initial force was 0.75 N, the puncture speed was 1 mm/s, the return speed was 1 mm/s, the puncture distance was 25 mm, and the maximum resistance recorded by the texture analyzer was the peel hardness. Peel viscosity was measured using a texture analyzer (Food Technology Corporation, Virginia, USA) and a clamp probe. Each fruit sample was cut into blocks with an average growth of about 40 mm and a width of about 20 mm on the equatorial plane. The short side of the peel was fixed by the clamp, and the pulp on one side was pulled down vertically with tweezers until the peel and pulp were completely separated. The texture analyzer began to count when the tweezers clamped the pulp. The maximum resistance recorded by the texture analyzer was the peel adhesion.

4.3. Fruit Quality Determination

The fruit longitudinal diameter, transverse diameter, peel thickness, and pulp diameter were measured by vernier caliper. The TA content was measured using a sugar–acid all-in-one machine (ATAGO, Tokyo, Japan). The contents of TS and ST were measured by anthrone ethyl acetate and concentrated sulfuric acid reagent. The sugar components in the tissue were extracted by ethanol solution, and the sugar components were measured. The acid components were determined by high performance liquid chromatography. The determination conditions of high performance liquid chromatography were as follows: column temperature 25 °C, injection volume 10 μL, mobile phase 4% methanol aqueous solution, and flow rate 0.8 mL/min.

4.4. Paraffin Section Preparation and Cell Microscopic Observation

Saffron solid green paraffin sections were made by Seville Biotechnology Co., Ltd. (Wuhan, Hubei). An Olympus BH-2 optical microscope (Olympus, Japan) was used to observe and photograph the appropriate field of view at 20 times magnification, and Image J was used to measure the perimeter and area of pericarp cells.

4.5. Determination of Cell Wall Polysaccharide Content and Related Degradation Enzyme Activity

The cell wall material (CWM) was extracted from 1 g of dried and ground peel samples with dimethyl sulfoxide and acetone reagents. Then, 0.05 g of the CWM was added to sodium acetate twice and Na_2CO_3 once. The contents of WSP, ISP, and CSP were determined by carbazole colorimetry. Then NaOH and concentrated sulfuric acid were added to the precipitate to determine the contents of CL and HCL by anthrone colorimetry. Cx, PG, PL, and PME activities were measured using commercial kits.

4.6. Determination of Mineral Element Content

First, 0.5 g of dried and grinded peel samples were digested with nitric acid on an adjustable electric heating plate, and the contents of P, K, S, P, Zn, Ca, Mg, Fe, Cu, and Mn were determined by an atomic absorption spectrophotometer.

4.7. Determination of Plant Hormone Content

The contents of IAA, CTK, GA, ABA, and ETH were determined by high performance liquid chromatography (HPLC) from Sanshu Biotechnology Co., Ltd. (Nantong, China). Then, 100 mg of sample was accurately weighed, and 1.0 mL of pre-cooled 50% acetonitrile (ACN) aqueous solution was added. Subsequently, the sample was sonicated at 4 °C for 3 min, and then extracted at 4 °C for 30 min. After extraction, the sample was centrifuged at 12,000 rpm for 10 min at 4 °C, and the supernatant was collected for further processing and analysis. The results were based on high resolution mass spectrometry, and a qualitative analysis was performed by the identification of secondary mass spectrometry fragments. A quantitative analysis was performed by the external standard method.

4.8. Quantitative Reverse Transcription Polymerase Chain Reaction

RNA was extracted from peel samples using the chloroform–isopropanol method, cDNA was obtained using abm's 5X All-In-One RT MasterMix reverse transcription kit, and a fluorescence quantitative polymerase chain reaction was performed using a Beijing Jumei Co., Ltd., commercial kit (Beijing Changping) and a fluorescence quantitative PCR instrument (Bio-Rad Laboratories, Inc, Hercules, CA, USA). Each reaction system used 10 μL and the reaction system included 5 μL 5 × M5 RT Super plus Mix, 3 μL ddH_2O, 1 μL cDNA, and 1 μL 10 mM primer. Actin was used as an internal control. The $2^{-\Delta\Delta CT}$ method was applied to calculate the relative gene expression.

4.9. Statistical Analysis

Statistical significance was evaluated using an analysis of variance (ANOVA) in SPSS 23.0 (IBM, Armonk, NY, USA), with a significance threshold set at $p < 0.05$. Origin 2021 (Origin Lab Corporation, Northampton, MA, USA) software was used to draw the figures.

5. Conclusions

The results of this study demonstrate that the peelability of QJ, MRJ, and CJ gradually decreases during fruit development and ripening, with all three cultivars exhibiting easier peeling characteristics at full maturity, following the order of QJ (most difficult to peel) > MRJ (moderate) > CJ (easiest). Throughout the phenological period, compared to MRJ and CJ, QJ showed several distinctive features: larger oil glands, more regularly shaped and uniformly sized spongy layer cells with a tighter arrangement, a lower accumulation of intracellular organic matter (including soluble sugars, titratable acids, and starch), and a higher content of cell wall polysaccharides (cellulose and hemicellulose) and calcium. These structural characteristics likely contribute to QJ's superior peel integrity and most-difficult peelability. At full maturity, the observed differences in mineral elements and phytohormones may lead to cultivar-specific variations in organic matter accumulation and cell wall degradation, ultimately affecting their cellular morphology and arrangement and resulting in different peelability characteristics among these citrus cultivars. These results provide a reference for future studies on citrus peelability.

Author Contributions: Y.Y.: formal analysis, visualization, writing—original draft, methodology, and software. Z.H. and Y.W.: formal analysis, visualization, and writing—original draft. L.D., T.W. and D.C.: formal analysis and visualization. L.L. and B.X.: data curation and investigation. M.T.: methodology. Z.W. and J.W.: conceptualization, project administration, visualization, and writing—review and editing. All authors have read and agreed to the published version of the manuscript.

Funding: This research was supported by the National Key R&D Program of China [Grant No. 2021YFD1600802-02], the 14th–fifth–plan of Breeding in Sichuan Province, China [Grant No. 2021YFYZ0023-14], and the Sichuan Provincial Science and Technology Plan Project [Grant No. 2024YFHZ0258].

Data Availability Statement: Data are contained within the article.

Conflicts of Interest: The authors declare no conflicts of interest.

References

1. Bayer, R.J.; Mabberley, D.J.; Morton, C.; Miller, C.H.; Sharma, I.K.; Pfeil, B.E.; Rich, S.; Hitchcock, R.; Sykes, S. A molecular phylogeny of the orange subfamily (Rutaceae: Aurantioideae) using nine cpDNA sequences. *Am. J. Bot.* **2009**, *96*, 668–685. [CrossRef] [PubMed]
2. Harker, F.R.; Hallett, I.C.; White, A.; Seal, A.G. Measurement of fruit peelability in the genus actinidia. *J. Texture Stud.* **2011**, *42*, 237–246. [CrossRef]
3. Goudeau, D.; Uratsu, S.L.; Inoue, K.; Dasilva, F.G.; Leslie, A.; Cook, D.; Reagan, R.L.; Dandekar, A.M. Tuning the orchestra: Selective gene regulation and orange fruit quality. *Plant Sci.* **2008**, *174*, 310–320. [CrossRef]
4. Goldenberg, L.; Yaniv, Y.; Porat, R.; Carmi, N. Mandarin fruit quality: A review. *J. Sci. Food Agric.* **2018**, *98*, 18–26. [CrossRef]
5. Goldenberg, L.; Yaniv, Y.; Kaplunov, T.; Doron-Faigenboim, A.; Porat, R.; Carmi, N. Genetic Diversity among Mandarins in Fruit-Quality Traits. *J. Agric. Food Chem.* **2014**, *62*, 4938–4946. [CrossRef]
6. Simons, T.; Sivertsen, H.; Guinard, J.-X. Mapping the Preferences of Adult and Child Consumers for California-grown Mandarins. *HortScience* **2018**, *53*, U1029–U1138. [CrossRef]
7. Yu, X.; Zhang, X.; Jiang, D.; Zhu, S.; Cao, L.; Liu, X.; Shen, W.; Zhao, W.; Zhao, X. Genetic diversity of the ease of peeling in mandarins. *Sci. Hortic.* **2021**, *278*, 109852. [CrossRef]
8. Sadka, A.; Shlizerman, L.; Kamara, I.; Blumwald, E. Primary Metabolism in Citrus Fruit as Affected by Its Unique Structure. *Front. Plant Sci.* **2019**, *10*, 1167–1180. [CrossRef]
9. Cháfer, M.; González-Martínez, C.; Chiralt, A.; Fito, P. Microstructure and vacuum impregnation response of citrus peels. *Food Res. Int.* **2003**, *36*, 35–41. [CrossRef]
10. Feng, G.; Wu, J.; Xu, Y.; Lu, L.; Yi, H. High-spatiotemporal-resolution transcriptomes provide insights into fruit development and ripening in Citrus sinensis. *Plant Biotechnol. J.* **2021**, *19*, 1337–1353. [CrossRef]
11. Ibáñez, A.M.; Martinelli, F.; Reagan, R.L.; Uratsu, S.L.; Vo, A.; Tinoco, M.A.; Phu, M.L.; Chen, Y.; Rocke, D.M.; Dandekar, A.M. Transcriptome and metabolome analysis of Citrus fruit to elucidate puffing disorder. *Plant Sci.* **2014**, *217*, 87–98. [CrossRef] [PubMed]
12. Krishna, K.S.; Reddy, B.S. Post-harvest physico-mechanical properties of orange peel and fruit. *J. Food Eng.* **2006**, *73*, 112–120.
13. Blaker, K.M.; Olmstead, J.W. Cell wall composition of the skin and flesh tissue of crisp and standard texture southern highbush blueberry genotypes. *J. Berry Res.* **2015**, *5*, 9–15. [CrossRef]
14. Waldron, K. Plant structure and fruit and vegetable texture. *Texture Food Solid Foods* **2004**, *2*, 241–258.

15. Reeve, R.M. Relationships of histological structure to texture of fresh and processed fruits and vegetables. *J. Texture Stud.* **1970**, *1*, 247–284. [CrossRef]
16. Wang, J.; Mujumdar, A.S.; Wang, H.; Fang, X.-M.; Xiao, H.-W.; Raghavan, V. Effect of drying method and cultivar on sensory attributes, textural profiles, and volatile characteristics of grape raisins. *Dry. Technol.* **2021**, *39*, 495–506. [CrossRef]
17. Huang, Y.; Qin, S.; He, J.; Lyu, D. Integration of cell wall fraction, organic matter content, and membrane to understand crispness changes in apples. *Sci. Hortic.* **2023**, *321*, 112309. [CrossRef]
18. Stitt, M.; Zeeman, S.C. Starch turnover: Pathways, regulation and role in growth. *Curr. Opin. Plant Biol.* **2012**, *15*, 282–292. [CrossRef]
19. Palmer, J.W.; Harker, F.R.; Tustin, D.S.; Johnston, J. Fruit dry matter concentration: A new quality metric for apples. *J. Sci. Food Agric.* **2010**, *90*, 2586–2594. [CrossRef]
20. Itai, A.; Tanahashi, T. Inhibition of sucrose loss during cold storage in Japanese pear (*Pyrus pyrifolia* Nakai) by 1-MCP. *Postharvest Biol. Technol.* **2008**, *48*, 355–363. [CrossRef]
21. Jia, K.; Zhang, Q.; Xing, Y.; Yan, J.; Liu, L.; Nie, K. A Development-Associated Decrease in Osmotic Potential Contributes to Fruit Ripening Initiation in Strawberry (*Fragaria ananassa*). *Front. Plant Sci.* **2020**, *11*, 1035. [CrossRef] [PubMed]
22. Videcoq, P.; Steenkeste, K.; Bonnin, E.; Garnier, C. A multi-scale study of enzyme diffusion in macromolecular solutions and physical gels of pectin polysaccharides. *Soft Matter* **2013**, *9*, 5110–5118.
23. Houben, K.; Jolie, R.P.; Fraeye, I.; Van Loey, A.M.; Hendrickx, M.E. Comparative study of the cell wall composition of broccoli, carrot, and tomato: Structural characterization of the extractable pectins and hemicelluloses. *Carbohydr. Res.* **2011**, *346*, 1105–1111. [CrossRef]
24. Goulao, L.; Oliveira, C. Cell wall modifications during fruit ripening: When a fruit is not the fruit. *Trends Food Sci. Technol.* **2008**, *19*, 4–25. [CrossRef]
25. Philip, J.H.; Bruce, A.S. Chemistry and Molecular Organization of Plant Cell Walls. In *Biomass Recalcitrance: Deconstructing the Plant Cell Wall for Bioenergy*; Blackwell Publishing Ltd.: Hoboken, NY, USA, 2009.
26. Gilbert, H.J. The Biochemistry and Structural Biology of Plant Cell Wall Deconstruction. *Plant Physiol.* **2010**, *153*, 444–455. [CrossRef]
27. Höfte, H.; Voxeur, A. Plant cell walls. *Curr. Biol.* **2017**, *27*, R865–R870. [CrossRef]
28. Gapper, N.E.; McQuinn, R.P.; Giovannoni, J.J. Molecular and genetic regulation of fruit ripening. *Plant Mol. Biol.* **2013**, *82*, 575–591. [CrossRef]
29. Brummell, D.A. Cell wall disassembly in ripening fruit. *Funct. Plant Biol.* **2006**, *33*, 103–119. [CrossRef]
30. White, P.J.; Broadley, M.R. Calcium in Plants. *Ann. Bot.* **2003**, *92*, 487–511. [CrossRef]
31. Alfaro, J.M.; Bermejo, A.; Navarro, P.; Quiñones, A.; Salvador, A. Effect of Rootstock on Citrus Fruit Quality: A Review. *Food Rev. Int.* **2023**, *39*, 2835–2853. [CrossRef]
32. Yu, X.; Zhang, X.; Liu, X.; Ren, Y.; Jiang, D.; Shen, W.; Zhao, X.; Cao, L. Comparative transcriptomic profile of two mandarin varieties during maturation reveals pectinase regulating peelability. *Sci. Hortic.* **2024**, *331*, 113148. [CrossRef]
33. Rahman, M.M.; Joardder, M.U.; Karim, A. Non-destructive investigation of cellular level moisture distribution and morphological changes during drying of a plant-based food material. *Biosyst. Eng.* **2018**, *169*, 126–138. [CrossRef]
34. Liu, D.; Guo, W.; Li, Q.; Xie, D. Relationship of the bulk optical properties in 950–1650 nm wavelength range with internal quality and microstructure of kiwifruit during maturation. *Biosyst. Eng.* **2019**, *184*, 45–54. [CrossRef]
35. Gwanpua, S.G.; Verlinden, B.E.; Hertog, M.L.; Nicolai, B.M.; Hendrickx, M.; Geeraerd, A. Slow softening of Kanzi apples (*Malus×domestica* L.) is associated with preservation of pectin integrity in middle lamella. *Food Chem.* **2016**, *211*, 883–891. [CrossRef]
36. Dong, Z.; Shi, X.; Liu, X.; Srivastava, A.K.; Shi, X.; Zhang, Y.; Hu, C.; Zhang, F. Calcium application regulates fruit cracking by cross-linking of fruit peel pectin during young fruit growth stage of citrus. *Sci. Hortic.* **2025**, *340*, 113922. [CrossRef]
37. Wada, H.; Matthews, M.A.; Shackel, K.A. Seasonal pattern of apoplastic solute accumulation and loss of cell turgor during ripening of Vitis vinifera fruit under field conditions. *J. Exp. Bot.* **2009**, *60*, 1773–1781. [CrossRef]
38. Naoki, S.; Ichiro, I.S.; Shoji, T.; Ryoichi, Y. Texture Evaluation of Cucumber by a New Acoustic Vibration Method. *J. Jpn. Soc. Hortic. Sci.* **2005**, *74*, 31–35.
39. Centeno, D.C.; Osorio, S.; Nunes-Nesi, A.; Bertolo, A.L.; Carneiro, R.T.; Araújo, W.L.; Steinhauser, M.-C.; Michalska, J.; Rohrmann, J.; Geigenberger, P.; et al. Malate Plays a Crucial Role in Starch Metabolism, Ripening, and Soluble Solid Content of Tomato Fruit and Affects Postharvest Softening. *Plant Cell* **2011**, *23*, 162–184. [CrossRef]
40. Zhang, Y.; Li, P.; Cheng, L. Developmental changes of carbohydrates, organic acids, amino acids, and phenolic compounds in 'Honeycrisp' apple flesh. *Food Chem.* **2010**, *123*, 1013–1018. [CrossRef]
41. Huang, X.-Y.; Wang, C.-K.; Zhao, Y.-W.; Sun, C.-H.; Hu, D.-G. Mechanisms and regulation of organic acid accumulation in plant vacuoles. *Hortic. Res.* **2021**, *8*, 227. [CrossRef]

42. Lahaye, M.; Falourd, X.; Laillet, B.; Le Gall, S. Cellulose, pectin and water in cell walls determine apple flesh viscoelastic mechanical properties. *Carbohydr. Polym.* **2020**, *232*, 115768. [CrossRef] [PubMed]
43. Pauly, M.; Gille, S.; Liu, L.; Mansoori, N.; de Souza, A.; Schultink, A.; Xiong, G. Hemicellulose biosynthesis. *Planta* **2013**, *238*, 627–642. [CrossRef]
44. Kumar, M.; Campbell, L.; Turner, S. Secondary cell walls: Biosynthesis and manipulation. *J. Exp. Bot.* **2016**, *67*, 515–531. [CrossRef]
45. Punumong, P.; Sangsuwan, J.; Kim, S.M.; Rattanapanone, N. Combined Effect of Calcium Chloride and Modified Atmosphere Packaging on Texture and Quality of Minimally-Processed Litchi Fruit. *Chiang Mai J. Sci.* **2016**, *43*, 556–569.
46. Braybrook, S.A.; Hofte, H.; Peaucelle, A. Probing the mechanical contributions of the pectin matrix. *Plant Signal. Behav.* **2012**, *7*, 1037–1041. [CrossRef]
47. Guo, J.; Wang, S.; Yu, X.; Dong, R.; Li, Y.; Mei, X.; Shen, Y. Polyamines Regulate Strawberry Fruit Ripening by Abscisic Acid, Auxin, and Ethylene. *Plant Physiol.* **2018**, *177*, 339–351. [CrossRef]
48. Zhou, Q.; Zhang, F.; Ji, S.; Dai, H.; Zhou, X.; Wei, B.; Cheng, S.; Wang, A. Abscisic acid accelerates postharvest blueberry fruit softening by promoting cell wall metabolism. *Sci. Hortic.* **2021**, *288*, 110325. [CrossRef]
49. Qin, Z.; Pan, J.; Li, J.; Sun, J.; Khoo, H.E.; Dong, X. Effects of 1-methylcyclopropene and abscisic acid treatments on texture properties and microstructures of postharvest tangerine (*Citrus reticulata* cv. Orah). *J. Food Process. Preserv.* **2022**, *46*, e16633. [CrossRef]
50. Li, R.; Wang, Y.; Li, W.; Shao, Y. Comparative Analyses of Ripening, Texture Properties and Cell Wall Composition in Three Tropical Fruits Treated with 1-Methylcyclopropene during Cold Storage. *Horticulturae* **2023**, *9*, 126. [CrossRef]
51. Wu, W.; Bao, Z.-Y.; Xiong, C.-X.; Shi, L.-Y.; Chen, W.; Yin, X.-R.; Yang, Z.-F. The Softening of Persimmon Fruit Was Inhibited by Gibberellin via DkDELLA1/2. *J. Agric. Food Chem.* **2024**, *73*, 1159–1166. [CrossRef]
52. Haji, T. Inheritance of Flesh Texture in Peach and Effects of Ethylene Treatment on Softening of the Stony Hard Peach. *Jpn. Agric. Res. Q.* **2014**, *48*, 57–61. [CrossRef]

plants

Article

A Novel Approach for Comparing Selected Metabolites in Citrus Leaves and Fruits Across Datasets

Ryan C. Traband [1,†], Xuesong Wang [2,†], Mariano Resendiz [1], Megan Meng [1], Yoko Hiraoka [1], Qiong Jia [2], Rendell Chang [2], Ethan Eurmsirilerd [2], Tracy Kahn [1], Peggy A. Mauk [1], Amancio De Souza [3], Anil Bhatia [3], Haiyan Ke [3], Donald Merhaut [1], Mikeal L. Roose [1], Zhenyu Jia [1,2,*] and John M. Chater [4,*]

1 Department of Botany and Plant Sciences, University of California, Riverside, CA 92521, USA; ryan.traband@email.ucr.edu (R.C.T.); peggym@ucr.edu (P.A.M.); donald.merhaut@ucr.edu (D.M.)
2 Genetics, Genomics, and Bioinformatics Program, University of California, Riverside, CA 92521, USA
3 Metabolomics Core Facility, University of California, Riverside, CA 92521, USA; haiyanke@ucr.edu (H.K.)
4 Department of Horticultural Sciences, University of Florida, Lake Alfred, FL 33850, USA
* Correspondence: zhenyuj@ucr.edu (Z.J.); jchater@ufl.edu (J.M.C.)
† These authors contributed equally to this work.

Abstract: Citrus fruits are valued not only for their nutritional benefits but also for their rich phytochemical content. Metabolomics has emerged as a comprehensive technique for assessing the chemical composition of fruits. The botanical connection between leaves, flowers, and fruits is reflected in both their structure and chemical composition, particularly in the flow of nutrients between plant organs. We introduced a new logarithm ratio-based approach to compare metabolite profiles between fruits and leaves. We hypothesize that this method allows for the analysis of multiple citrus metabolomic profiles to reveal known and novel correlation patterns, reflecting the dynamic connections between metabolic sources. To test this hypothesis, we leveraged comprehensive leaf metabolomic profiles from over 200 accessions in the Givaudan Citrus Variety Collection and reviewed published metabolomics data for fruits and juices of matching citrus types. By employing logarithm-transformed metabolic ratios within each dataset, we accounted for systematic differences across metabolomic platforms, achieving an unbiased analysis.

Keywords: metabolomics; method; citrus; comparison; datasets

1. Introduction

Citrus production holds significant importance in the agricultural sector on a global scale, with annual production at 158.5 million tons. As of 2021, the top citrus-producing countries include China, Brazil, India, and Mexico [1]. In the United States, citrus fruits account for 14% of the total fresh fruit volume, with oranges ranking fourth in terms of per capita consumption, behind bananas, apples, and melons. Highly valued for their flavor, vitamin C, lycopene, and high phenolic content, citrus fruits are also used for their distinct scents and oils, derived from the rind, in a variety of applications [2]. In light of these remarkable attributes, citrus, along with other fruits, serves as a veritable treasure trove of diverse phytochemicals, all synthesized within a single organism. In fact, plants outshine all other kingdoms on Earth by possessing the highest number and the most diverse set of secondary metabolites distributed across different parts of their bodies [3]. In the fruit industry, various metrics are employed to assess the chemical composition of fruits, including total soluble solids, total acidity, flavor, texture, and water content, using a variety of assays. However, metabolomics is a powerful tool for comprehensively analyzing plant biochemical diversity when used in sync with genetics or transcriptomics. Metabolomics is the study of all small molecules, known as metabolites, detected within a biological system. It provides insights into biochemical processes by analyzing the composition, concentration, and interactions of metabolites, reflecting the organism's physiological state. This powerful but costly method uses techniques like mass spectrometry (MS) and nuclear magnetic resonance (NMR). Metabolites are either detected through targeted analysis where specific metabolites and their isomers can be quantified in terms such as mg/L and µg/g or untargeted analysis often measured in arbitrary units (a.u). Metabolomics helps uncover metabolic pathways, identify biomarkers, and understand how genetics and environmental factors influence metabolism.

Metabolomics offers valuable insights into metabolic pathways that contribute to species differentiation and bioactive compound accumulation. In citrus, metabolomics has been instrumental in identifying key genetic factors influencing flavonoid and coumarin biosynthesis. Recent large-scale studies, such as that by Xiao Liang (2024) [4], have mapped metabolic variation across diverse citrus accessions, linking thousands of genetic markers to metabolite accumulation patterns. Their findings highlight how distinct citrus species have undergone metabolic specialization, particularly in pathways governing flavonoid glycosylation and coumarin accumulation, which influence both the nutritional properties and potential pharmacological effects of citrus fruits. By integrating metabolomic and genomic data, researchers can better understand how selective breeding and evolutionary pressures have shaped citrus metabolite profiles, providing a foundation for developing varieties optimized for health benefits and reduced adverse effects [4]. This approach, using genetics and transcriptomics in conjunction with metabolomics, has been widely adopted in plant research. For instance, Wang et al. analyzed metabolomic profiles in various tissues of a single plant [5]. Yazici et al. compared the metabolic profiles of identical tissues among different genotypes [6]. Additionally, Asai et al. explored the variability in metabolite abundance within a specific tissue type across different species of citrus subjected to tissue wounding [7]. In this paper, we introduce a new log-ratio-based approach to compare metabolite profiles between fruits and leaves and across major citrus groups.

2. Materials and Methods

2.1. Fruit Datasets and Selection Criteria Used

Fruit datasets were chosen in terms of their ability to provide raw metabolomic data for Citrus fruit flesh, specifically juice or pulp. Data needed to be displayed in units such as (a.u.) for untargeted work and ng/mL, µg/mL, mg/L, µg/g, nmol/g, or a.u. for targeted work. Publications providing data exclusively in the form of principal components or in formats inaccessible unless through paid or outdated software were also excluded. Shi Feng et al. (2018) focused on the metabolite profiles responsible for how preharvest factors, including genotype, rootstock, and grove location, influence the quality of mandarin fruit [8]. Volatile analysis was carried out on a Clarus 680 gas chromatograph (GC) (PerkinElmer, Inc., Waltham, MA, USA) equipped with a Clarus SQ 8T mass spectrometry (MS) detector. The publication provided a broader list of metabolites but focused exclusively on only four varieties of mandarin hybrids [8]. The four varieties studied were Clementines (*Citrus clementina* hort. ex Tanaka), Page (*Citrus tangelo*), Daisy (*Citrus reticulata* Blanco), and Satsumas (*Citrus unshiu* Marcovitch). Each variety had fruit collected from three different locations: Clementines were sourced from Los Angeles, CA, USA, Pasadena, CA, USA, and Fowler, CA, USA, Page mandarins were sourced from Fowler, CA, USA, and Daisy Mandarins were from Lindcove, CA, USA. For each variety, three fruits were sampled to prepare the juice, flesh, and other tissues. Further details on replicates were not specified.

Wang S (2016) analyzed 47 different cultivars of citrus using a non-targeted metabolomics approach [5]. They detected more than 2000 metabolite signals, from which more than 54 metabolites, including amino acids, flavonoids, and limonoids, were identified/annotated to understand the spatial and temporal distribution of metabolites in different species and different tissues. The 47 accessions were collected from the following species: lemon (*Citrus lemon* [L.] Burm f.), pummelo (*Citrus grandis* (L.) Osbeck), grapefruit (*Citrus paradisi* Macf), sweet orange (*C. sinensis* [L.] Osbeck), and mandarin (*C. reticulata* Marcf.). Cultivars were harvested randomly from trees in three positions and pooled for each biological replicate. The fruits were washed and the tissues of the flavedo, albedo, segment membrane (SM), and juice sacs (JS) were separated.

In a paper published by Shouchuang Wang in 2024, a comprehensive metabolic profiling of various citrus species from different locations was performed, and metabolic profiles were compared among the species, with a focus on the phenylpropanoid metabolic pathway [9]. In this study, a total of 189 citrus germplasm resources of different types were collected, mainly including 154 pummelos, as well as 15 mandarins, 15 oranges, 1 lemon, 1 citron, and 1 papeda. For the reasonableness of sampling, five to ten mature fruits around the crown of a tree were picked from three random positions per tree as biological replicates. Two biological replicates per cultivar were collected. The structures of 360 metabolites were identified by comparing the mass spectral fragments and retention times of standards, by the manual annotation of mass spectral profiles, and by searching metabolic databases; these included primary metabolites of amino acids and vitamins, as well as secondary metabolites such as lipids and phenylpropanoids. A multiple reaction monitoring method based on a triple quadrupole–linear ion trap mass spectrometer (API 4000, AB Sciex, Framingham, MA, USA) was used for the quantitative analysis of metabolites and for accurate time-of-flight mass spectrometry. More specific information on these datasets can be found in their respective publications.

Wang (2017) [10] analyzed 62 Citrus accessions from a collection of popular/local cultivars in China. Twelve to twenty-one healthy fruits true to their cultivars and at commercial maturity were randomly collected from the peripheral canopies of at least three trees and were randomly divided into three biological replicates [10]. The washed material was separated into four tissues, including flavedo, albedo, SM, and JSs, and immediately placed in liquid nitrogen. Qualitative metabolic analysis via HPLC-DAD-ESI-QqTOF-MS/MS (6520B, Agilent, Santa Clara, CA, USA) was performed in the targeted MS2 mode. The UV spectra (DAD) were recorded from 270 to 380 nm. The raw data were analyzed using MassHunter software (https://www.agilent.com/en/promotions/masshunter-mass-spec) and the processing method was the same as previously described 32. Quantitative analysis of metabolites was carried out in the multiple reaction monitoring (MRM) mode by LC-ESI-Q TRAP-MS/MS (4000Q TRAP, ABI, Los Angeles, CA, USA). For a list of samples used from all datasets, please refer to Supplementary Table S1.

2.2. *Leaf Dataset*

Leaf material was collected from the Givaudan Citrus Variety Collection on a single day in both 2020 and 2022, with sampling from different parts of the canopy in all cardinal directions. In 2020, due to limited research activity during the COVID-19 lockdown, leaves were only collected from 60 varieties, with 2 trees sampled per variety, totaling 120 trees. This included the control varieties Clementina de Nules and Parent Washington, sampled at 9 am, 10 am, and 11 am. In 2022, approximately 230 varieties were sampled using the same strategy, including the same controls. Collected leaves were immediately frozen in liquid nitrogen and stored at −80 °C until analysis. Sample preparation involved freeze-drying and coarse grinding of the leaves with a mortar and pestle, followed by fine grinding using a bead beater. Approximately 10 mg of the finely ground powder was extracted using a monophasic solvent mixture. The extracts were then sonicated, vortexed, and centrifuged, and the supernatants were collected for further analysis.

For polar metabolite analysis, we used a TQ-XS triple quadrupole mass spectrometer coupled with an Acquity I-class UPLC system fitted with a ZIC-pHILIC column (made by Waters Corporation, Milford, MA, USA), following the protocols of Vliet et al. (2019) [11]. The system operated in MRM mode with optimized source and desolvation temperatures, gas flows, and capillary voltage. Quality control samples, prepared by pooling aliquots from all study samples, were used to ensure system stability.

Secondary metabolite analysis was performed using a Synapt G2-Si quadrupole time-of-flight mass spectrometer coupled with an Acquity I-class UPLC system, equipped with a CSH phenyl–hexyl column, as described in established protocols [10]. The mass spectrometer operated in a data-dependent acquisition mode, scanning a mass range of 50 to 1200 *m/z*. Quality control was ensured through the periodic analysis of pooled samples and the use of Leucine enkephalin for mass correction. All samples were analyzed in a randomized order to maintain consistency in data acquisition.

2.3. Log-Ratios of Two Metabolites Under Analysis

One advantage of this method is the ability to draw insights from multiple studies in which the abundance of metabolites was profiled for citrus fruits and leaves. However, a significant challenge lies in the potential systematic differences between samples or studies, including variations in assay platforms and other uncontrollable variables. To tackle this issue, we calculated the logarithm base 2 ration, R_{AB}, of two selected metabolites within each sample, as specified in Equation (1) below:

$$R_{AB} = log_2\left(\frac{M_A}{M_B}\right) = log_2(M_A) - log_2(M_B), \quad (1)$$

where M_A and M_B are the absolute abundances of metabolites A and B, respectively, and R_{AB} is the log_2 ratio of these two abundance values. Logarithm transformation is necessary due to the non-Gaussian distribution of the absolute abundance values for the metabolites, and using the ratio of two metabolite abundances is a special form of normalization that eliminates sample-specific or study-specific biases. Indeed, ratios of molecules have been widely utilized as metric scales in fruit research. For example, the ratio of total phenolics to antioxidant capacity is essential for determining what portion of the phenolics is responsible for the antioxidant activities of fruits [12]. Additionally, the ratio of malic acid to citric acid is commonly used to numerically determine a fruit's flavor profile, ripeness, its value in fermentation processes, and more [13]. Through this bias correction step using log-ratio, the R_{AB} value of these two selected metabolites can be calculated across samples and datasets for subsequent analysis and comparison. This approach is being used for the first time to investigate profiles in distinct plant organs across various accession groups ("accession groups" in this case meaning heavily related varieties and species that do not bear the same name or direct parentage), leveraging different data sources.

2.4. Statistical Methods

2.4.1. Standard Quantitative Methods

Bar plots were used to illustrate the comparison of single metabolite ratios between leaf and fruit datasets, determined using Equation (1). Two-sided t-tests were used to compare the log ratios of metabolite pairs. To visualize metabolite profile comparisons across accession groups, datasets, and organs (leaf vs. fruit) using the log-ratio approach, we developed two volcano plot-based visualizations, as described below.

2.4.2. Single Log-Ratio Analysis

This display only investigates the log-ratio of two selected metabolites A and B (a potential log-ratio signature) and surveys the mean log-ratio, i.e., $\overline{log_2\left(\frac{M_A}{M_B}\right)}$, of replicated samples for each accession group in each dataset. Each data point represents the statistics calculated from the replicates of an accession group in a dataset (representing leaf or fruit samples). For each data point, a one-sample t-test was performed to determine whether the mean $log_2(M_A)$ is significantly different from the mean $log_2(M_B)$. The x-axis shows the mean of the log-ratios (or the difference between the mean $log_2(M_A)$ and the mean $log_2(M_B)$), while the y-axis indicates the -log_{10} (p-value), where the p-value is calculated from the one-sample t-test. This analysis can be used to check (1) the repeatability of log-ratios in the same tissue type of the same accession groups across datasets, (2) the variation in or correlation of log-ratios between different tissues (leaf versus fruit) of the same accession groups across datasets, and (3) the general pattern of log-ratios between tissue types across accession groups and datasets.

2.4.3. Analysis of Multiple Log-Ratios Between Leaves and Fruits

This display compares the log-ratios of multiple selected metabolite pairs (potential log-ratio signatures) and the means of the corresponding mean log-ratios between organs (leaf versus fruit in this study) for each accession group. Note that there are multiple fruit datasets but only one leaf dataset (our data) in this study. We treated the fruit datasets as replicates and determined whether the mean of the mean log ratios for these replicates is significantly different from the mean log ratio of the single leaf dataset using a one-sample t-test. Each data point represents the statistics for a metabolite pair (a potential log-ratio signature) calculated from the replicates of fruit datasets for an accession group, compared to the leaf dataset. The x-axis shows the difference between the mean of the mean log-ratio of the metabolite pair in fruit samples across datasets and the mean log-ratio in the leaf samples in our dataset, as shown in Equation (2):

$$\frac{1}{n}\sum_{i=1}^{n}\overline{log_{10}\left(\frac{M_{Ai}}{M_{Bi}}\right)} - \overline{log_{10}\left(\frac{M'_A}{M'_B}\right)} \quad (2)$$

where n is the number of replicates of fruit datasets, M_{Ai} and M_{Bi} are the abundances of metabolites A and B in the ith replicate fruit dataset (with $i = 1 \ldots n$), and M'_A and M'_B are the abundances of metabolites A and B in the leaf dataset. The y-axis indicates the -log_{10} (p-value), where the p-value is calculated from the one-sample t-test. This analysis can be used to identify metabolite ratio signatures that are either similarly or differentially represented between plant organs, such as fruit versus leaf, in specific citrus accession groups or across accession groups.

3. Results and Discussion

3.1. Expected Similar Metabolites

Feng [8] analyzed the fruit metabolomes of four mandarin varieties using a targeted panel. In our study, we extracted metabolite profiles from the leaves of four similar varieties for comparison. Figure 1 illustrates the ratios of citric and malic acids, showing higher levels of citric acid in the fruit and higher levels of malic acid in the leaves. This difference may stem from the fruit's need for more energy, as citric acid in the Krebs cycle generates guanosine triphosphate (GTP) and provides intermediates for amino acids and fatty acids. Citrus fruits also produce excess citric acid in the vacuoles of their juice sac cells. The vacuoles are relatively enlarged, and the tonoplast allocates citrate into the tonoplast through vacuolar H-ATPase and H-pyrophosphatases. The movement and maintenance of pH in these vacuoles creates the rapid acidification we taste in citrus [14,15]. In leaves, the lower levels of citric acid could be due to two main factors. First, citric and malic acids are more balanced in leaves, with malic acid playing a critical role in the Calvin cycle during photosynthesis, including the CAM, C4, and C3 pathways. Second, malic acid is essential for the water stress responses [16].

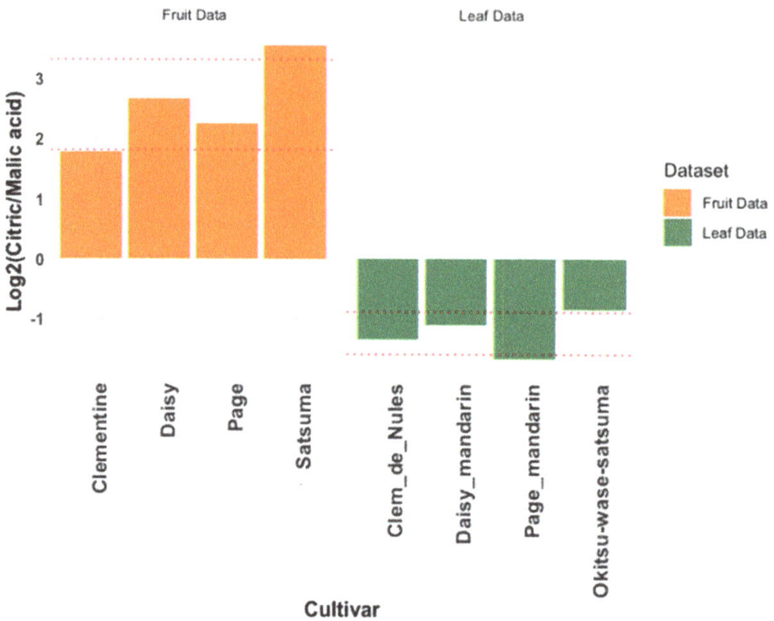

Figure 1. Citric-acid-to-malic-acid ratio in citrus fruit and leaves. This figure compares the citric-acid-to-malic-acid-ratio in the fruits and leaves of four mandarin varieties. Fruit data were sourced from Feng et al. (2018) [8], while leaf data were collected from the Givaudan Citrus Variety Collection at UC Riverside, representing similar but not identical mandarin varieties. Because each sample lacked replicates, error bars could not be calculated. The red dotted lines represent the standard error of the existing samples displayed on the chart.

Naringenin, a bitter molecule present in citrus and various plants, shares a biosynthetic pathway with hesperidin, another flavonoid. Despite their similarity, naringenin is notably more concentrated in pummelo and grapefruit than other citrus varieties and lines. Hesperidin is abundant in citrus fruits, particularly those referred to as "hesperidium fruits." Their flavonoid profile, including the naringenin-to-hesperidin ratio, serves as a valuable chemotaxonomic marker for classifying and distinguishing citrus species and cultivars. Understanding the flavonoid composition of these molecules is crucial for assessing their potential nutritional benefits to humans and for guiding breeding programs and cultivar identification. As anticipated, pummelo and grapefruit cultivars consistently exhibit significantly higher naringenin than hesperidin concentrations across all datasets (Figure 2). Interestingly, certain mandarin varieties have been observed to contain higher quantities of hesperidin compared to other cultivars [17,18].

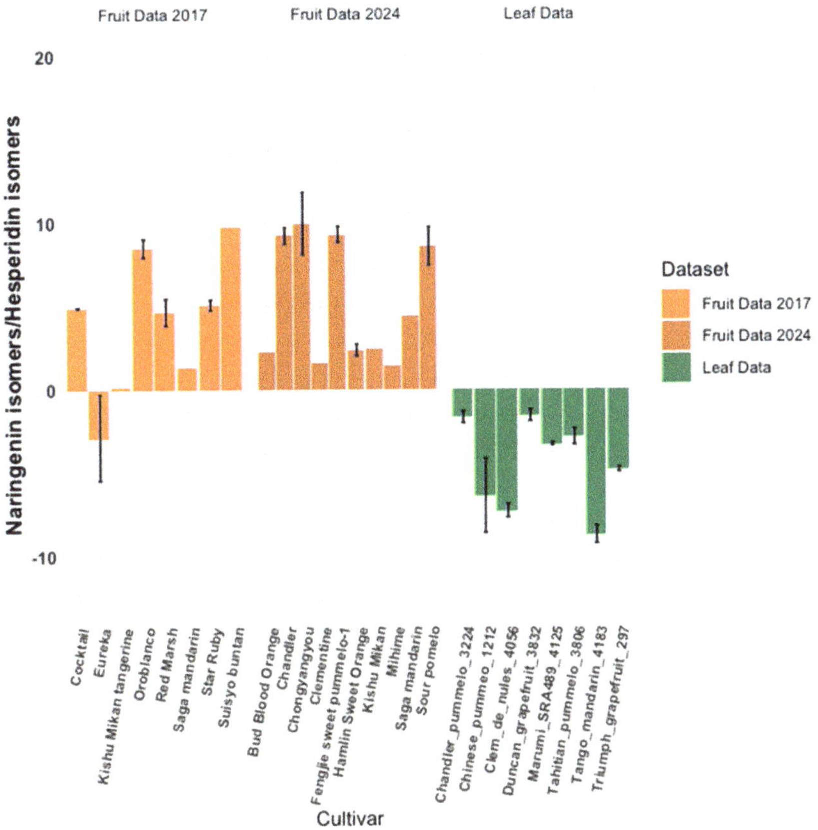

Figure 2. Naringenin-to-hesperidin ratio comparison. This figure compares naringenin-to-hesperidin ratios between a leaf metabolomics dataset and two fruit datasets, calculated using Equation (1) (Section 2). Error bars are unique to each sample. Samples without error bars lacked sufficient replicates.

Unlike mammals, plants synthesize their own aromatic amino acids through the shikimate pathway. Analyzing amino acid ratios can provide insights into a plant's metabolic state at the time of collection or reveal differences between treatment groups or related individuals. For instance, the proline-to-valine ratio serves as an indicator of abiotic stress. Proline accumulates in plants under drought, salinity, and extreme temperatures due to its roles as an osmoprotectant, protein and membrane stabilizer, and free radical scavenger. Figure 3 shows a distinct difference in the proline-to-valine ratio between leaves and fruit, suggesting a potential developmental response related to late fruit maturation. This observation is consistent with findings in Hayet et al. [17], which examined proline responses in grapevines. It is noteworthy that the proline in fruit likely results from maturation signals involving ornithine rather than stress-induced accumulation through glutamine [19].

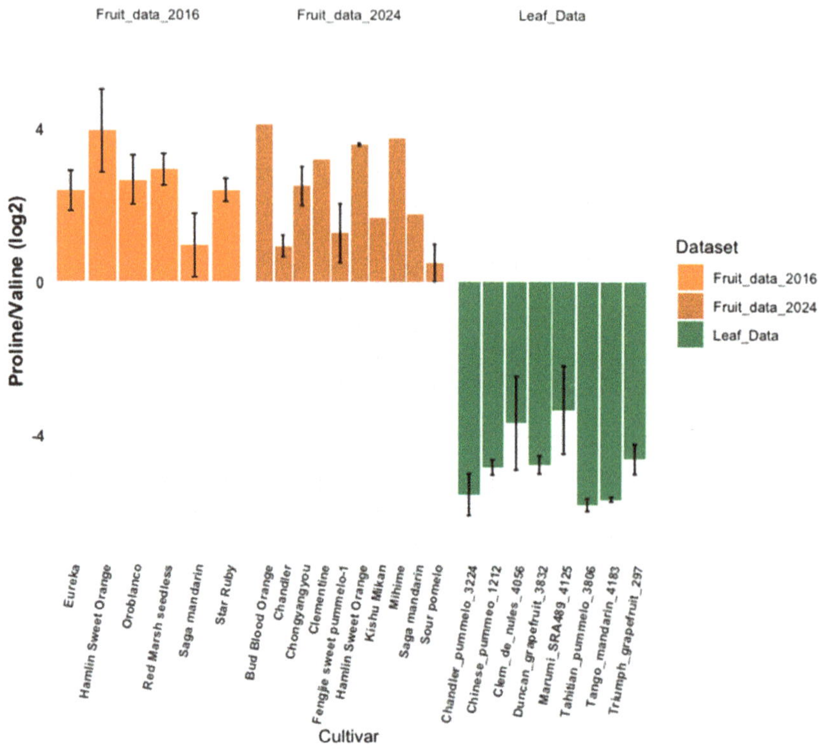

Figure 3. Proline-to-valine ratio comparison. This figure compares proline-to-valine ratios between a leaf metabolomics dataset and two fruit datasets, calculated using Equation (1) (Section 2). Error bars are unique to each sample. Samples without error bars lacked sufficient replicates.

3.2. Expected Similar Metabolite Ratios

Essential building blocks like the 20 amino acids and ubiquitous plant compounds such as quercetin [20] are consistently present across organisms. The stability of these common metabolites within the ratio model highlights the method's robustness, even in the face of dataset variability. Glutamine and glutamic acid, closely related compounds differing by a single amide group, are key players in plant stress responses and amino acid synthesis. These amino acids are vital for protein synthesis, influencing plant growth and development, and their abundance and interconversion affect the overall amino acid pool. Figure 4 shows the glutamine-to-glutamate ratio across three databases from a subset of mandarins and pummelos. While some variance and outliers exist, a general trend of similar ratios appears across both fruit juice and leaf tissue datasets, reflecting relative stability. Ideally, log ratios should center around zero. This is in stark contrast to the naringenin-to-hesperidin ratios (Figure 2) and citric-to-malic-acid ratios (Figure 1), which exhibit more distinct behaviors. Overall, the variance in Figure 4 is relatively low, with glutamine-to-glutamate ratios ranging from approximately −2 to 2, likely due to environmental factors and genetic differences among citrus varieties. Given their structural similarity, these amino acids are expected to be consistently abundant across plant tissues. However, as demonstrated in Figure 2, stronger differences appear when different species and cultivar groups are compared.

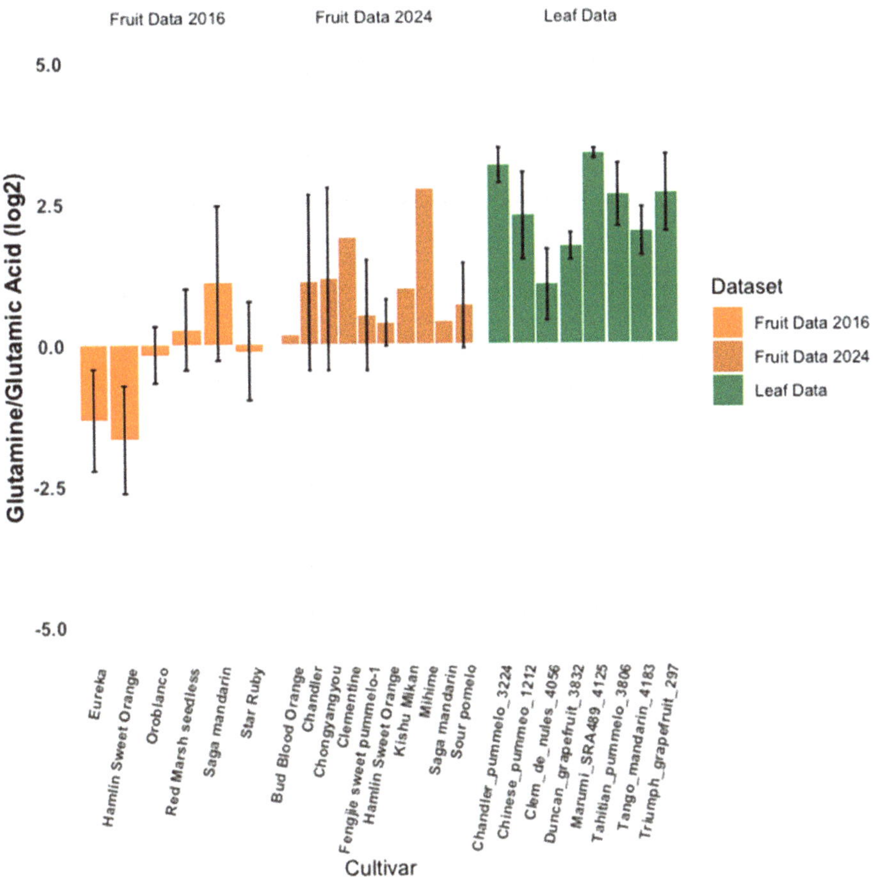

Figure 4. Glutamate-to-glutamic-acid ratio comparison. This figure compares glutamate-to-glutamic-acid ratios between a leaf metabolomics dataset and two fruit datasets, calculated using Equation (1) (Section 2). Error bars are unique to each sample. Samples without error bars lacked sufficient replicates.

3.3. *Metabolite Ratios: Intra-Cultivar-Group Comparison*

We adapted the volcano plot, typically used in differential gene expression analysis, to compare metabolite ratios between fruits and leaves across major citrus cultivar groups. Figure 5 illustrates six distinct metabolite ratios, derived from three fruit-based and one leaf-based dataset. This plot integrates the significance and association of different tissues, species, and datasets into a single chart. Two of the selected ratios, glutamate to glutamic acid and proline to valine, represent homeostatic relationships, reflecting the structural similarity of these metabolites. The figure highlights both expected similarities and unique characteristics of each metabolite ratio.

The naringenin-to-hesperidin ratio, as previously observed, is visualized with a broader range of varieties across samples (Figure 5a). Pummelo and grapefruit fruit samples consistently exhibit higher levels of naringenin compared to hesperidin, while pummelo leaf data demonstrate lower levels, distinct from other leaf and fruit data points, confirming a known trend. The glutamate-to-glutamine and phenylalanine-to-tyrosine ratios (Figure 5b) exhibit remarkable stability, with data points clustering according to their respective datasets, indicating minimal variation beyond data source differences.

This is expected, as phenylalanine is a precursor to tyrosine, and glutaminase catalyzes the conversion of glutamine to glutamate. Although cultivar groups show no discernible trends, comparing leaf and fruit data reveals higher glutamine levels in leaves and higher glutamic acid levels in fruit datasets.

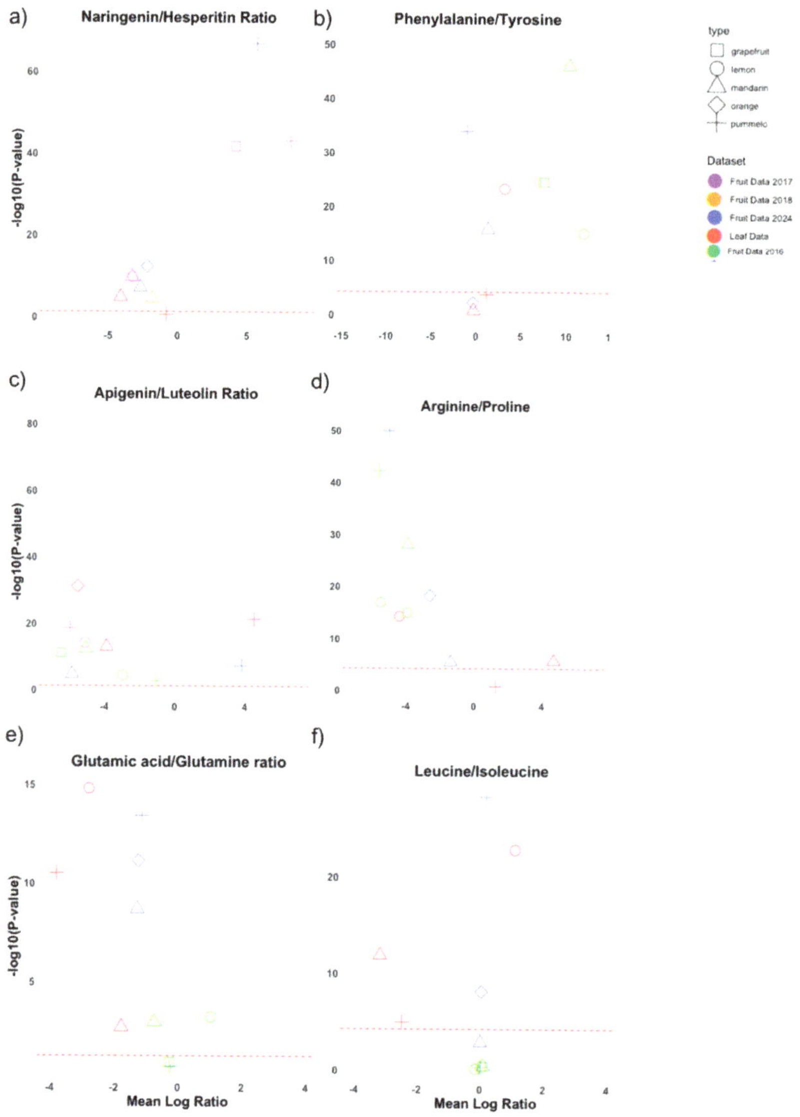

Figure 5. Volcano plot comparing metabolite ratios across multiple datasets (including fruit and leaf tissues) and citrus categories. The x-axis represents the mean log ratio of two selected metabolites, with negative values indicating a higher abundance of the denominator metabolite and positive values indicating a higher abundance of the numerator metabolite. The y-axis represents the significance level for the difference in the mean log ratio from 0, with a threshold set at $-log_{10}(0.05)$ (one-sample *t*-test). (**a**) Examines the ratio between naringenin hesperitin. (**b**) Examines the ratio between phenylalanine and tryosine. (**c**) Examines the ratio between apigenin and luteolin. (**d**) Examines the ratio between arginine and proline. (**e**) Examines the ratio between glutamic acid and glutamine. (**f**) Examines the ratio between leucine and isoleucine.

The apigenin-to-luteolin ratio (Figure 5c), a comparison of two flavonoids, can reflect a plant's response to various stresses, including UV radiation, drought, or pathogen attack. Flavonoids serve as antioxidants and UV protectants, and their balance may indicate a plant's resource allocation for different stress responses. In Figure 5a, mandarins generally exhibit a higher luteolin-to-apigenin ratio, suggesting greater activity of flavone synthase (FNS) and possibly other enzymes like flavonoid 3′-hydroxylase (F3′H), which catalyze the hydroxylation step necessary for luteolin production. Conversely, pummelos tend to produce more apigenin, indicating higher FNS activity but lower F3′H activity, resulting in apigenin, which lacks the additional hydroxyl group found in luteolin. Figure 5d indicates that the arginine-to-proline ratio (Figure 5d) generally favors proline in fruit but arginine in leaves. Leaves might prioritize arginine for protein synthesis and nitrogen assimilation, essential for growth and metabolism. Fruit's preference for proline could indicate an adaptation to osmotic stress or a nitrogen storage mechanism readily convertible to arginine [21,22]. The glutamine-to-glutamic-acid ratio (Figure 5e) has all datasets clustering within their respective colors (dataset), meaning that there is a distinct difference between datasets but not between these different cultivar groups. Glutamine and glutamic acid often show relatively consistent levels across various conditions due to their crucial functions in nitrogen assimilation and transport [23,24]. Figure 5f, covering the leucine-to-isoleucine ratio, has a large clustering in one area, with some variability. Leucine is clearly favored because it tends to accumulate better than isoleucine [25,26], which follows what is found in most studies in the literature covering these amino acids in organisms. The ratio of leucine to isoleucine varies based on species, environmental conditions, and growth stages, with leucine often present in higher concentrations due to its role in growth regulation and biomass accumulation. This balance is crucial for protein synthesis, the stress response, and metabolic stability, as both amino acids contribute to energy metabolism and resilience against environmental stress.

3.4. Metabolite Ratios: Inter-Cultivar Group Comparison

We then compared the selected metabolite ratios of the combined fruit datasets to our leaf dataset. The volcano plots in Figure 6 indicate which metabolite is more abundant in which plant organ. For example, Figure 6a shows a significant preference for the citric-to-malic-acid ratio in fruit. The citric-to-aspartic-acid ratio, in this context, is used as a placeholder to compare citric acid to another common organic acid to confirm its abundance. More common protein ratios (Figure 6a), such as proline to alanine, exhibit lower significance and are more balanced, hovering close to zero on the x-axis. The proline vs. glutamate ratio indicated higher proline levels in citrus fruit, which often result from environmental stress. Proline acts as an osmoprotectant, helping cells manage drought, salt, and temperature extremes. Proline stabilizes proteins and provides energy for recovery, while glutamine focuses on nitrogen transport, making it less responsive to immediate stress. In Figure 6b, several phenolic ratios are compared to apigenin as a reference, with each ratio ascending to a higher level of significance. Only apigenin and quercetin exceeded the significance threshold in Figure 6b. Apigenin and quercetin are flavonoids, which are often associated with plant stress responses and defense. The higher, significant apigenin/quercetin ratio in leaves may reflect the leaves' role in synthesizing or storing such compounds to protect the plant [27]. While Figure 6b lacks significant findings in this series, it is intentionally designed to demonstrate how to analyze the relationship between a specific metabolite (like apigenin) and a list of other metabolites. With specific targets and congruent datasets, entire pathways can be plotted in this same manner.

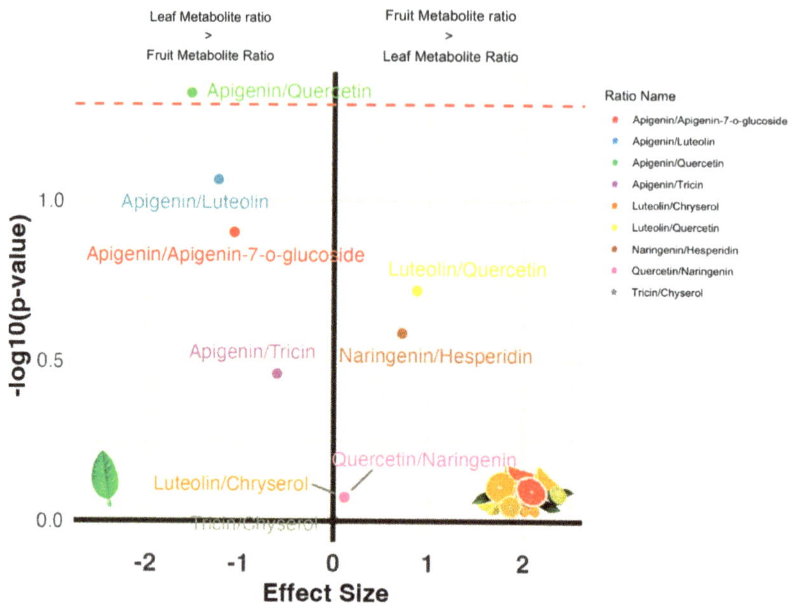

(a)

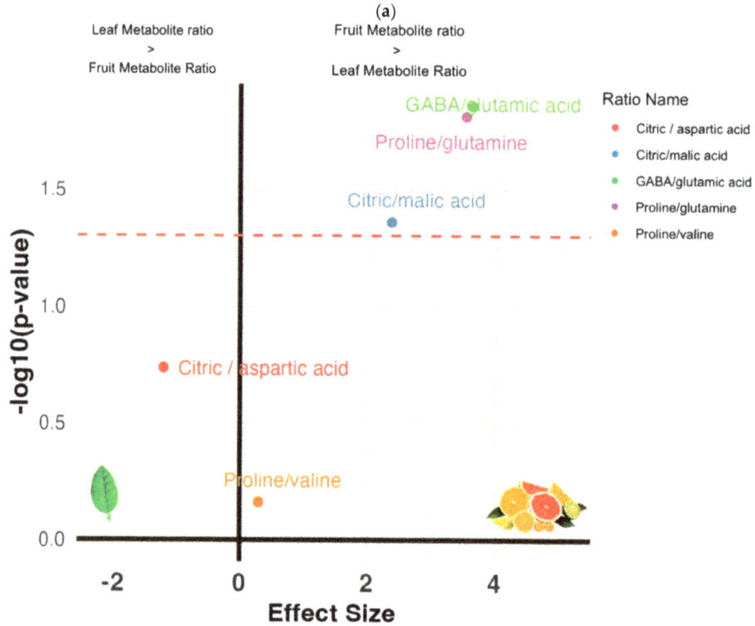

(b)

Figure 6. (**a**) Volcano plot of untargeted metabolites (flavonoids): This plot identifies significant differences in flavonoid ratios between fruit and leaf datasets. The x-axis represents the difference in average log ratios between fruits and leaves, with positive values indicating higher log ratios in fruits and negative values indicating higher log ratios in leaves. The y-axis represents the significance level, with a threshold of $-\log_{10}(0.05)$. Each data point represents a specific flavonoid ratio. (**b**) Volcano plot of targeted metabolites (amino acids and common acids): This plot identifies significant differences in ratios of targeted metabolites (amino acids and common acids) between fruit and leaf datasets. The x-axis and y-axis interpretations are the same as in Figure 6a. See Equation (2) for the calculation [28].

Genomics and transcriptomic data are often favored over metabolomics data, largely due to the larger sample sizes and greater potential for extensive biological analysis and collaboration. Metabolomics, however, is inherently dynamic—its variability across tissues, organisms, and environments, along with differences in analytical platforms, presents challenges for large-scale data collection and analysis. Consequently, many metabolomics studies are limited by a small number of replicates and a narrow focus on specific metabolites, and it is challenging to combine data from different studies for joint analysis [29]. Despite these obstacles, we have developed an approach to assess citrus metabolomics databases, integrating numerous samples across different organs, diverse varieties, and multiple studies, representing a significant advancement in the field. We propose a logarithmic ratio of selected metabolites to mitigate systematic differences across studies, enabling the joint analysis of data focused on different crop parts to test relevant hypotheses. Although the datasets (from different studies) included in this analysis were originally designed for specific objectives, they offer valuable data for broader comparative analyses. The strength of this approach lies in identifying large-scale trends across multiple datasets, allowing informed decision-making before embarking on costly new metabolomic experiments. While comparing logarithmic ratios is effective, it has limitations—ratios can amplify trends, and log transformations may distort the biological relevance of small differences. Nonetheless, this method offers a normalized and statistically manageable framework to analysis of metabolites.

Comparing metabolite profiles between different plant organs, such as citrus fruits and leaves, provides insights into molecular transport dynamics and the reasons for specific molecular enrichment in certain organs [30,31]. Genetic factors likely drive the observed variation in metabolite enrichment across cultivars. Identifying these genetic determinants will clarify the functions of associated genes and pathways and their roles in different plant tissues. Chemotaxonomic markers and their genetic counterparts can be developed to help identify promising lines for breeding and transformation [32].

Large databases, like the one created in the publication by Xiao Liang (2024) [4], are an excellent example of comprehensive reference databases that can be established to support future metabolomic comparisons, potentially transforming citrus breeding by enabling the prediction of fruit traits based on metabolite or genetic profiles from young leaf samples. In the publication, through a metabolite-based genome-wide association study (mGWAS) and strong validation, they discovered two loci key to UDP-glycosyltransferase (UGT) and glycosylation. This kind of discovery allows breeders to select for desirable traits and optimal parental lines, expediting cultivar development. However, methods like log ratio transformation could be the first step in connecting these databases to others for the discovery of even more loci. In this study, we demonstrated an approach using a few selected metabolite ratios; future research could expand this by using Procrustes analysis to explore other important ratios between fruits, leaves, and additional organs, such as flowers, given the available data. For example, comparing citric-to-malic-acid or glucose-to-fructose ratios could offer insights into sugar and acid levels [33,34], while tannin or saponin ratios might serve as indicators of bitterness. Establishing standardized practices for data collection, storage, and metabolite profiling across institutions is also essential to foster collaboration, data sharing, and robust analysis in the field.

4. Conclusions

We introduced a novel log-ratio-based approach to compare metabolite profiles between fruits and leaves across major citrus groups. This method revealed both distinguishing relationships—such as those involving citric and malic acids—and minimal differences, as seen with glutamic acid and glutamine. We used known metabolite relationships, like naringenin and hesperidin, to show the approach can come to similar conclusions as the scientific literature. Further, we demonstrated that the method is able to compare multiple ratios at once across just two datasets. This method has the advantage of using metabolite log-ratios for the reduction in inter-study bias, allowing smaller datasets from various sources to be combined for analyses and thereby enhancing the statistical power to test diverse hypotheses. This method holds promise for standardizing metabolite profiles in different studies with similar research directions across other crops or organisms, including humans.

Supplementary Materials: The following supporting information can be downloaded at: https://www.mdpi.com/article/10.3390/plants14101406/s1.

Author Contributions: R.C.T.: conceptualization, methodology, data curation, formal analysis, writing—review and editing. X.W.: formal analysis, writing—review and editing. Z.J. and J.M.C.: conceptualization, methodology, writing—review and editing. M.L.R.: formal analysis, writing—review and editing. M.M., Y.H., Q.J., R.C., E.E., T.K., P.A.M., D.M., and M.R.: writing—review and editing. A.D.S., A.B., and H.K.: metabolomic profiling, review and editing. All authors have read and agreed to the published version of the manuscript.

Funding: This project was supported by the U.S. Department of Agriculture National Institute of Food and Agriculture, USDA NIFA FACT grant 2019-67022-29930 to Z.J. and M.L.R.

Data Availability Statement: The original contributions presented in this study are included in the article/Supplementary Materials. Further inquiries can be directed to the corresponding author(s).

Conflicts of Interest: The authors declare no conflicts of interest.

References

1. Statistical Bulletin FRESH AND PROCESSED. Available online: https://www.fao.org/3/cb6492en/cb6492en.pdf (accessed on 15 November 2023).
2. Maqbool, Z.; Khalid, W.; Atiq, H.T.; Koraqi, H.; Javaid, Z.; Alhag, S.K.; Al-Shuraym, L.A.; Bader, D.M.D.; Almarzuq, M.; Afifi, M.; et al. Citrus Waste as Source of Bioactive Compounds: Extraction and Utilization in Health and Food Industry. *Molecules* **2023**, *28*, 1636. [CrossRef] [PubMed]
3. Kessler, A.; Kalske, A. Plant Secondary Metabolite Diversity and Species Interactions. *Annu. Rev. Ecol. Evol. Syst.* **2018**, *49*, 115–138. [CrossRef]
4. Liang, X.; Wang, Y.; Shen, W.; Liao, B.; Liu, X.; Yang, Z.; Chen, J.; Zhao, C.; Liao, Z.; Cao, J.; et al. Genomic and Metabolomic Insights into the Selection and Differentiation of Bioactive Compounds in Citrus. *Mol. Plant* **2024**, *17*, 1753–1772. [CrossRef]
5. Wang, S.; Tu, H.; Wan, J.; Chen, W.; Liu, X.; Luo, J.; Xu, J.; Zhang, H. Spatio-Temporal Distribution and Natural Variation of Metabolites in Citrus Fruits. *Food Chem.* **2016**, *199*, 8–17. [CrossRef] [PubMed]
6. Yazici, K.; Balijagic, J.; Goksu, B.; Bilgin, O.F.; Ercisli, S. Comparison of Some Fruit Quality Parameters of Selected 12 Mandarin Genotypes from Black Sea Region in Turkey. *ACS Omega* **2023**, *8*, 19719–19727. [CrossRef]
7. Asai, T.; Matsukawa, T.; Kajiyama, S. ichiro Metabolomic Analysis of Primary Metabolites in Citrus Leaf during Defense Responses. *J. Biosci. Bioeng.* **2017**, *123*, 376–381. [CrossRef] [PubMed]
8. Feng, S.; Niu, L.; Suh, J.H.; Hung, W.-L.; Wang, Y. Comprehensive Metabolomics Analysis of Mandarins (Citrus Reticulata) as a Tool for Variety, Rootstock, and Grove Discrimination. *J. Agric. Food Chem.* **2018**, *66*, 10317–10326. [CrossRef]
9. Wang, S.; Shen, S.; Wang, C.; Wang, X.; Yang, C.; Zhou, S.; Zhang, R.; Zhou, Q.; Yu, H.; Guo, H.; et al. A Metabolomics Study in Citrus Provides Insight into Bioactive Phenylpropanoid Metabolism. *Hortic. Res.* **2024**, *11*, uhad267. [CrossRef]
10. Wang, S.; Yang, C.; Tu, H.; Zhou, J.; Liu, X.; Cheng, Y.; Luo, J.; Deng, X.; Zhang, H.; Xu, J. Characterization and Metabolic Diversity of Flavonoids in Citrus Species. *Sci. Rep.* **2017**, *7*, 10549. [CrossRef]
11. Vliet, S.M.F.; Dasgupta, S.; Sparks, N.R.L.; Kirkwood, J.S.; Vollaro, A.; Hur, M.; Zur Nieden, N.I.; Volz, D.C. Maternal-to-Zygotic Transition as a Potential Target for Niclosamide during Early Embryogenesis. *Toxicol. Appl. Pharmacol.* **2019**, *380*, 114699. [CrossRef]
12. Babbar, N.; Oberoi, H.S.; Uppal, D.S.; Patil, R.T. Total Phenolic Content and Antioxidant Capacity of Extracts Obtained from Six Important Fruit Residues. *Food Res. Int.* **2011**, *44*, 391–396. [CrossRef]
13. Romero Rodriguez, M.A.; Vazquez Oderiz, M.L.; Lopez Hernandez, J.; Simal Lozano, J. Determination of Vitamin C and Organic Acids in Various Fruits by HPLC. *J. Chromatogr. Sci.* **1992**, *30*, 433–437. [CrossRef]

14. Brune, A.; Müller, M.; Taiz, L.; Gonzalez, P.; Etxeberria, E. Vacuolar Acidification in Citrus Fruit: Comparison between Acid Lime (Citrus Aurantifolia) and Sweet Lime (Citrus Limmetioides) Juice Cells. *J. Am. Soc. Hortic. Sci.* **2002**, *127*, 171–177. [CrossRef]
15. Müller, M.L.; Jensen, M.; Taiz, L. The Vacuolar H+-ATPase of Lemon Fruits Is Regulated by Variable H+/ATP Coupling and Slip. *J. Biol. Chem.* **1999**, *274*, 10706–10716. [CrossRef]
16. Sun, X.; Han, G.; Meng, Z.; Lin, L.; Sui, N. Roles of Malic Enzymes in Plant Development and Stress Responses. *Plant Signal. Behav.* **2019**, *14*, e1644596. [CrossRef]
17. Zhou, P.; Zheng, M.; Li, X.; Zhou, J.; Shang, Y.; Li, Z.; Qu, L. A Consecutive Extraction of Pectin and Hesperidin from Citrus Aurantium L.:Process Optimization, Extract Mechanism, Characterization and Bio-Activity Analysis. *Ind. Crops Prod.* **2022**, *182*, 114849. [CrossRef]
18. Zhang, M.; Duan, C.; Zang, Y.; Huang, Z.; Liu, G. The Flavonoid Composition of Flavedo and Juice from the Pummelo Cultivar (Citrus Grandis (L.) Osbeck) and the Grapefruit Cultivar (Citrus Paradisi) from China. *Food Chem.* **2011**, *129*, 1530–1536. [CrossRef]
19. Hayat, S.; Hayat, Q.; Alyemeni, M.N.; Wani, A.S.; Pichtel, J.; Ahmad, A. Role of Proline under Changing Environments: A Review. *Plant Signal. Behav.* **2012**, *7*, 1456–1466. [CrossRef]
20. Aghababaei, F.; Hadidi, M. Recent Advances in Potential Health Benefits of Quercetin. *Pharmaceuticals* **2023**, *16*, 1020. [CrossRef]
21. Winter, G.; Todd, C.D.; Trovato, M.; Forlani, G.; Funck, D. Physiological Implications of Arginine Metabolism in Plants. *Front. Plant Sci.* **2015**, *6*, 534. [CrossRef]
22. Ingrisano, R.; Tosato, E.; Trost, P.; Gurrieri, L.; Sparla, F. Proline, Cysteine and Branched-Chain Amino Acids in Abiotic Stress Response of Land Plants and Microalgae. *Plants* **2023**, *12*, 3410. [CrossRef] [PubMed]
23. Cánovas, F.M.; Avila, C.; Cantón, F.R.; Cañas, R.A.; de la Torre, F. Ammonium Assimilation and Amino Acid Metabolism in Conifers. *J. Exp. Bot.* **2007**, *58*, 2307–2318. [CrossRef] [PubMed]
24. Miflin, B.J.; Habash, D.Z. The Role of Glutamine Synthetase and Glutamate Dehydrogenase in Nitrogen Assimilation and Possibilities for Improvement in the Nitrogen Utilization of Crops. *J. Exp. Bot.* **2002**, *53*, 979–987. [CrossRef]
25. Binder, S. Branched-Chain Amino Acid Metabolism in Arabidopsis Thaliana. *Arab. Book* **2010**, *8*, e0137. [CrossRef]
26. Joshi, V.; Joung, J.-G.; Fei, Z.; Jander, G. Interdependence of Threonine, Methionine and Isoleucine Metabolism in Plants: Accumulation and Transcriptional Regulation under Abiotic Stress. *Amino Acids* **2010**, *39*, 933–947. [CrossRef] [PubMed]
27. Shomali, A.; Das, S.; Arif, N.; Sarraf, M.; Zahra, N.; Yadav, V.; Aliniaeifard, S.; Chauhan, D.K.; Hasanuzzaman, M. Diverse Physiological Roles of Flavonoids in Plant Environmental Stress Responses and Tolerance. *Plants* **2022**, *11*, 3158. [CrossRef]
28. Corliss, J. Fruit of the Month: Citrus Fruits. Available online: https://www.health.harvard.edu/heart-health/fruit-of-the-month-citrus-fruits (accessed on 28 February 2025).
29. Broadhurst, D.I.; Kell, D.B. Statistical Strategies for Avoiding False Discoveries in Metabolomics and Related Experiments. *Metabolomics* **2007**, *2*, 171–196. [CrossRef]
30. Li, D.; Heiling, S.; Baldwin, I.T.; Gaquerel, E. Illuminating a Plant's Tissue-Specific Metabolic Diversity Using Computational Metabolomics and Information Theory. *Proc. Natl. Acad. Sci. USA* **2016**, *113*, E7610–E7618. [CrossRef]
31. Ma, X.; Tang, K.; Tang, Z.; Dong, A.; Meng, Y.; Wang, P. Organ-Specific, Integrated Omics Data-Based Study on the Metabolic Pathways of the Medicinal Plant Bletilla Striata (Orchidaceae). *BMC Plant Biol.* **2021**, *21*, 504. [CrossRef]
32. Peters, K.; Blatt-Janmaat, K.L.; Tkach, N.; van Dam, N.V.; Neumann, S. Untargeted Metabolomics for Integrative Taxonomy: Metabolomics, DNA Marker-Based Sequencing, and Phenotype Bioimaging. *Plants* **2023**, *12*, 881. [CrossRef]
33. Li, J.; Zhang, C.; Liu, H.; Liu, J.; Jiao, Z. Profiles of Sugar and Organic Acid of Fruit Juices: A Comparative Study and Implication for Authentication. *J. Food Qual.* **2020**, *2020*, 1–11. [CrossRef]
34. Bae, H.; Yun, S.; Yoon, I.; Nam, E.; Kwon, J.; Jun, J. Assessment of Organic Acid and Sugar Composition in Apricot, Plumcot, Plum, and Peach during Fruit Development. *J. Appl. Bot. Food Qual.* **2014**, *87*, 24–29.

plants

MDPI

Article

Proposal of Critical Nutrient Levels in Soil and Citrus Leaves Using the Boundary Line Method

Antonio João de Lima Neto [1], Amanda Veridiana Krug [2], Jean Michel Moura-Bueno [2,*], Danilo Eduardo Rozane [3], William Natale [4], Jacson Hindersmann [2], Ana Luiza Lima Marques [2], Lincon Oliveira Stefanello [5], Daniéle Gonçalves Papalia [2] and Gustavo Brunetto [2]

[1] Department of Plant Science, Federal University of Ceará (UFC), Fortaleza 60356-000, CE, Brazil; antonio.joao@ufc.br
[2] Department of Soil Science, Federal University of Santa Maria (UFSM), Santa Maria 97105-900, RS, Brazil; krug.amanda@hotmail.com (A.V.K.); jacsonjh7@gmail.com (J.H.); marquesluizalima@gmail.com (A.L.L.M.); danipapalia@hotmail.com (D.G.P.); brunetto.gustavo@gmail.com (G.B.)
[3] Department of Agronomy and Natural Resources, São Paulo State University (UNESP), Registro 11900-000, SP, Brazil; danilo.rozane@unesp.br
[4] Brazilian Agricultural Research Corporation (EMBRAPA—Agroindustry Tropical), Fortaleza 60511-110, CE, Brazil; natale@ufc.br
[5] Department of Agronomy, Federal Technological University of Parana, Santa Helena 85892-000, PR, Brazil; lincono@utfpr.edu.br
* Correspondence: bueno.jean1@gmail.com

Abstract: Establishing critical levels (CLs) and sufficiency ranges (SRs) for nutrients improves fertilizer recommendations and supports citrus yield and fruit quality. The objective of this study was to establish CLs, soil fertility classes, and leaf nutrient SRs for citrus. This study used data on the yield and nutrients of the soil and leaves, collected from 2016 to 2021, of commercial orange (*Citrus sinensis*) and tangerine (*Citrus deliciosa*) orchards in the Southwest and Metropolitan regions of the state of Rio Grande do Sul, in southern Brazil. The yield data were related to the soil attributes/leaf nutrient contents. From the models obtained from this relationship, soil fertility classes and leaf sufficiency ranges were established using the boundary line (BL) method. The appropriate classes are 5.1–5.6 for pH, 1.0–1.4% for OM, 65.8–129.0 mg dm^{-3} for P, 161.4–326.0 mg dm^{-3} for K, 0.9–1.4 cmol$_c$ dm^{-3} for Ca, 0.22–0.34, cmol$_c$ dm^{-3} for Mg, 1.9–2.9 cmol$_c$ dm^{-3} for SB, 4.5–5.8 cmol$_c$ dm^{-3} for CEC, and 40.6–53.2% for V. The appropriate ranges of leaf contents were as follows: 19.1–22.7 g kg^{-1} of N, 0.8–1.3 g kg^{-1} of P, 7.8–11.3 g kg^{-1} of K, 20.9–28.4 g kg^{-1} of Ca, 2.0–3.3 g kg^{-1} of Mg, 2.0–3.0 g kg^{-1} of S, 88.8–127.5 mg kg^{-1} of B, 28.3–73.6 mg kg^{-1} of Cu, 74.3–122.5 mg kg^{-1} of Fe, 55.7–89.3 mg kg^{-1} of Mn, and 10.9–15.6 mg kg^{-1} of Zn. The BL method made it possible to establish nutrient CLs using data from commercial orchards, which is not possible when using conventional approaches. The established norms will allow for a more precise definition of the real need for fertilizer application in citrus orchards.

Keywords: fertilization in orchards; soil fertility in orchards; leaf nutrient sufficiency ranges; rational use of fertilizers

1. Introduction

Citrus (*Citrus* spp.) is commercially grown in more than 140 countries in an area of approximately 10.6 million hectares, with a world production of 169.4 million tons, playing a key role in the agricultural economy. Brazil is the second largest producer in the

https://doi.org/10.3390/plants14121764

world with an area of approximately 701.8 thousand hectares and a production volume of 20.5 million tons [1].

Citrus plants are predominantly cultivated in tropical and subtropical regions around the world, where soils, in general, are characterized by high acidity and low organic matter content, which are the main factors in limiting nutrients [2,3]. Because of this, the soil does not provide sufficient quantities of nutrients to meet the demand of citrus [4–6]. Therefore, it is necessary to use soil acidity correctives and fertilizers to meet the high nutritional demand of citrus plants, maintain high levels of yield, and produce fruits with excellent quality [7,8].

Liming and fertilization recommendations for citrus are based on the results of soil and leaf chemical analyses carried out periodically [8,9]. The efficient use of these tools requires the establishment of interpretation standards, which have been established in calibration studies carried out over several years and at various locations under a wide variation of soil and climate conditions [10–12]. However, the use of new scion and rootstock varieties in citriculture, as well as changes in the nutritional management of orchards, requires an update of these reference tables to interpret the results of the analyses [13,14].

Establishing diagnostic norms for soil fertility and the nutritional status of fruit-bearing plants through calibration studies takes a lot of time and is expensive [15]. As an alternative to classical calibration studies, parameters for interpreting soil fertility and plant nutritional status can be established using the boundary line (BL) method, which is a solution to the limitations of regional diagnosis [15–17]. This method has the advantage of using data from nutritional monitoring carried out by producers and consists of relating yield (dependent factor) to soil attributes/leaf content, fitting mathematical models to the points of the upper boundary that make it possible to establish the optimal value of the independent factor [15–18]. This method has been successfully used to establish adequate nutrient levels in soil and leaf tissues for various crops [15,17–23].

The objective of this study was to establish the critical and optimal levels of nutrients in the soil and leaf tissues of citrus for production areas, using the BL method, while showing why the traditional calibration methods are inadequate in this region.

2. Materials and Methods

2.1. Study Area

This study was carried out using a database with information on yield, soil fertility attributes, and leaf nutrient contents. These data were collected in non-irrigated commercial orchards of oranges (*Citrus sinensis*) and (*Citrus deliciosa*) tangerines, whose production is destined for fresh consumption, distributed in the Southwest and Metropolitan regions of Rio Grande do Sul, in the municipalities of Rosário do Sul, Pareci Novo, and Montenegro.

The data totaled 495 soil observations and 486 leaf observations, collected in the 2016/2017, 2017/2018, 2019/2020, and 2020/2021 seasons in citrus orchards, with ages ranging from 11 to 35 years, and planting density between 500 and 667 plants ha^{-1} (3 × 5 m; 3 × 6 m; 4 × 5 m; 5 × 3 m), consisting of three rootstocks (*Citrange troyer*, *Citrumelo swingle*, and *Porcirus trifoliata*) and thirteen varieties of scion (Caí, Cara Cara, Ellendale, Lane Late, Midnight, Montenegrina, Murcot, Nadorcott, Navelina, Ortanique, Pareci, Ponkan, and Salustiana). The orchards are planted in soils predominantly classified as Alfisols and Ultisols [24].

2.2. Soil, Leaf and Production Analyses

The soil was randomly collected in the 0–20 cm layer in homogeneous plots, in a total of twenty single subsamples to form a composite sample. The samples were collected in the crop row, where the fertilizers were applied. Then, the soil samples were air-dried, passed through a sieve (2 mm mesh), and chemically analyzed according to the methodology proposed by [25]. Soil pH was determined in water (soil: water ratio of 1:1), and available concentrations of P and K were extracted by Mehlich-1; exchangeable Ca, Mg, and Al were extracted with 1 mol L^{-1} KCl. P was determined by spectrophotometry (Bell Photonics, 1105, São Paulo, Brazil), K by flame photometry (Digimed, DM-62, São Paulo, Brazil), Ca and Mg by atomic absorption spectrometry (PerkinElmer—Analyst 200, Norwalk, CT, USA), and Al was titrated with 0.0125 mol L^{-1} NaOH. The sum of bases (SBs), cation exchange capacity at pH 7.0 (CEC $pH_{7.0}$), aluminum saturation (Al), base saturation (V), and potential acidity (H + Al) were calculated. Total organic carbon content was determined by oxidation (Walkley–Black) and multiplied by 1.724 to estimate the soil organic matter (OM) contents.

Leaf collection was carried out on the same plants used for soil collection; the 3rd or 4th leaf (approximately six months of age) of the branch with terminal fruit of 2 to 4 cm in diameter was collected, for a total of four healthy and undamaged leaves per tree, with one in each quadrant and at the medium height of the plant [26] collected at the same phenological stage, according to the fruiting period of each variety.

After collection, the leaves were washed in distilled water, a detergent solution (0.1%), a hydrochloric acid solution (0.3%), and deionized water. Then, the plant material was dried in an oven at 65 °C for 48 to 96 h, and ground and analyzed for total nutrient contents according to the methodology described by [25]. A subsample of the tissue was subjected to sulfuric digestion to determine N content using a Kjeldahl steam-drag distiller. Another subsample was subjected to nitric-perchloric digestion and Ca, Mg, Cu, Zn, Fe, and Mn contents were determined in the extract by atomic absorption spectrophotometry (PerkinElmer—Analyst 200, Norwalk, CT, USA). P contents were determined by colorimetry [27] in a spectrophotometer (Bell Photonics, 1105, Brazil), and K contents were determined in a flame photometer (Digimed, DM-62, São Paulo, Brazil). B contents were extracted by burning 0.5 g of a plant tissue sample in a muffle furnace (600 °C for 1 h) and determined in a spectrophotometer (Bell Photonics, 1105, São Paulo, Brazil) [28,29].

The trees where soil and leaf samples were collected were marked and later used to obtain yield. To calculate the production, all the fruits of each tree were collected at the time of harvest for each variety, and the fruit boxes were weighed and multiplied by the number of plants per hectare.

2.3. Establishing Diagnostic Norms

The critical levels (CLs), soil fertility classes (FCs), and leaf nutrient sufficiency ranges (SRs) for citrus were established using the BL method as proposed by [17]. Before establishing the BLs, outliers were removed from the dataset using box-and-whisker diagrams in Statistica software version 12.5. Scatter plots were constructed relating the relative fruit yield (RFY) of each plot (y-axis) to the attributes of the soil fertility/leaf nutrient contents (x-axis) of the same plot [18]. Next, the x-axis was divided into 10–15 intervals and the highest point of each interval of the upper boundary was selected [22].

The data pairs (y,x) selected at the upper boundary (10 to 15 points) of the data cloud were used to establish mathematical models, chosen based on the best fit, significance, biological significance, and coefficient of determination (R^2). These models were used to establish the critical levels, soil fertility classes, and leaf sufficiency ranges: Low (RFY < 70%), Medium (70% ≤ RFY < 90%), Adequate (90% ≤ RFY ≤ 100%), and High (RFY > 100%).

3. Results

3.1. Soil Fertility Attributes

The soils of commercial citrus plots, in general, are heterogeneous in terms of fertility attributes, with a coefficient of variation (CV) ranging between 12.6% and

91.1% (Table 1). The high CV values indicate that the soil fertility data of the evaluated areas present high variability. The soils of the orchards have a pH ranging from 4.1 to 7.1 and OM concentrations from 0.7 to 2.0%. The available P and K concentrations ranged from 1.2 to 2.66.6 mg dm^{-3} and from 16.0 to 590 mg dm^{-3}, respectively. The Ca and Mg concentrations in the soil ranged from 0.2 to 4.2 cmol$_c$ dm^{-3} and from 0.1 to 0.9 cmol$_c$ dm^{-3}, respectively. The sum of bases (SBs), cation exchange capacity (CEC), and base saturation (V) ranged from 0.5 to 5.2 cmol$_c$ dm^{-3}, from 1.7 to 9.2 cmol$_c$ dm^{-3}, and from 18.8 to 84.2%, respectively (Table 1).

Table 1. Descriptive statistics of soil fertility attributes (0–20 cm) of commercial citrus plots.

Soil Fertility Attributes	Minimum	Maximum	Mean	SD	CV (%)
pH in H_2O	4.1	7.1	5.5	0.7	12.6
OM (%)	0.7	2.0	1.2	0.3	21.6
P (mg dm^{-3})	1.2	266.6	78.2	61.1	78.1
K (mg dm^{-3})	16.0	590.0	156.6	142.1	90.7
Ca (cmol$_c$ dm^{-3})	0.2	4.2	1.7	0.8	47.8
Mg (cmol$_c$ dm^{-3})	0.1	0.9	0.4	0.2	39.2
SB (cmol$_c$ dm^{-3})	0.5	5.2	2.6	0.9	35.3
CEC (cmol$_c$ dm^{-3})	1.7	9.2	5.1	1.6	31.1
V (%)	18.8	84.2	56.5	16.0	28.3

pH in water (1:2.5); OM—organic matter obtained by organic carbon × 1.724 (Walkley–Black); P and K available—Mehlich-1 extractant; Ca and Mg exchangeable—1 mol L^{-1} KCl extractant; SB—sum of exchangeable bases; CEC—cation exchange capacity at pH$_{7.0}$; V—base saturation index; SD—Standard deviation; CV—coefficient of variation.

3.2. Critical Levels and Soil Fertility Classes

The scatter plots show the relationship between the relative fruit yield of citrus and the attributes of soil fertility (Figure 1). These relationships allowed the fitting of mathematical models with coefficients of determination (R^2) ranging from 0.75 to 0.96. The response curves obtained by the BL method indicate that the increase in pH and nutrient availability in the soil promoted yield gains up to a plateau, beyond which the excessive concentrations of nutrients are related to nutritional imbalances reflected in low yields.

From the equations obtained from the relationship between the relative fruit yield of citrus and the fertility attributes, the critical levels and soil fertility classes were established (Table 2). The adequate soil pH for crop growth and development ranged from 5.1 to 5.6 and the OM concentration ranged from 1.0 to 1.4. The P and K concentrations related to higher yields ranged from 65.8 to 129.0 mg dm^{-3} and from 161.4 to 326.0 mg dm^{-3}, respectively. The Ca and Mg concentrations should be 0.9–1.4 cmol$_c$ dm^{-3} and 0.22–0.34 cmol$_c$ dm^{-3}. Also, the soil must have a SBs ranging from 1.9 to 2.9 cmol$_c$ dm^{-3}, a CEC ranging from 4.5 to 5.8 cmol$_c$ dm^{-3}, and a V ranging from 40.6 to 53.2% (Table 2).

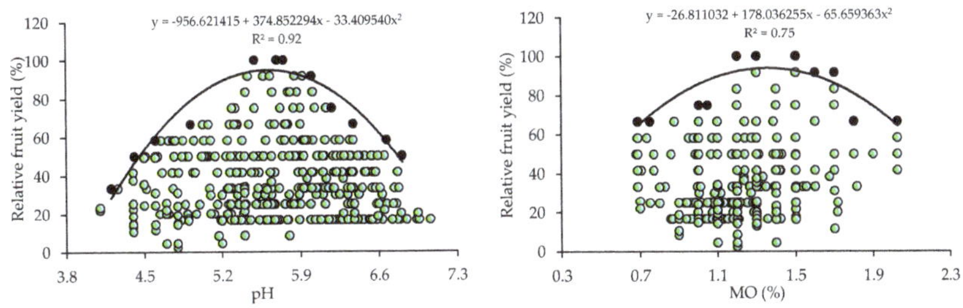

Figure 1. *Cont.*

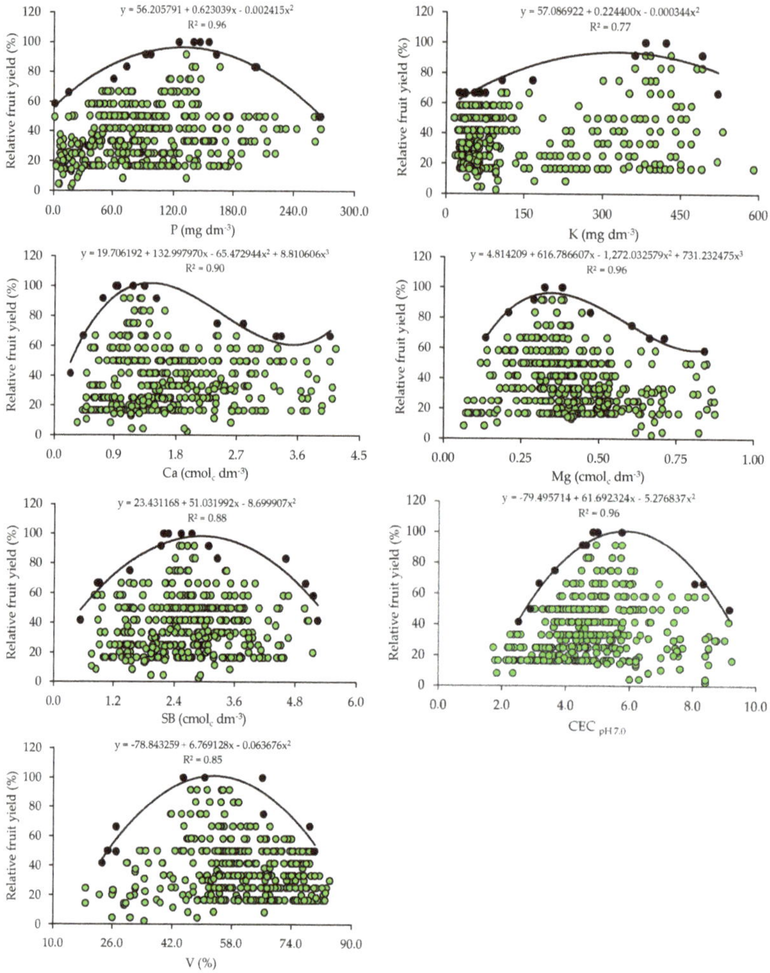

Figure 1. Scatter plots and boundary lines of the relationship between the relative fruit yield of citrus and soil hydrogen potential (pH), organic matter (OM), available phosphorus (P), available potassium (K), exchangeable calcium (Ca), exchangeable magnesium (Mg), sum of exchangeable bases (SB), cation exchange capacity (CEC), and base saturation (V) (0–20 cm).

Table 2. Classes for interpretation of soil fertility (0–20 cm) for citrus, established by the boundary line method.

Relative Fruit Yield (RFY)	pH (H_2O)	OM	P	K	Ca	Fertility Classes
%		%	----- mg dm^{-3} -----		cmol$_c$ dm^{-3}	
RFY < 70	<4.7	<0.7	<19.6	<40.5	<0.5	Low
70 ≤ RFY < 90	4.7–5.1	0.7–1.0	19.6–65.8	40.5–161.4	0.5–0.9	Medium
90 ≤ RFY ≤ 100	5.1–5.6	1.0–1.4	65.8–129.0	161.4–326.0	0.9–1.4	Adequate [1]
RFY > 100	>5.6	>1.4	>129.0	>326.0	>1.4	High

Relative fruit yield (RFY)	Mg	SB	CEC	V	Fertility Classes
%	------------- cmol$_c$ dm^{-3}	-------------		--- % ---	
RFY < 70	<0.13	<1.1	<3.5	<31.3	Low

Table 2. *Cont.*

Relative Fruit Yield (RFY)	pH (H_2O)	OM	P	K	Ca	Fertility Classes
70 ≤ RFY < 90	0.13–0.22	1.1–1.9	3.5–4.5		31.3–40.6	Medium
90 ≤ RFY ≤ 100	0.22–0.34	1.9–2.9	4.5–5.8		40.6–53.2	Adequate [1]
RFY > 100	>0.34	>2.9	>5.8		>53.2	High

[1] The lower and upper limits of the range correspond to the critical and optimal levels (90 and 100% of yield); RFY < 70% (FY < 171.4 kg plant^{-1}); 70% ≤ RFY < 90% (≤171.4 kg plant^{-1} FY < 220.3 kg plant^{-1}); 90% ≤ FY ≤ 100% (220.3 kg plant^{-1} ≤ FY ≤ 244.8 kg plant^{-1}); RFY > 100% (FY > 244.8 kg plant^{-1}); and pH in water (1:2.5). OM—organic matter obtained by organic carbon × 1.724 (Walkley & Black); P and K—Mehlich-1 extractant; Ca and Mg—1 mol L^{-1} KCl extractant; SB—sum of exchangeable bases; CEC—cation exchange capacity at $pH_{7.0}$; V—base saturation index.

3.3. Leaf Nutrient Contents

Descriptive statistics of yield and the nutrient contents in citrus leaf tissues indicate that there is wide variation in the nutritional status of the plants in the orchards (Table 3). The yield of the orchards ranged from 6.0 to 244.8 kg plant^{-1}, with an average of 85.9 kg plant^{-1}. Leaf nitrogen (N) contents ranged from 14.9 to 32.2 g kg^{-1}, P contents from 0.5 to 2.5 g kg^{-1}, K contents from 4.2 to 19.6 g kg^{-1}, Ca contents from 13.0 to 43.9 g kg^{-1}, Mg contents from 0.8 to 6.5 g kg^{-1}, and S contents from 0.7 to 5.8 g kg^{-1} (Table 3). The micronutrients B, Cu, Fe, Mn and Zn showed contents in the leaf tissues ranging from 13.4 to 224.1 g kg^{-1}, 6.8 to 177.7 g kg^{-1}, 36.7 to 243.7 g kg^{-1}, 7.1 to 153.6 g kg^{-1}, and 6.6 to 39.7 g kg^{-1}, respectively.

Table 3. Descriptive statistics of yield and leaf nutrient contents in citrus.

Nutrients	Minimum	Maximum	Mean	SD	CV (%)
Fruit yield (kg plant^{-1})	6.0	244.8	85.9	47.6	55.4
N (g kg^{-1})	14.9	32.2	22.9	3.1	13.3
P (g kg^{-1})	0.5	2.5	1.3	0.5	34.4
K (g kg^{-1})	4.2	19.6	9.9	3.2	32.0
Ca (g kg^{-1})	13.0	43.9	28.3	5.6	19.8
Mg (g kg^{-1})	0.8	6.5	3.3	1.3	38.3
S (g kg^{-1})	0.7	5.8	3.2	1.1	35.0
B (mg kg^{-1})	13.4	224.1	115.8	48.1	41.5
Cu (mg kg^{-1})	6.8	177.7	58.3	39.4	67.7
Fe (mg kg^{-1})	36.7	243.7	98.4	44.6	45.4
Mn (mg kg^{-1})	7.1	153.6	63.0	31.8	50.4
Zn (mg kg^{-1})	6.6	39.7	17.7	7.6	42.9

SD—Standard deviation; CV—coefficient of variation.

3.4. Critical Levels and Leaf Sufficiency Ranges

The relationships between relative fruit yield and leaf nutrient contents are presented below (Figures 2 and 3). Based on the relationships, the points of the upper boundary were selected, which allowed for the fitting of mathematical models with high coefficients of determination (R^2), ranging from 0.87 to 0.97. These models were used to establish the critical levels and fertility classes of the nutrients in leaf tissue.

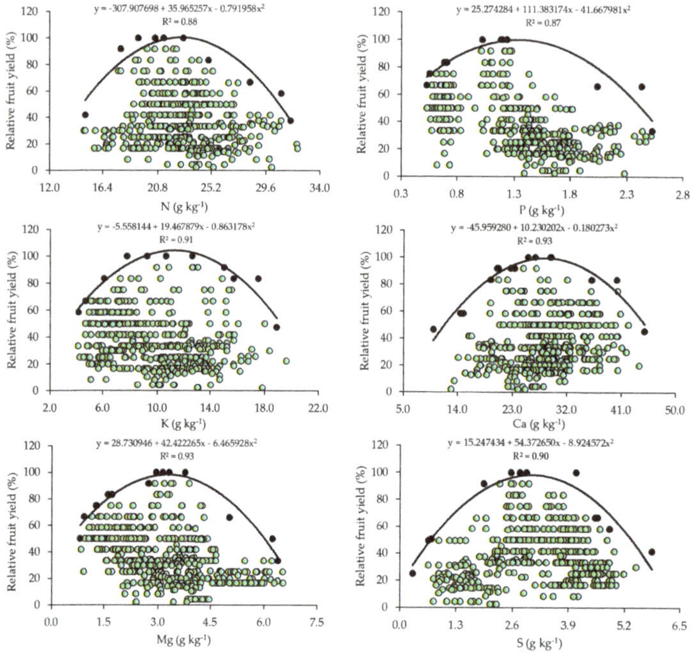

Figure 2. Scatter plots and boundary lines of the relationship between the relative fruit yield of citrus and the leaf contents of nitrogen (N), phosphorus (P), potassium (K), calcium (Ca), magnesium (Mg), and sulfur (S).

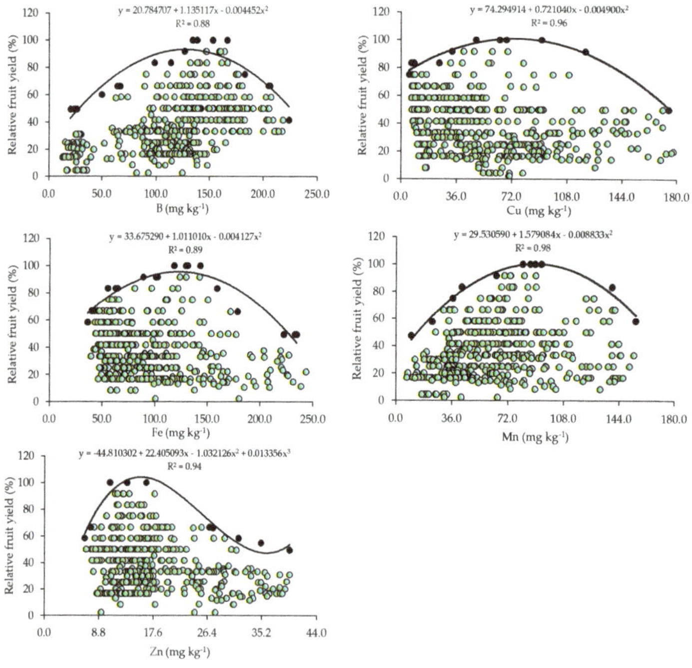

Figure 3. Scatter plots and boundary lines of the relationship between the relative fruit yield of citrus and the leaf contents of boron (B), copper (Cu), iron (Fe), manganese (Mn), and zinc (Zn).

The critical levels and leaf sufficiency ranges of the nutrients were obtained from the response curves of the relationship between yield and leaf contents (Table 4). The appropriate contents for citrus are 19.1–22.7 g kg^{-1} of N, 0.8–1.3 g kg^{-1} of P, 7.8–11.3 g kg^{-1} of K, 20.9–28.4 g kg^{-1} of Ca, 2.0–3.3 g kg^{-1} of Mg, and 2.0–3.0 g kg^{-1} of S (Table 4). In the case of micronutrients, the ranges of adequate contents of B, Cu, Fe, Mn, and Zn are 88.8–127.5 mg kg^{-1}, 28.3–73.6 mg kg^{-1}, 74.3–122.5 mg kg^{-1}, 55.7–89.3 mg kg^{-1}, and 10.9–15.6 mg kg^{-1}, respectively (Table 4).

Table 4. Leaf sufficiency ranges of nutrients for citrus, established by the boundary line method.

Relative Fruit Yield (RFY)	N	P	K	Ca	Mg	S	Sufficiency Ranges
%			g kg^{-1}				
RFY < 70	<16.5	<0.5	<5.2	<15.5	<1.1	<1.2	Low
70 ≤ RFY < 90	16.5–19.1	0.5–0.8	5.2–7.8	15.5–20.9	1.1–2.0	1.2–2.0	Medium
90 ≤ RFY ≤ 100	19.1–22.7	0.8–1.3	7.8–11.3	20.9–28.4	2.0–3.3	2.0–3.0	Adequate [1]
RFY > 100	>22.7	>1.3	>11.3	>28.4	>3.3	>3.0	High

Relative fruit yield (RFY)	B	Cu	Fe	Mn	Zn	Sufficiency ranges
%			mg kg^{-1}			
RFY < 70	<48.3	-	<39.1	<31.1	<7.8	Low
70 ≤ RFY < 90	48.3–88.8	-	39.1–74.3	31.1–55.7	7.8–10.9	Medium
90 ≤ RFY ≤ 100	88.8–127.5	28.3–73.6	74.3–122.5	55.7–89.3	10.9–15.6	Adequate [1]
RFY > 100	>127.5	>73.6	>122.5	>89.3	>15.6	High

[1] The lower and upper limits of the range correspond to the critical and optimal levels (90 and 100% of yield); RFY < 70% (FY < 171.4 kg $plant^{-1}$); 70% ≤ RFY < 90% (≤ 171.4 kg $plant^{-1}$ FY < 220.3 kg $plant^{-1}$); 90% ≤ FY ≤ 100% (220.3 kg $plant^{-1}$ ≤ FY ≤ 244.8 kg $plant^{-1}$); and RFY > 100% (FY > 244.8 kg $plant^{-1}$).

4. Discussion

4.1. Relationships Between Soil Fertility and Yield

The citrus orchards evaluated show a wide variability of attributes related to soil fertility (Table 1). According to the classification proposed by [30], considering the coefficient of variation (CV), the variability is very high (CV > 30%) for most attributes, except OM and V, which showed high variability (20% < CV ≤ 30%), and pH, which showed medium variability (10% < CV ≤ 20%). The areas have a pH ranging from acidic to neutral, with low levels (≤2.5%) of OM, very low to very high concentrations (≤10.0 mg dm^{-3} to > 60.0 mg dm^{-3}) of P, very low to very high concentrations (≤20.0 mg dm^{-3} to > 120.0 mg dm^{-3}) of K, low to high concentrations (<2.0 $cmol_c$ dm^{-3} to >4.0 $cmol_c$ dm^{-3}) of Ca, and low to medium concentrations (<0.5 $cmol_c$ dm^{-3} to >1.0 $cmol_c$ dm^{-3}) of Mg, with CEC ranging from low to medium (≤7.5 $cmol_c$ dm^{-3} to 7.6–15.0 $cmol_c$ dm^{-3}) [31] and V ranging from very low to high (<45% to >80%) [32]. The low values of pH, Ca, Mg, and V observed in the diagnostic layer (0–20 cm) in most soils in citrus orchards in southern Brazil may occur because the doses of corrective materials were not established following the technical recommendation or the method of application was not adequate [33,34]. Farmers often define doses of corrective materials by considering empirical information, and limestone is applied to the soil surface, without being homogeneously incorporated in the 0–20 cm layer, which makes it difficult for the corrective to descend into the soil profile due to its very low mobility, not being efficient in correcting acidity, and in increasing concentrations of Ca and Mg [33,34]. High concentrations of P and K in the soil may be a consequence of annual applications of doses above the plants' needs [33,34].

The variability of soil chemical attributes allowed for the fitting of mathematical models using the BL method (Figure 1). These models confirm the correlation of pH and nutrient availability in the soil with citrus fruit yield [35,36]. According to [16], BLs represent the limiting effect of the independent variable on the dependent variable and correspond to the maximum yields that can be attained in each location [15,18]. The data pairs located below the BL correspond to the plots whose yields were reduced by the influence of another independent variable (soil, climate, pests, diseases, nutrients, etc.) or by the interactions of other independent variables [17,37].

Soil pH determines the solubility and availability of nutrients in soil [8]. The adequate pH values for the maximum yield of citrus (5.1–5.6) obtained by the BL method (Table 2) are slightly below the reference pH (6.0) proposed for citrus in the states of RS and SC [31]. However, the soil pH values are within the ideal range (5.0 to 6.5) for citrus growth [38]. It should be noted that citrus are sensitive to acidity and, when grown in soils with a pH lower than 5.0, the greater availability of Al can cause toxicity in their root systems [2]. In areas where acidity is a restrictive factor for citrus growth and development, it is necessary to apply acidity correctives to raise the soil pH and nutrient availability to adequate levels, promoting yield gains and improvements in fruit quality [39,40].

The OM concentrations that promoted the maximum citrus yield (1.0–1.4%) are low (≤2.5%), according to [31]. In this study region, the soils, in general, are highly weathered, with low OM contents [33,34,41]. Therefore, farmers use organic sources of nutrients, such as organic compost [33,34], to increase the availability of nutrients, especially mineral forms of N, which can be absorbed by plants, increasing fruit yield [40–42].

The critical and optimal (90 and 100%) P levels in the soil (65.8–129.0 mg dm^{-3}) obtained for citrus using BL (Table 2) are higher than those proposed in the regional fertilization recommendation (18.1–36.0 mg dm^{-3} and 30.1–60.0 mg dm^{-3}) [31] for soils with clay contents lower than 20% (class 4) and between 21 and 40% (class 3), which predominate in the areas studied. In the case of K, the reference values (161.4–326.0 mg dm^{-3}) are higher than those recommended in the regional recommendation (61–120 mg dm^{-3}) [31] for soils with a CEC ≤ 7.5 $cmol_c$ dm^{-3}, which are predominant in the evaluated orchards (Table 1). High concentrations of P and K in the soil have already been reported in the citrus orchards in the main producing regions of southern Brazil and may be related to the frequent and excessive use of industrialized fertilizers and organic residues [33,34]. Excess K in soils can compete with Ca and Mg, inducing a deficiency of these nutrients in leaf tissues, which can cause yield losses [3,14]. Excess P in soils can increase the more labile forms of P, enhancing transfer, especially through the runoff solution, which increases the probability of surface water contamination, making it necessary to quantify the risk of P runoff with current practices [43,44]. The standards for interpreting soil fertility attributes currently used were established for fruit trees in general, considering a very wide region (RS and SC) with great variation in soil and climate conditions, and were not specific to citrus [31]. Divergences in the critical levels of nutrients in the soil and leaf tissues as compared to those in the manuals have also been found in other studies and have been attributed to the use of more demanding cultivars and changes in management [15,18,21].

The adequate concentrations of Ca (0.9–1.4 $cmol_c$ dm^{-3}) and Mg (0.22–0.34 $cmol_c$ dm^{-3}) proposed for citrus using the BL (Table 2) method are lower than the adequate concentrations of Ca and Mg (2.0–4.0 $cmol_c$ dm^{-3} and 0.5–1.0 $cmol_c$ dm^{-3}) proposed for fruit trees in general, according to the regional recommendation [31]. These regional recommendation standards were established for a group of plants of different species (fruit trees), and are not specific to citrus fruits, which justifies the differences observed. The lack of pH correction during the implementation of the orchard through liming, associated with the non-homogeneous incorporation of the soil corrective, may be directly related to the lower critical levels of Ca and Mg. However, research results have shown positive correlations between citrus yield and soil pH and exchangeable Ca and Mg concentrations [35,38,40]. This is because an increase in pH contributes to a greater availability of nutrients, and the supply of Ca and Mg, provided by liming, promotes a greater development of the citrus root system, with positive effects on the absorption of other nutrients, plant nutrition, and fruit production [3,36].

The adequate CEC class for citrus (4.5–5.8 $cmol_c$ dm^{-3}) falls into the low class ($\leq$ 7.5 cmolc dm^{-3}), according to the regional fertilization recommendation [31]. These lower critical levels for soil CEC are related to the low values of clay and also of OM in most orchard soils. A positive correlation was even observed between these soil attributes. In general, soils cultivated with citrus in southern Brazil have low OM and CEC values [33,34].

The ideal base saturation (V) for citrus (53.2%) obtained by the BL method (Table 2) is classified as low (45–64%), according to the regional fertilization recommendation [32], and is below the recommendation for citrus (V = 70%) in the state of São Paulo [2]. However, ref. [45], when evaluating the effect of liming (limestone applied on the surface) in a 'Pêra' orange orchard in the state of São Paulo, concluded that the maximum production was obtained with a V around 50% and suggested that the base saturation indicated for the crop may be lower than that recommended by the official recommendation.

4.2. Relationships Between Leaf Nutrient Contents and Yield

The wide variability observed in soil chemical attributes (Table 2) resulted in a wide variation in yield levels and leaf nutrient contents in the orchards (Table 3). This variability was very high for yield and nutrient contents (CV > 30%), except for N and K, which showed medium variability (10% < CV $\leq$ 20%), according to [30]. This high variability in yield and nutrient contents in leaf tissues is directly related to the variability of nutrient concentrations in the soil (reflecting fertilization management) and may also be due to the variation in the age of the orchards and monitoring across multiple seasons.

The appropriate ranges of leaf nutrient contents for citrus, established for the producing region of Rio Grande do Sul using the BL method, were compared with those available in the literature (Table 5). It is possible to observe that the ranges vary between the producing regions in Brazil and that, although they are in agreement with those proposed by [4] for the same cultivation region using the Compositional Nutrient Diagnosis (CND) method, they have a smaller amplitude (Table 5). When the contents of these nutrients in the leaves fall below the optimal ranges, it results in yield losses and the need to be corrected through adjustments in the fertilization program [9].

Table 5. Sufficiency ranges and adequate nutrient contents for citrus, established by the boundary method and those found in the literature.

Reference	N	P	K	Ca	Mg	S
			g kg^{-1}			
BL-Citrus	19.1–22.7	0.8–1.3	7.8–11.3	20.9–28.4	2.0–3.3	2.0–3.0
[4]	21–26	0.9–1.5	7–11	25–33	1.9–3.7	2.3–3.8
[31]	23–27	1.2–1.6	10–15	35–45	3.0–4.0	-
[2]	25–30	1.2–1.6	12–16	35–50	3.5–5.0	2.0–3.0
Reference	**B**	**Cu**	**Fe**		**Mn**	**Zn**
			mg kg^{-1}			
BL-Citrus	88.8–127.5	28.3–73.6	74.3–122.5		55.7–89.3	10.9–15.6
[4]	85–149	7–83	50–145		38–94	8–29
[31]	50–100	4.1–10	50–120		35–50	35–50
[2]	75–150	10–20	50–150		35–70	50–75

The ranges of adequate contents of N, K, Ca, Mg, and Zn in citrus leaf tissues, obtained using BL, in addition to the lower amplitude, have lower limits than the ranges currently used to interpret the nutritional status of the crops in RS and SC [31]. This pattern was also observed in relation to the ranges proposed for the interpretation of leaf contents in citrus in the state of São Paulo [2]. These lower ranges and with lower amplitudes indicate a lower demand for this micronutrient by citrus [4]. In addition, the doses of these micronutrients based on regional standards [31] are being overestimated. It should be highlighted that, although Cu, Fe, and Mn are required in smaller quantities, their deficiencies impair the photosynthesis of citrus trees, while B and Zn deficiencies affect meristematic growth [2]. It should be noted that the norms established for the south region of the country [31] and for the state of São Paulo [2] were obtained for a wide producing region, encompassing soils and climates that diverged from the current conditions, which justifies the differences.

Foliar and soil analyses are useful in the nutritional management of citrus and should be complementary and never individualized. This is because it is not always possible to establish good correlations between the nutrients in soil and leaf tissues [3,8]. This was observed for P and K, whose critical and optimal levels in the soil (Table 2), despite being much higher than those proposed by the current fertilization recommendation system [31], did not result in high levels of these nutrients in the leaf tissues (Tables 4 and 5). This can occur because of the low availability of water in the soil, which reduces the probability of nutrients approaching the outer surface of the root, soil compaction, or even ion interactions in the soil, which is very common for P and can decrease the free chemical species of the nutrients in a soil solution [46]. In addition, in the specific case of N, a foliar analysis of this nutrient is used to evaluate its availability in soil and to recommend the doses of nitrogen fertilizer to be applied [10].

The differences observed between the diagnostic norms of soil fertility and plant nutrition between the producing regions reinforce the need to define current and regionalized diagnostic norms for specific climate, soil, and management conditions [3,7,13,14]. It is important that these patterns of interpretation of soil fertility and leaf nutrient content are validated in the field for conditions similar to those in this study [15].

5. Conclusions

The soil fertility classes and leaf sufficiency ranges of nutrients were established for citrus-producing regions in southern Brazil, using the BL method. This method made it possible to establish interpretation standards using data from commercial plantations, which present wide variability in soil, climate, and management conditions. These data reflect the best interactions between soil and climate variables and the management of areas that influence productivity, and thus the standards can be more widely applicable. These norms diverged from those used in other citrus-producing regions and reinforce the need to establish regional or local standards to interpret soil fertility and the nutritional status of plants. However, to ensure the accuracy of the established nutritional standards, their validation under field conditions is necessary through fertilization experiments. The norms obtained using the BL method contemplate data from several years and under current cultivation conditions; they will contribute to defining the real need for nutrient application in citrus orchards, greater efficiency in the use of correctives and fertilizers, and lowering environmental impacts.

Author Contributions: Conceptualization, A.J.d.L.N., D.E.R., W.N. and G.B.; methodology, A.J.d.L.N., D.E.R. and W.N.; validation, A.V.K., J.H., A.L.L.M., D.G.P. and L.O.S.; resources, A.V.K. and G.B.; data curation, A.V.K., J.M.M.-B., J.H., A.L.L.M. and D.G.P.; writing—original draft preparation, A.J.d.L.N., D.E.R. and W.N.; writing—review and editing, A.J.d.L.N., D.E.R., W.N., J.M.M.-B. and G.B.; visualization, A.V.K., J.H., A.L.L.M., D.G.P. and L.O.S.; supervision, G.B.; project administration, A.V.K. and G.B.; funding acquisition, A.V.K. and G.B. All authors have read and agreed to the published version of the manuscript.

Funding: This study was partly funded by the Coordenação de Aperfeiçoamento de Pessoal de Nível Superior—Brazil (Capes)—Financial Code 001, the National Council of Scientific and Technological

Development (CNPq—Process:302023/2019-4), and the Foundation for Research Support of Rio Grande do Sul (FAPERGS—Process:21/2551-0001952-9).

Data Availability Statement: Data are contained within the article.

Acknowledgments: The authors acknowledge the company Citrusul and the producers from the municipalities of Montenegro and Pareci Novo who welcomed us and made their orchards available for collection and carrying out this study.

Conflicts of Interest: The remaining authors declare that the research was conducted in the absence of any commercial or financial relationships that could be construed as a potential conflict of interest.

References

1. FAOSTAT—Food and Agriculture Organization of the United Nations. Crops and Livestock Products 2023. Available online: http://www.fao.org/faostat/en/#data/QCL (accessed on 20 January 2025).
2. Mattos, D., Jr.; Kadyampakeni, D.M.; Oliver, A.Q.; Boaretto, R.M.; Morgan, K.T.; Quaggio, J.A. Soil and nutrition interactions. In *The Genus Citrus*, 1st ed.; Talon, M., Caruso, M., Gmitter, F.G., Jr., Eds.; Woodhead Publishing: Cambridge, UK, 2020; pp. 311–331. [CrossRef]
3. Mousavi, S.M.; Srivastava, A.K.; Raiesi, T. Citrus nutrition in Iran: Lessons from calcareous soils. *J. Plant Nutr.* **2024**, *47*, 3367–3392. [CrossRef]
4. Krug, A.V.; Papalia, D.G.; Marques, A.L.L.; Hindersmann, J.; Soares, V.M.; Grando, D.L.; Moura-Bueno, J.M.; Trapp, T.; Rozane, D.E.; Natale, W.; et al. Proposition of critical levels of nutrients in citrus leaves, grown in a subtropical climate, for fresh market fruit production. *Sci. Hortic.* **2023**, *317*, 112047. [CrossRef]
5. Stefanello, L.; Schwalbert, R.; Schwalbert, R.; Tassinari, A.; Garlet, L.; De Conti, L.; Ciotta, M.; Ceretta, C.; Ciampitti, I.; Brunetto, G. Phosphorus critical levels in soil and grapevine leaves for South Brazil vineyards: A Bayesian approach. *Eur. J. Agron.* **2023**, *144*, 126752. [CrossRef]
6. Brunetto, G.; Ceretta, C.A.; Kaminski, J.; Melo, G.W.B.; Welter, P.D.; Girotto, E.; Lorenzi, C.R.; Vieira, R.C.B.; De Conti, L.; Tiecher, T.L. Annual urea nitrogen contribution to the nutrition of Cabernet Sauvignon grapevine grown in sandy and clayey soil. *Agronomy* **2024**, *14*, 101. [CrossRef]
7. Labaied, M.B.; Serra, A.P.; Mimoun, M.B. Establishment of nutrients optimal range for nutritional diagnosis of mandarins based on DRIS and CND methods. *Commun. Soil Sci. Plant Anal.* **2018**, *49*, 2557–2570. [CrossRef]
8. Chen, Y.W.; Li, F.F.; Wu, Y.C.; Zhou, T.; Chang, Y.Y.; Lian, X.F.; Yin, T.; Ye, L.; Li, Y.S.; Lu, X.P. Profiles of citrus orchard nutrition and fruit quality in Hunan Province, China. *Int. J. Fruit Sci.* **2022**, *22*, 779–793. [CrossRef]
9. Ahmad, N.; Hussain, S.; Ali, M.A.; Minhas, A.; Waheed, W.; Danish, S.; Fahad, S.; Ghafoor, U.; Baig, K.S.; Sultan, H.; et al. Correlation of soil characteristics and citrus leaf nutrients contents in current scenario of Layyah district. *Horticulturae* **2022**, *8*, 61. [CrossRef]
10. Quaggio, J.A.; Cantarella, H.; Van Raij, B. Phosphorus and potassium soil test and nitrogen leaf analysis as a base for citrus fertilization. *Nutr. Cycl. Agroecosyst.* **1998**, *52*, 67–74. [CrossRef]
11. Tassinari, A.; Stefanello, L.O.; Schwalbert, R.A.; Vitto, B.B.; Kulmann, M.S.S.; Santos, J.P.J.; Arruda, W.S.; Schwalbert, R.; Tiecher, T.L.; Ceretta, C.A.; et al. Nitrogen critical level in leaves in 'Chardonnay' and 'Pinot Noir' grapevines to adequate yield and quality must. *Agronomy* **2022**, *12*, 1132. [CrossRef]
12. Andrade, C.B.; Comin, J.J.; Moura-Bueno, J.M.; Brunetto, G. Obtaining reference values for nutrients in vineyard soils through boundary line approach using Bayesian segmented quantile regression on commercial farm data. *Eur. J. Agron.* **2023**, *150*, 126928. [CrossRef]
13. Rozane, D.E.; Mattos, D., Jr.; Parent, S.É.; Natale, W.; Parent, L.E. Meta-Analysis in the selection of groups in varieties of citrus. *Commun. Soil Sci. Plant Anal.* **2015**, *46*, 1948–1959. [CrossRef]
14. Yamane, D.R.; Parent, S.-É.; Natale, W.; Cecílio Filho, A.B.; Rozane, D.E.; Nowaki, R.H.D.; Mattos Junior, D.; Parent, L.E. Site-Specific Nutrient Diagnosis of Orange Groves. *Horticulturae* **2022**, *8*, 1126. [CrossRef]
15. Lima Neto, A.J.; Natale, W.; Deus, J.A.L.; Rozane, D.E. Establishment of critical nutrient levels in the soil and leaf of 'Prata' banana using the boundary line. *Sci. Hortic.* **2024**, *328*, 112923. [CrossRef]
16. Webb, R. Use of the boundary line in the analysis of biological data. *J. Hortic. Sci.* **1972**, *47*, 309–319. [CrossRef]
17. Walworth, J.L.; Letzsch, W.S.; Sumner, M.E. Use of boundary lines in establishing diagnostic norms. *Soil Sci. Soc. Am. J.* **1986**, *50*, 123–128. [CrossRef]
18. Lima Neto, A.J.; Neves, J.C.L.; Martinez, H.E.P.; Sousa, J.S.; Fernandes, L.V. Establishment of critical nutrient levels in soil and plant for eucalyptus. *Rev. Bras. Ciênc. Solo* **2020**, *44*, e0190150. [CrossRef]

19. Evanylo, G.K.; Sumner, M.E. Utilization of the boundary line approach in the development of soil nutrient norms for soybean production. *Commun. Soil Sci. Plant Anal.* **1987**, *18*, 1379–1401. [CrossRef]
20. Manorama, K.; Behera, S.K.; Suresh, K. Establishing optimal nutrient norms in leaf and soil for oil palm in India. *Ind. Crops Prod.* **2021**, *174*, 114223. [CrossRef]
21. Guimarães, G.G.F.; Deus, J.A.L.; Lima Neto, A.J. Boundary line method to update critical soil phosphorus and potassium levels in banana plantations in Santa Catarina. *Rev. Bras. Frutic.* **2023**, *45*, e-979. [CrossRef]
22. Ali, A.M. Nutrient sufficiency ranges in mango using boundary-line approach and compositional nutrient diagnosis norms in El-Salhiya, Egypt. *Commun. Soil Sci.Plant Anal.* **2018**, *49*, 188–201. [CrossRef]
23. Ali, A.M. Establishment of nutrient sufficiency ranges in olive using boundary-line approach. *J. Plant Nutr.* **2023**, *46*, 453–461. [CrossRef]
24. Santos, H.G.; Jacomine, P.K.T.; Anjos, L.H.C.; Oliveira, V.A.; Lumbreras, J.F.; Coelho, M.R.; Almeida, J.A.; Araujo Filho, J.C.; Oliveira, J.B.; Cunha, T.J.F. *Sistema Brasileiro de Classificação de Solos*, 5th ed.; Embrapa: Brasília, Brazil, 2018.
25. Tedesco, M.J.; Gianello, C.; Bissani, C.A.; Bohnen, H. *Análises de solo, Plantas e Outros Materiais*, 2nd ed.; Departamento de solo da UFRGS: Porto Alegre, Brazil, 1995.
26. Quaggio, J.A.; Mattos, D., Jr.; Boaretto, R.M. Citros. In *Boas Práticas Para o uso Eficiente de Fertilizantes*; Prochnow, L.I., Casarin, W., Stipp, S.R., Eds.; International Plant Nutrition Institute: Piracicaba, Brazil, 2010; pp. 371–409.
27. Murphy, J.; Riley, J. A modified single solution method for the determination of phosphate in natural waters. *Anal. Chim. Acta* **1962**, *27*, 31–36. [CrossRef]
28. Krug, F.J.; Mortatti, J.; Pessenda, L.; Zagatto, E.A.G.; Bergamin, H. Flow injection spectrophotometric determination of boron in plant material with azomethine-H. *Anal. Chim. Acta* **1981**, *125*, 29–35. [CrossRef]
29. Gaines, T.P.; Mitchell, G.A. Boron determination in plant tissue by the Azomethine H method. *Commun. Soil Sci. Plant Anal.* **1979**, *10*, 1099–1108. [CrossRef]
30. Pimentel-Gomes, F. *Curso de Estatística Experimental*, 15th ed.; Fealq: Piracicaba, Brazil, 2009.
31. CQFS-RS/SC—Comissão de Química e Fertilidade do Solo RS e SC. *Manual de Calagem e Adubação Para os Estados do Rio Grande do Sul e Santa Catarina*, 11th ed.; Sociedade Brasileira de Ciência do Solo, Núcleo Regional Sul: Porto Alegre, Brazil, 2016.
32. CQFS-RS/SC—Comissão de Química e Fertilidade do Solo RS e SC. *Manual de Calagem e Adubação Para os Estados do Rio Grande do Sul e Santa Catarina*, 10th ed.; Sociedade Brasileira de Ciência do Solo, Núcleo Regional Sul: Porto Alegre, Brazil, 2004.
33. Griebeler, S.R.; Gonzatto, M.P.; Scivittaro, W.B.; Oliveira, R.P.; Schwarz, S.F. Diagnóstico nutricional de pomares de laranjeiras da Fronteira Oeste do Rio Grande do Sul. *Pesqui. Agropecu. Gaúch.* **2020**, *26*, 114–130. [CrossRef]
34. Griebeler, S.R.; Gonzatto, M.P.; Schwarz, S.F.; Böettcher, G.N.; Pauletti, G.F.; Rota, L.D. Nutritional diagnosis of 'Montenegrina' mandarin orchards at the Southern Brazil. *Rev. Ceres* **2023**, *70*, 69–77. [CrossRef]
35. Fidalski, J.; Auler, P.A.M. Alterações químicas temporais nas faixas de adubação e entrelinhas do pomar, nutrição e produção de laranja após calagem superficial. *Rev. Bras. Cienc. Solo* **2008**, *32*, 689–696. [CrossRef]
36. Auler, P.A.M.; Neves, C.S.V.J.; Fidalski, J.; Pavan, M.A. Calagem e desenvolvimento radicular, nutrição e produção de laranja "Valência" sobre porta-enxertos e sistemas de preparo do solo. *Pesqui. Agropecu. Bras.* **2011**, *46*, 254–261. [CrossRef]
37. Smith, J.F.N.; Hardie, A.G. Determination of foliar nutrient sufficiency ranges in cultivated rooibos tea using the boundary-line approach. *S. Afr. J. Plant Soil* **2022**, *39*, 226–233. [CrossRef]
38. Wang, Y.; Long, Q.; Li, Y.; Kang, F.; Fan, Z.; Xiong, H.; Zhao, H.; Luo, Y.; Guo, R.; He, X.; et al. Mitigating magnesium deficiency for sustainable citrus production: A case study in Southwest China. *Sci. Hortic.* **2022**, *295*, 1083215. [CrossRef]
39. Zhang, S.; Yang, W.; Muneer, M.A.; Ji, Z.; Tong, L.; Zhang, X.; Li, X.; Wang, W.; Zhang, F.; Wu, L. Integrated use of lime with Mg fertilizer significantly improves the pomelo yield, quality, economic returns and soil physicochemical properties under acidic soil of southern China. *Sci. Hortic.* **2021**, *290*, 110502. [CrossRef]
40. Wu, L.; Qin, M.; Muneer, M.A.; Bao, J.; Chen, X.; Yang, Y.; Huang, J.; Zhang, S.; Su, D.; Yan, X. Soil pH and organic matter: Key edaphic factors in sustaining optimum yield and quality of pomelo fruit. *Sci. Hortic.* **2024**, *337*, 113524. [CrossRef]
41. Stefanello, L.O.; Schwalbert, R.; Schwalbert, R.A.; De Conti, L.; Kulmann, M.S.; Garlet, L.P.; Silveira, M.L.R.; Sautter, C.K.; Melo, G.W.B.; Rozane, D.E.; et al. Nitrogen supply method affects growth, yield and must composition of young grape vines (*Vitis vinifera* L. cv Alicante Bouschet) in southern Brazil. *Sci. Hortic.* **2020**, *261*, 108910. [CrossRef]
42. Stefanello, L.O.; Schwalbert, R.; Schwalbert, R.A.; Drescher, G.L.; De Conti, L.; Pott, L.P.; Tassinari, A.; Kulmann, M.S.D.S.; Silva, I.C.B.; Brunetto, G. Ideal nitrogen concentration in leaves for the production of high-quality grapes cv 'Alicante Bouschet' (*Vitis vinifera* L.) subjected to modes of application and nitrogen doses. *Eur. J. Agron.* **2021**, *123*, 126200. [CrossRef]
43. Grando, D.L.; Gatiboni, L.C.; Mumbach, G.L.; Dall'orsoletta, D.J.; Souza Junior, A.A.; Schmitt, D.E. Phosphorus in the runoff of soils with contrasting textures influenced by soil slope and pig slurry application. *Agric. Water Manag.* **2021**, *258*, 107178. [CrossRef]

44. Lourenzi, C.R.; Ceretta, C.A.; Ciancio, N.H.R.; Tiecher, T.L.; Silva, L.O.S.; De Conti, L.; Girotto, E.; Ferreira, P.A.A.; Vidal, R.F.; Scopel, G.; et al. Forms of nitrogen and phosphorus transfer by runoff in soil under no-tillage with successive organic waste and mineral fertilizers applications. *Agric. Water Manag.* **2021**, *248*, 106779. [CrossRef]
45. Silva, M.A.C.; Natale, W.; Prado, R.M.; Corrêa, M.C.M.; Stuchi, E.S.; Andrioli, I. Aplicação superficial de calcário em pomar de laranjeira 'Pêra' em produção. *Rev. Bras. Frutic.* **2007**, *29*, 606–612. [CrossRef]
46. De Conti, L.; Ceretta, C.A.; Ferreira, P.A.A.; Lorensini, F.; Lourenzi, C.R.; Vidal, R.F.; Tassinari, A.; Brunetto, G. Effects of pig slurry application and crops on phosphorus content in soil and the chemical species in solution. *Rev. Bras. Cienc. Solo* **2015**, *39*, 774–787. [CrossRef]

plants

MDPI

Article

Soil pH Determining the Assembly Processes of Abundant and Rare Bacterial Communities in Response to Cultivation Modes in Lemon Farmlands

Hao-Qiang Liu [1,2,*], Si-Chen Li [1,2], Hong-Jun Li [1,2] and Zhu-Chun Peng [1,2]

1 Citrus Research Institute, Southwest University, Beipei District, Chongqing 400715, China; lisichencitrus@126.com (S.-C.L.); lihongjun@cric.cn (H.-J.L.); pengzhuchun@cric.cn (Z.-C.P.)
2 National Engineering Research Center for Citrus, Chinese Academy of Agricultural Sciences, Beipei District, Chongqing 400712, China
* Correspondence: 113883823816@126.com

Abstract: Here, the biogeographic patterns of abundant and rare bacterial taxa in lemon farmlands with different cultivation modes were examined using the dataset obtained from high-throughput sequencing. The abundant sub-communities exhibited a lower richness, a similar abundance proportion, and lower compositional variations than rare taxa. With regard to different cultivation modes, a lower richness but higher beta-diversity distance was observed in abundant bacterial taxa from greenhouse soils compared to other open field farmlands. In addition, some potential indicators, including Proteobacteria, Chloroflexi, and Bacteroidota, were found to be enriched in the abundant sub-communities in greenhouse soils. Moreover, a stronger environmental-related distance–decay of similarity was observed in abundant taxa from greenhouse soils, but in hilly-converted farmlands for rare taxa. The abundant sub-communities were more sensitive to environmental changes and more tightly phylogenetically clustered. In contrast, homogeneous selection dominated the assembly of rare taxa, which was insensitive to dispersal limitations. Soil pH was identified as the key factor to driving the assembly of soil bacterial communities, with a more deterministic and stochastic assembly for abundant and rare taxa, respectively, at the neutral environments.

Keywords: environmental adaptation; distance–decay of similarity; greenhouse; phylogenetic signal; stochastic assembly

1. Introduction

The microbial community is typically composed of a small number of highly abundant taxa and a large number of rare taxa with extremely high diversity [1]. Differentiating between abundant and rare sub-communities is crucial for understanding the structure and function of microbial communities [2]. Abundant and rare microbial taxa exhibit distinct patterns in community diversity and taxonomic composition [3]. Abundant taxa account for a small proportion of the total number of taxa, but represent a large proportion of the overall community abundance [4]. In contrast, rare microbial taxa contribute significantly to the biodiversity of ecosystems with a low abundance [5]. Abundant and rare taxa also differ markedly in their ecological roles in the relationships between biodiversity and ecosystem functions [6]. Abundant taxa are often considered the most important for core ecosystem functions due to their high abundances [7], and changes in the abundance of these taxa can drive variations in these broad functional measures [8]. For rare taxa, studies have shown that these taxa, despite their low abundance, can have disproportionately large effects on ecosystem processes and stability [9]. In addition, rare taxa generally provide functional redundancy, which is crucial for maintaining ecosystem processes under changing environmental conditions, as rare taxa can potentially replace functions lost if dominant taxa are diminished due to environmental stresses [10]. By comparing them with abundant taxa, researchers can better understand the overall biodiversity and how these different components contribute to ecosystem resilience and function [11].

In addition, the dynamics between abundant and rare taxa can determine the resilience of ecosystems to environmental changes [12]. Understanding how these groups respond differently to environmental stresses can inform the conservation strategies and management practices of ecosystems [13]. Abundant taxa tend to have wider niche breadths and can better adapt to environmental changes compared to rare taxa, allowing them to thrive under changing conditions [14]. Understanding how abundant taxa respond and adapt to environmental changes is crucial for predicting ecosystem-level responses [15]. On the other hand, rare taxa often show different responses to environmental changes compared to abundant taxa [16]. Rare taxa might serve as a reservoir of functional capabilities that become crucial under specific environmental conditions [17]. For instance, in response to environmental stressors or changes, these rare taxa can increase in abundance and take over functions necessary for ecosystem recovery and maintenance [18]. Moreover, rare microbial taxa can act as indicators of ecosystem health and shifts in environmental conditions [19]. Their presence or absence can provide early warning signs of ecological changes that might not be detectable by examining only the abundant taxa [20]. By understanding these contrasting responses, researchers can better predict how microbial communities as a whole will respond to environmental perturbations, and how the balance between abundant and rare taxa may shift [21]. This knowledge is crucial for managing and conserving ecosystem functions in the face of global change.

Moreover, comparing rare and abundant taxa helps refine ecological theories and models about community assembly within ecosystems [22]. The assembly of microbial communities is influenced by various processes such as diversification, dispersal, selection, and drift [23]. These processes could be divided into deterministic and stochastic processes, and their balance drive the biogeography of microorganisms [24]. Environmental filtering, which structures communities based on abiotic factors, and trait clustering due to competitive ability differences or ecological interactions, are key mechanisms of deterministic processes influencing community assembly [25]. These processes are positively related to community alpha diversity, microbial interaction degree, and bacterial predatory-specific gene abundance [26]. On the other hand, stochastic processes are governed by neutral theory and involve random assembly through birth–death, drift, and speciation [27]. Among them, drift is most important when selection is weak, alpha diversity is low, and the total number of community members is small [28]. Accordingly, previous studies indicated that abundant taxa were dominantly shaped by deterministic processes, such as environmental filtering, while the assembly of rare taxa was governed by stochastic processes, such as dispersal limitation [29–31]. As their relative importance can vary depending on the environment and the specific microbial community [32–34], distinguishing the assembly mechanisms of abundant and rare taxa is crucial for accurately interpreting microbial community dynamics, interactions, and functional potential [35]. This knowledge is valuable for understanding microbial community ecology and its applications.

Agricultural ecosystems provided humans with food, forage, bioenergy, and pharmaceuticals, which are essential for human well-being [36]. As one of the most valuable fruits, lemon cultivation has rapidly developed in China, especially for Chongqing and Sichuan Province in southwest area [37]. In the present study, we aimed to (I) evaluate the effects of cultivation modes on the diversity and composition of microbial communities, (II) explore the differences of environmental adaptation, and (III) recognize the major factors influencing the assembly of abundant and rare bacteria in soils of lemon farmlands. These questions were addressed using the high-throughput sequencing datasets of soil bacteria along with 15 environmental factors in greenhouse and open-field lemon farmlands. Given the unique environmental conditions of greenhouses, we hypothesized that the biogeographic patterns and assembly mechanisms of soil bacterial communities could be different in this cultivation mode compared to other open farmlands in Chongqing, China. Moreover, we also hypothesized that the responses of assembly processes governing abundant and rare bacterial sub-communities could be distinct under environmental changes. Insights gained from this study contribute to the development of ecological theories related to microbial succession, community assembly, and the impact of microbial diversity on ecosystem processes.

2. Results

2.1. General Patterns of Abundant and Rare Taxa

According to the sequencing results, a total of 2,634,985 high-quality reads were obtained from the 108 soil samples, which were clustered into 36,680 ASVs belonging to 48 bacterial phyla and 857 genera. As expected, most of them were identified as rare taxa (34,949, 95.28%) and only 115 ASVs (0.31%) were recognized as abundant taxa. The average richness proportion of abundant and rare taxa in the studied soil samples was 6.67% and 64.91%, respectively, and the average abundance proportion was 24.61% and 32.81%, respectively (Figure 1a). For soils from different cultivation modes, the richness of abundant taxa was significantly higher in soils from hilly- and paddy-converted farmlands compared to those from the greenhouses (Tukey's HSD test, $p < 0.05$, Figure 1b). The richness of rare taxa was also the highest in hilly-converted samples but significantly decreased in paddy-converted farmlands (Tukey's HSD test, $p < 0.05$, Figure 1b). The total abundance of abundant taxa in soils was the highest in greenhouse but lowest in hilly-converted farmlands (Figure 1c). In addition, a significantly higher total abundance of rare taxa was found in soils from greenhouse and hilly-converted farmlands than those in paddy-converted farmlands (Tukey's HSD test, $p < 0.05$, Figure 1c).

2.2. Effects of Cultivation Modes on Abundant and Rare Taxa

PCoA revealed the differences in the community structure of abundant and rare taxa among different cultivation modes. The distribution of abundant sub-communities among different cultivation modes had a large of overlap (Figure 2a); in contrast, the rare sub-communities were totally separately clustered (Figure 2b). The results of the adonis test also revealed the stronger influences of cultivation modes on rare taxa than that to abundant sub-communities (0.209 vs. 0.153), although both of them were significant ($p < 0.05$, Figure 2a,b). The Bray–Curtis distance of abundant and rare sub-communities between different soils from each cultivation mode were further compared, respectively. Both of them showed the highest intra-variation in greenhouses, followed by paddy-converted farmlands, and the lowest in hilly-converted farmlands (Tukey's HSD test, $p < 0.05$, Figure 2c,d).

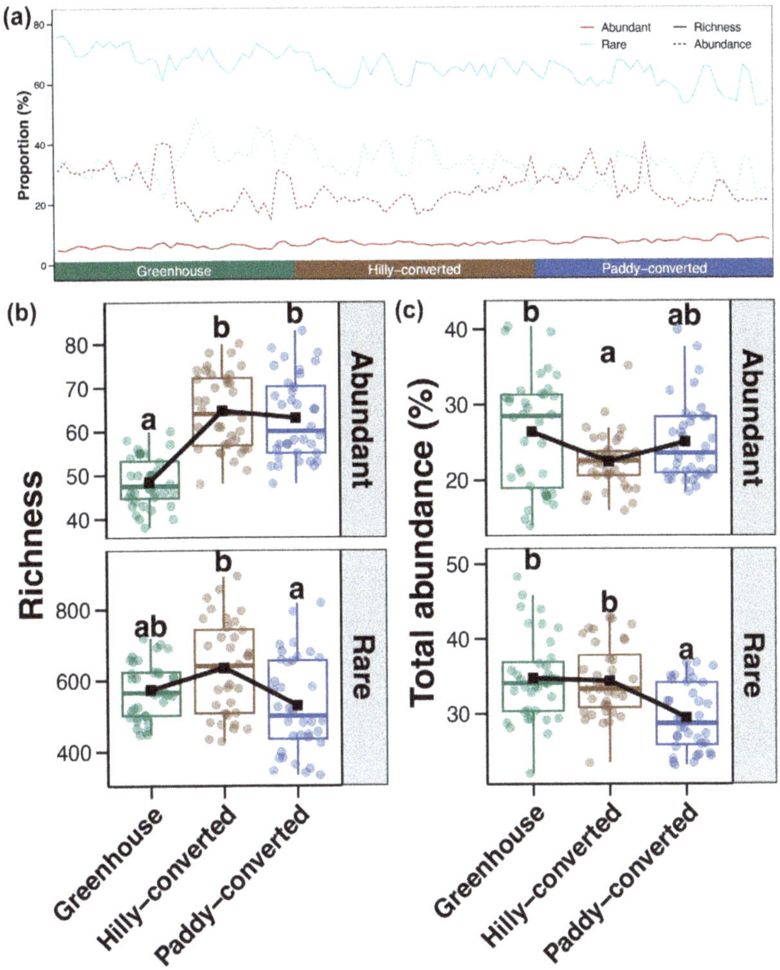

Figure 1. (**a**) The proportions (percentage of whole community) of abundant and rare taxa in richness and abundance, respectively, across studied soil samples. The red and blue lines represent abundant and rare taxa, respectively. The solid and dotted lines represent the proportion based on richness and relative abundance, respectively. (**b**) Differences in the richness of abundant and rare sub-communities among different cultivation modes. (**c**) Variations in the total abundance of abundant and rare taxa among different cultivation modes. Different lowercase letters above each box in the same sub-figure represent significant differences among samples from different cultivation modes (Tukey's HSD test, $p < 0.05$).

For both abundant and rare taxa, Proteobacteria was the most dominant bacterial phyla, following by Actinobacteriota, Acidobacteriota, and Chloroflexi, but their proportion in abundant or rare taxa were totally distinct (Figure 2e). Abundant sub-communities contained a higher proportion of Proteobacteria, Actinobacteria, Cyanobacteria, and Nitrospirota, while the proportion of Acidobacteriota, Chloroflexi, Gemmatimonadota, and Myxococcota were higher in rare taxa (Wilcox rank-sum test, $p < 0.05$, Figure 2e). In addition, their variations among different cultivation modes were also compared. The results showed higher abundances of Proteobacteira, Chloroflexi, and Bacteroidota in abundant sub-communities from greenhouse soils, with more abundant Acidobacteriota and Actinobacteriota in hilly- and paddy-converted farmlands, respectively (Tukey's HSD test, $p < 0.05$, Figure 2f). In contrast, only Gemmatimonadota was found to be enriched in the rare sub-communities of hilly- and paddy-converted farmlands compared to those in greenhouses (Tukey's HSD test, $p < 0.05$, Figure 2f).

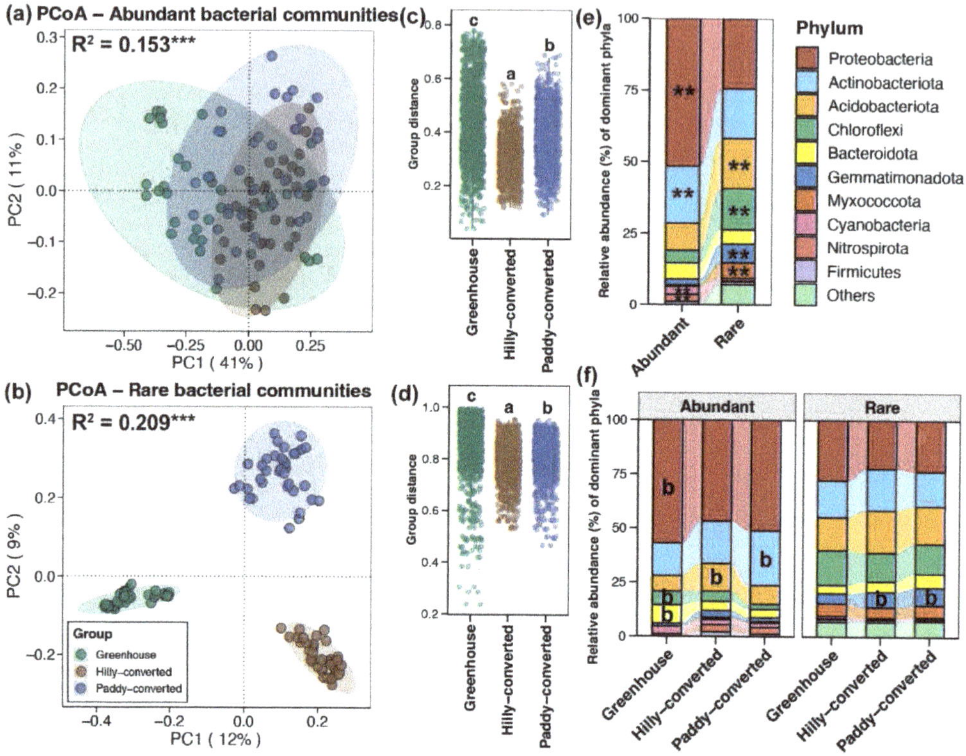

Figure 2. PCoA and adonis test for abundant (**a**) and rare (**b**) sub-communities in soils among different cultivation modes. *** represents the p-value of adonis test lower than 0.001. Differences in the Bray–Curtis distance between of abundant (**c**) and rare (**d**) sub-communities, respectively, between different soils from each cultivation mode. Different lowercase letters above each box in the same sub-figure represent significant differences among samples from different cultivation modes (Tukey's HSD test, $p < 0.05$). (**e**) The average relative abundances of dominant bacterial phyla for abundant and rare sub-communities in all studied soils. ** represents significantly higher relative abundance of bacterial phylum in abundant or rare sub-communities (Wilcox rank-sum test, $p < 0.05$). (**f**) Variations in the relative abundances of dominant bacterial phyla in abundant or rare sub-communities from soils among different cultivation modes. The letter "b" represents significantly higher relative abundance of bacterial phylum in corresponding cultivation mode compared to others (Tukey's HSD test, $p < 0.05$).

2.3. *Environmental Adaptation of Abundant and Rare Sub-Communities*

To explore the potential association between environmental conditions with the abundant and rare sub-communities in lemon farmlands, the DDCS was first performed. A significant DDCS with environmental factors was obtained for both abundant and rare sub-communities in all three cultivation modes ($p < 0.05$, Figure 3a). For abundant taxa, the strength of the DDCS was the strongest in greenhouses, followed by paddy-converted farmlands, and the weakest in hilly-converted farmlands, whereas this trend was just the reverse for rare sub-communities (Figure 3a). In addition to the DDCS based on taxonomic distance, the DDCS based on phylogenetic distance was further carried out. Similar to the results based on taxonomic distance, the DDCS with environmental factors based on phylogenetic distance was all significant for both abundant and rare sub-communities in all three cultivation modes ($p < 0.05$, Figure 3b). The trend of the DDCS strength among different cultivation modes was also consistent to the results of the DDCS based on taxonomic distance (Figure 3b). Moreover, variations in the phylogenetic distance of abundant and rare sub-communities between different soils from each cultivation mode were also compared. For both abundant and rare sub-communities, the phylogenetic distance between soils was the highest in greenhouse and the lowest in hilly-converted farmlands (Figure 3c), also consistent with the results based on taxonomic distance (Figure 2c).

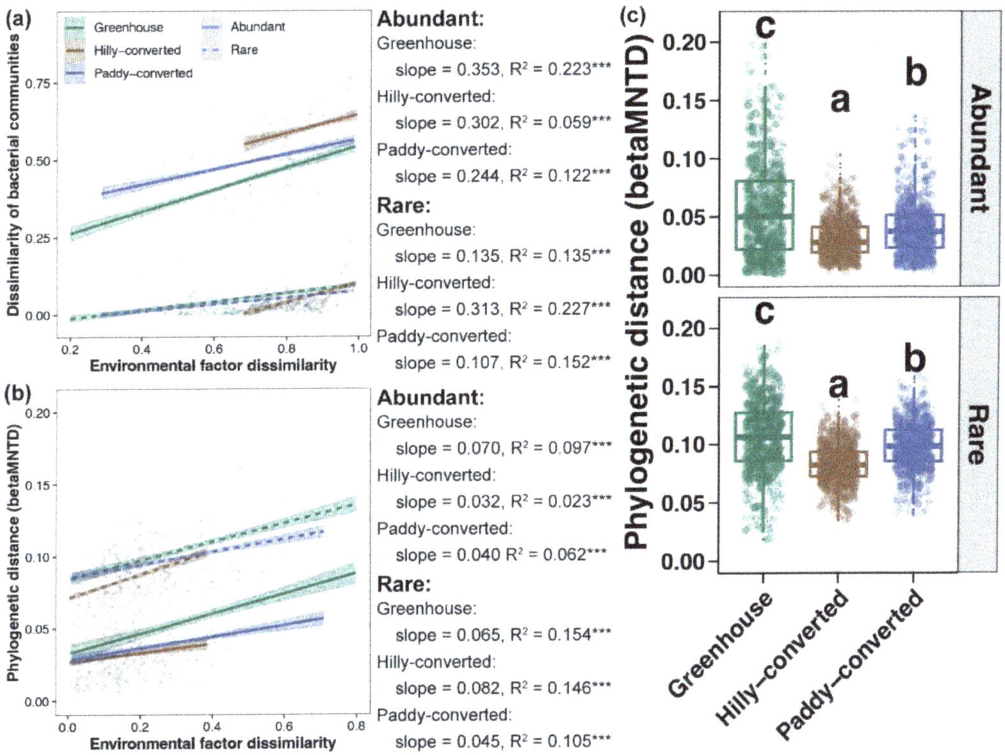

Figure 3. DDCS with environmental factors for the abundant and rare sub-communities among different cultivation modes based on the taxonomic (**a**) and phylogenetic (**b**) distances, respectively. *** represents the p-value of adonis test lower than 0.001. Lines in each sub-figure represents the fitted curve and gray shadow represents the 95% confidence interval. (**c**) Differences in the phylogenetic distance between abundant and rare sub-communities, respectively, between different soils from each cultivation mode. Different lowercase letters above each box in the same sub-figure represent significant differences among samples from different cultivation modes (Tukey's HSD test, $p < 0.05$).

The environmental threshold analysis was further employed to compare the differences of abundant and rare sub-communities in response to each of the studied environmental factors. In all three studied cultivation modes, rare taxa exhibited a broader environmental breadth compared to abundant taxa for almost all measured environmental factors (Figure 4a). Moreover, relationships between the phylogeny of abundant or rare taxa with their environmental preferences were further measured. Stronger phylogenetic signals for most environmental factors were found in abundant taxa compared to those in rare taxa in all three cultivation modes (Figure 4b). This suggested that the narrow ecological preferences within the abundant sub-community could be due to their phylogenetic conservatism.

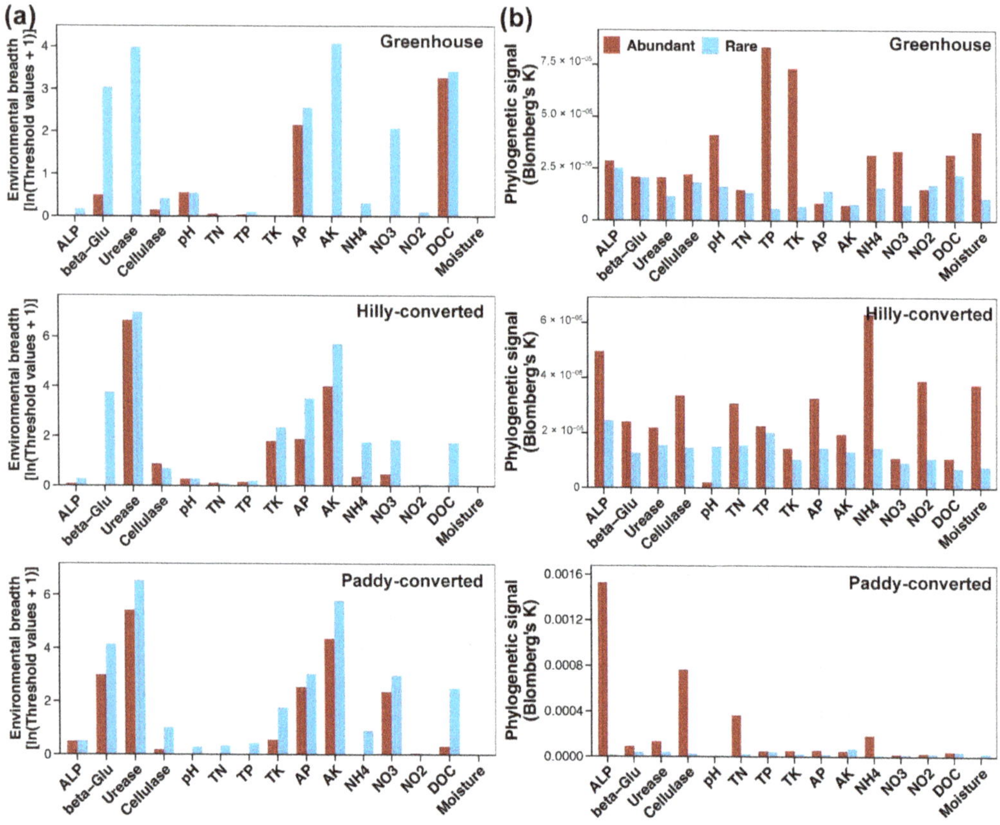

Figure 4. Environmental breadths (**a**) and phylogenetic signals (**b**) of abundant and rare taxa in response to environmental factors among different cultivation modes, respectively.

2.4. Community Assembly Mechanisms of Abundant and Rare Taxa

Based on the results of the null model, the relative importance of stochastic and deterministic processes for the assembly of abundant and rare sub-communities among different cultivation modes was investigated. The median of betaNTI for abundant sub-communities in all three cultivation modes were between −2 to 2, indicating that stochastic processes governed their assembly (Figure 5a). In contrast, the betaNTI for rare sub-communities were significantly lower than that for abundant taxa in all three cultivation modes (Wilcox rank-sum test, $p < 0.05$), with the median lower than −2 in hilly- and paddy-converted farmlands (Figure 5a). These results suggested that deterministic processes contributed more to the assembly of rare sub-communities than those for abundant taxa. Dispersal limiting, a stochastic process, was the dominant ecological process for the assembly of abundant sub-communities in all three cultivation modes, while it was, instead of homogeneous selection, one of the deterministic processes in rare taxa (Figure 5b).

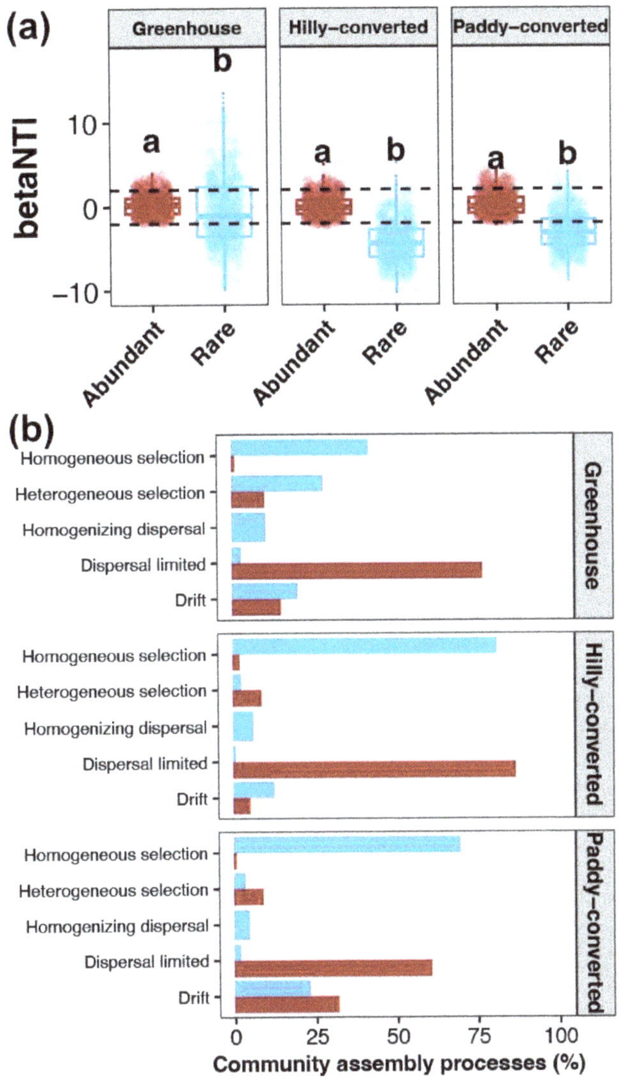

Figure 5. (**a**) Variations in betaNTI between the abundant and rare sub-communities in soils from each cultivation mode. Different lowercase letters above each box in the same sub-figure represent significant differences between abundant and rare sub-communities (Wilcox rank-sum test, $p < 0.05$). (**b**) Contributions of different ecological processes for assembly of abundant and rare sub-communities in soils from each cultivation mode. Red and blue colors represent the abundant and rare sub-communities, respectively.

2.5. Associations of Environmental Conditions and Bacterial Community Assembly

Relationships between betaNTI and measured environmental factors were evaluated in order to explore the potential drivers for soil bacterial community assembly in lemon farmlands. Linear regression results showed that soil pH was the factor with the strongest relationships with assembly processes in both the abundant and rare sub-communities (Table 1). Although both of them were significant (linear regression, $p < 0.05$), the association between soil pH and betaNTI was stronger for rare sub-communities that that for abundant taxa (Figure 6a). These relationships were further explored in each cultivation mode, and significant associations were found in soil pH with abundant taxa in hilly-converted farmlands, while rare taxa were found in greenhouse and paddy-converted farmlands (Linear regression, $p < 0.05$, Figure 6b). Furthermore, the samples were separated into sub-groups based on the soil pH and correlated with betaNTI. With an increasing soil pH, the relative importance of stochastic assembly first decreased and then increased in the abundant sub-communities, while it first increased and then decreased in the rare-communities (Figure 6c).

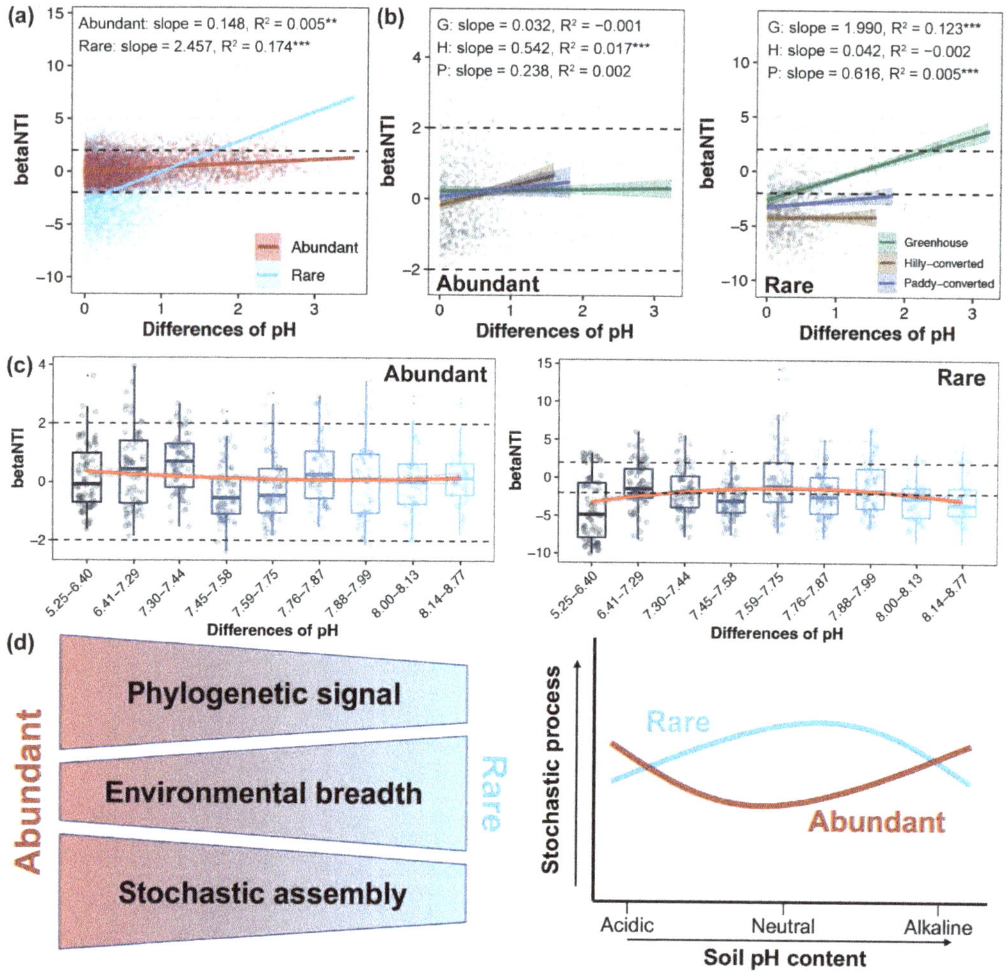

Figure 6. Relationships between betaNTI and differences in soil pH for abundant and rare sub-communities in all studied samples (**a**) and samples from different cultivation modes (**b**), respectively. ** and *** represent the p-value lower than 0.01 and 0.001, respectively. (**c**) Patterns of betaNTI across different categories in soil pH for the abundant and rare sub-communities. We ranked all 108 samples based on their pH from low to high; then, we sub-grouped them into 9 groups with 12 samples of each. Then, the pH intervals were determined according to the pH value belonging to each sub-group. (**d**) A conceptual paradigm showing environmental responses and stochastic processes in the assembly of abundant and rare bacterial sub-communities under the influence of soil pH in lemon farmlands.

Table 1. Linear regression between differences in environmental factors with betaNTI of abundant or rare sub-communities.

Environmental Factors	Abundant Taxa			Rare Taxa		
	Slope	R^2	*p*-Value	Slope	R^2	*p*-Value
ALP	0.1603	0.0060	0.0005	0.5656	0.0088	2.64×10^{-5}
beta-Glu	0.0028	0.0125	6.28×10^{-7}	0.0008	−0.0004	0.6260
Urease	5.72×10^{-6}	−0.0005	0.9250	9.87×10^{-5}	−0.0004	0.5809
Cellulase	0.0966	0.0080	5.97×10^{-5}	0.0272	−0.0005	0.7014
pH	0.1484	0.0050	0.0012	2.4570	0.1745	6.72×10^{-81}
TN	0.0770	−0.0003	0.5100	2.0898	0.0190	1.03×10^{-9}
TP	−0.1068	0.0007	0.1348	−0.6283	0.0042	0.0028
TK	0.0015	−0.0005	0.8607	−0.0173	−0.0003	0.4852
AP	−0.0009	6.94×10^{-5}	0.2877	0.0055	0.0022	0.0241
AK	−0.0006	0.0060	0.0004	−0.0008	0.0008	0.1168
NH4	0.0465	0.0038	0.0043	0.2238	0.0110	2.80×10^{-6}
NO3	0.0037	0.0006	0.1378	−0.0393	0.0145	8.98×10^{-8}
NO2	0.9222	0.0009	0.0960	18.1326	0.0650	1.19×10^{-29}
DOC	0.0038	0.0006	0.1463	0.0472	0.0195	6.64×10^{-10}
Moisture	3.0480	0.0024	7.38×10^{-7}	32.0664	0.1645	5.54×10^{-76}

3. Discussion

Comparing abundant and rare microbial taxa is essential for a deeper understanding of ecosystem complexity, and resilience, and the functional roles of microbes in environmental sustainability [38]. This comparison not only enhances our understanding of microbial ecology but also informs broader ecological and environmental management practices [39,40]. Based on the survey of soil bacteria in lemon farmlands with different cultivation modes, the present study provides evidence of the potential effects of the greenhouse on the soil bacteria community, indicating the lower richness but higher DDCS strength of abundant taxa. The results of this study also uncover disparate assembly processes controlling the abundant and rare sub-communities, which are mainly mediated by soil pH.

3.1. *Effects of Cultivation Modes on Abundant and Rare Bacterial Sub-Communities*

Soil bacterial communities play critical roles in agricultural productivity by decomposing organic compounds, recycling nutrients, and improving plant tolerance to stress [41]. Understanding the biogeography and distribution patterns of these soil bacterial communities can yield valuable insights underpinning sustainable agriculture. Moreover, the biogeographic patterns of different subdivisions of soil bacteria, rather than just at the whole community level, can provide a more nuanced view of microbial diversity and the response to environmental changes [42]. Previous studies have indicated that different agricultural land use practices (e.g., greenhouses, orchards, and paddy fields) drove distinct soil bacterial community structures and compositions [43]. The results of the present study also showed significant variations in the richness, proportion, and compositions of both abundant and rare bacterial sub-communities among soils of lemon farmlands with different cultivation modes (Figures 1 and 2). The greenhouse was found to decrease the richness of soil bacterial communities compared to open-field cultivation due to the agricultural intensification and higher rates of nitrogen fertilization [44,45]. Our results further suggested that the decreased richness of soil bacterial communities in greenhouses was more obviously seen in abundant taxa (Figure 1b). In addition, a previous study indicated that the proportion of abundant taxa tended to increase with increasing environmental disturbance [29]. Generally, the disturbance from human activities is stronger in greenhouses compared to those open-field cultivation [46]. The highest total abundance of abundant bacterial taxa found in greenhouse soils in the present study (Figure 1c) consists of the previous discovery. Moreover, compared to rare taxa, abundant taxa generally have a lower beta-diversity—less variation in community composition compared to rare taxa [47]. A similar phenomenon was also observed in the present study with more close distribution patterns in abundant taxa compared to rare sub-communities (Figure 2a,b). These differences identified here could help monitor the impact of agricultural management on soil microbial communities.

Certain bacterial taxa are enriched in specific agricultural modes, which can serve as indicator species for monitoring soil health [48]. Preteobacteria, Chloroflexi, and Bacteroidota were found to be enriched in the abundant sub-communities of soils from lemon greenhouses compared to those open-field farmlands (Figure 2f). Proteobacteria have been associated with the production and absorption of greenhouse gases [49], making them an important component in understanding soil–climate interactions and the impacts of environmental change. Chloroflexi have been found to play important roles in the composting process and soil biota in sustainable agriculture [50]. Controlling the relative abundance of Chloroflexi has been shown to be closely related to reducing greenhouse gas emissions [51]. Bacteroidota are key degraders of complex organic matter in greenhouse soils, playing crucial roles in carbon cycling and nutrient turnover [52], and the application of biofertilizers can significantly decrease its relative abundance [53]. These enriched abundant bacterial phyla in soils of lemon greenhouses are valuable for developing biological monitoring tools for assessing the impacts of human activities.

Moreover, understanding the DDCS can help in predicting the distribution of microbes and assessing the impacts of human activities on microbial communities [54]. The overall similarity and rates of decay are primarily influenced by species abundances [55]. Abundant species are those that are common and widespread, while rare species are those that are infrequent and have limited distribution [56]. Thus, abundant species tend to show a more gradual DDCS due to their widespread distribution, while rare species may exhibit a steeper decay at short distances but have a weaker overall influence on the distance–decay relationship [57]. However, a steeper environmental-related DDCS was uncovered for abundant bacterial taxa in greenhouse soils compared to that for rare sub-communities in the present study (Figure 3a,b). The contradicting results between the previous and present studies could be due to the difference in geographic scales among the distinct investigations [58,59]. This study only focused on the lemon farmlands at a local scale and greenhouse buildings obviously limited the turnover of abundant soil bacteria due to their high abundance and wide distribution [60]. In contrast, the dispersal potential of rare taxa between different soil samples could be high based on their low frequency [61]. Understanding these patterns is crucial for biodiversity conservation, as it informs strategies for preserving species across different geographic scales.

3.2. Stronger Environmental Adaptations of Rare Bacterial Taxa

Microbial communities can exhibit resilience and resistance to disturbances, but the specific response depends on the nature and chronicity of the disturbance [62]. These responses can include shifts in community composition and function, which are critical for predicting how ecosystems adapt to long-term environmental changes [63]. Abundant and rare taxa have distinct ecological functions and environmental adaptations: comparing them can provide insights into the overall stability and functioning of microbial communities under environmental stresses or disturbances [64,65]. Generally, rare taxa are more sensitive to environmental changes and disturbances compared to abundant taxa, which have broader niches and can better adapt [66]. However, we observed a narrow environmental breadth of abundant bacteria taxa in soils of lemon farmlands compared to that of rare taxa (Figure 6d). The definition of abundant taxa having wider niches is gleaned more from previous studies on large geographic scales and spanning diverse ecosystems [67–69]. In this study, bacteria only need to have a high competitiveness in lemon farmlands to become abundant species, and a considerable portion of them do not have a broad environmental adaptability, so they will be classified as rare or intermediate taxa in larger scale studies. This inference is supported by the comparison of phylogenetic signals between abundant and rare sub-communities (Figure 6d). Lower phylogenetic signals indicated more closely related species in the abundant sub-communities, which generally exhibited more similar ecological preferences across environmental gradients [70,71]. Thus, the abundant taxa identified in this study were more specialized and sensitive to environmental changes. In contrast, rare taxa with lower phylogenetic signals are weaker in phylogenetic niche conservatism, which provide functional redundancy and contribute significantly to ecosystem multifunctionality [72,73]. This might explain the broader environmental breadths and distinct biogeographic patterns of rare taxa compared to abundant sub-communities among lemon farmlands with different cultivation modes.

3.3. Stochastic and Deterministic Processes Respectively Governing the Assembly of Abundant and Rare Bacterial Sub-Communities

Understanding the microbial assembly mechanisms of microbial communities is crucial for advancing our fundamental knowledge of microbial ecology [74]. Elucidating the relative importances of deterministic and stochastic processes in shaping microbial communities can guide strategies for managing and conserving microbial diversity in natural and engineered ecosystems [75]. Moreover, comparing the assembly processes between abundant and rare taxa provides insights into the distinct strategies they employ to adapt to environmental stresses [76]. Bacterial communities in greenhouses have been previously found to be more influenced by stochastic processes like dispersal limitation and less by deterministic processes like homogeneous selection, leading to a lower bacterial diversity and less closely associated taxa [77]. The findings of the present study further explored this perspective of stochastic processes contributing more to shaping the rare sub-communities in greenhouses (Figure 5a). In addition, similar to previous findings [10,30,31], the present study also revealed that abundant sub-communities are more affected by stochastic processes, including random events and dispersal limitations (Figure 6d). In contrast, rare sub-communities are generally more influenced by deterministic processes, which are driven by environmental selection pressures (Figure 6d). Dispersal involves the movement of microorganisms across different habitats, and selection is the process by which environmental pressures favor certain microbial traits [78]. This suggested that abundant taxa might be more restricted to turnover in soils of lemon farmlands and rare taxa are subjected to a greater convergence under environmental selection pressures.

3.4. Soil pH Mediating the Assembly Processes of Soil Bacterial Communities in Lemon Farmlands

On the basis of the above findings, determining the key factors that regulate the assembly processes of abundant and rare sub-communities in the soils of lemon farmlands is essential for stability and the functioning of soil ecosystems [79]. We determined that soil pH played a leading role in regulating soil bacterial community assembly with the stochastic ratio increased and decreased at the neutral pH for rare and abundant sub-communities, respectively (Figure 6d). Soil pH has also been recognized as the primary factor driving the distribution and function of microorganisms in farmland soils [57]. A study that collected arable soils with a broad pH range (4.26–8.43) revealed that soil pH was an important factor shaping the bacterial community composition, with a higher diversity in neutral samples and a lower one in acidic samples [53]. In addition, soil acidification, which was more pronounced in the greenhouse soils, was identified as a key driver mediating the changes in bacterial community assembly processes under agricultural intensification [77]. Moreover, bacterial growth is highly influenced by pH, with 50% of maximum growth at pH opt within an interval of ± 1.7 pH units [80]. A soil pH closer to neutral might decrease the selective pressures and, consequently, reduce the effect of pH [81]. In this study, the decreased role of stochasticity in neutral environments may mean that adapted lineages of abundant taxa are reduced in neutral soils. The dominance of homogeneous selection in acid or alkaline soils suggests that rare taxa are more sensitive to an extreme pH; by contrast, the dominance of stochastic assembly in neutral soils implies the weaker niche-based exclusion and increased arrival of rare taxa lineages. Due to rare taxa being important contributors to microbial diversity [6], a changing soil pH might have a strong effect on microbial diversity.

Some agricultural management through targeted pH adjustment has been observed to optimize soil health and ecosystem functioning by leveraging pH–bacteria relationships. Liming acidic soils remains one of the most effective interventions for restoring bacterial equilibrium [82]. A longitudinal study demonstrated that lime applications increased soil microbial diversity and enhanced some bacteria to accelerate nutrient cycling [83]. In addition, field trials in China's North China Plain revealed that maintaining a soil pH between 6.2–6.8 optimized the bacterial richness, enabling functional redundancy during environmental perturbations [84]. Moreover, alleviating soil acidification suppressed bacterial wilt disease by inhibiting *Ralstonia solanacearum* by recruiting potentially beneficial rhizobacteria [85]. However, it should be noted that the nonlinear response of bacterial diversity to pH necessitates region-specific management [86]. By embracing pH as an ecological lever rather than a mere chemical metric, farmers could unlock the full potential of soil as a living, dynamic resource.

4. Materials and Methods

4.1. Sample Collection

Soil samples were collected from farmlands with more than five years of lemon cultivation at the Chongqing Tongnan Seedling Breeding Center located in the Guopo Village, Baizi Town, Chongqing, China. One-hundred and eight soil samples were collected from nine farmlands with twelve samples in each. A multi-point sampling method was used to collect soils with surface bulk soils at three random points within a 20 cm radius around a lemon trunk, and the collected soils were mixed into one sample. Each soil sample was divided into two parts: one was for the measurement of the soil properties, and another part was treated with liquid nitrogen and then cryopreserved at -80 °C for sequencing of soil bacterial communities. Among the nine farmlands, three of them were transformed from hilly areas (Hilly-converted), three of them were transformed from paddy fields (Paddy-converted), and the last three are orchards with greenhouse (Greenhouse). This sampling strategy allowed us to explore whether agro-ecosystems with different backgrounds exhibited similar trends.

4.2. Measurements of Environmental Factors

Edaphic variables, including moisture, pH, ammonium-nitrogen, nitrite-nitrogen, nitrate-nitrogen, total nitrogen (TN), available phosphorus (AP), total phosphorus (TP), available potassium (AK), total potassium (TK), and dissolved organic carbon (DOC), were measured using standard analytical methods [87,88]. Meanwhile, the soil enzyme activities, including urease, cellulase, alkaline phosphatase (ALP), and beta-glucosidase (beta-Glu), were determined using respective ELISA kits (Shanghai Jiwei Biological Technology Co.,

Ltd., Shanghai, China) using an RT-6100 spectrophotometer (Rayto Life and Analytical Sciences Co., Ltd., Shenzhen, China).

4.3. High-Throughput Sequencing and Data Processing

After DNA was extracted by the FastDNA® SPIN kit for soil (MP Biomedicals, Santa Ana, CA, USA), bacterial community was determined using high-throughput sequencing of the V3–V4 region of 16S rRNA gene with the primer pair 341F (GCCTCCCTCGCGC-CATCAGCAGTAGACGT) and 806R (GCCTTGCCAGCCCGCTCAG) [89] on a NovaSeq 6000 platform (Illumina Inc., San Diego, CA, USA). A negative control was used throughout the whole-sequencing process and no obvious microbial contamination was found. Based on the unique barcode at the end of reverse primer for each gene, sequenced reads were assigned to the corresponding samples. Then, sequences were filtered for quality (reads with average Phred scores > 20, no ambiguous bases or mismatches in the primers, homopolymer runs > 8, and sequence lengths > 250 bp), paired-end reads were assembled, and chimera was removed using the QIIME2 (Quantitative Insights Into Microbial Ecology 2) program [90]. The remained reads were clustered into amplicon sequence variants (ASVs) based on the DADA2 algorithm [91] and the representative sequences were classified within the SILVA database (release 138) [92]. By a randomly selected subset of 24,168 sequences from each sample to standardize the sequencing effort across samples, the ASV abundance table was obtained. ASVs with average relative abundance above 0.1% and below 0.01% across all samples were recognized as "abundant" and "rare" taxa [67].

4.4. Statistics Analysis

All statistics analyses were performed in R v4.4.2 and the results were visualized by the "ggplot2" package (v3.5.1). Richness (Chao1 index) of abundant or rare sub-communities in each sample and the Bray–Curtis distance of them between different samples were calculated by the "vegan" package (v2.6-10) [93]. Tukey's HSD test ("multcomp" package) was used to assess differences in the richness, total abundance, and Bray–Curtis distance of abundant or rare sub-communities, respectively, among farmlands with different cultivation modes [94]. Principal coordinates analysis (PCoA) and adonis test based on the Bray–Curtis distance of abundant or rare sub-communities were also achieved, respectively, by the "vegan" package. Differences in the relative abundances of dominant bacterial phyla between the abundant and rare sub-communities and variations of them in abundant or rare sub-communities among different cultivation modes were further compared by the Tukey's HSD test.

Beta mean nearest taxon distance (betaMNTD) metric of abundant and rare sub-communities was, respectively, calculated by the "picante" package (v1.8.2) [95], and their differences in samples among different cultivation modes were evaluated by Tukey's HSD test. Euclidean distance of measured environmental factors between each of two samples were calculated ("vegan" package) and the distance–decay community similarity (DDCS) was assessed as the slope of an ordinary least-squares regression between it with the Bray–Curtis distance or betaMNTD of abundant or rare sub-communities. In addition, the environmental breadth of abundant or rare sub-communities in response to each of the environmental factors was estimated by the threshold indicator taxa analysis ("TITAN2" package) (v2.4.3) [96]. Then, the phylogenetic signals of abundant and rare sub-communities to measured environmental factors were appraised by the Blomberg's K statistic ("picante" package) (v1.8.2) [95].

A null model method reported by Stegen et al. [97] was used to investigate the assembly processes of abundant and rare sub-communities. Based on the value of beta nearest taxon index (betaNTI) and Bray–Curtis-based Raup–Crick metric obtained from the null model, the relative importance of ecological processes for community assembly were estimated [98]. For betaNTI, values < -2 indicate homogeneous selection, whereas values > 2 indicate heterogeneous selection. The Raup–Crick metric was used to partition the remaining parts with $|\text{betaNTI}| \leq 2$. Here, Raup–Crick < -0.95 represents homogenous dispersal, Raup–Crick > 0.95 represents dispersal limitation, and $|\text{Raup–Crick}| \leq 0.95$ represents ecological drift. Then, the major factors that influenced the assembly processes of abundant and rare sub-communities were assessed using the linear regression comparing betaNTI values with the Euclidean distance matrices of each of the environmental factors.

5. Conclusions

Based on the investigation of soil bacterial communities in diverse lemon farmlands, we observed the distinct diversity, composition, and assembly of abundant and rare sub-communities. Meanwhile, significant effects of cultivation modes, especially for greenhouses, on both abundant and rare bacterial sub-communities were observed. A lower richness, higher beta diversity distance, and stronger DDCS of abundant sub-communities were found in greenhouses compared to other open-field farmlands. Moreover, we proposed a conceptual paradigm describing the assembly of abundant and rare bacterial sub-communities influenced by environmental factors in lemon farmlands (Figure 6d). The abundant sub-communities were more phylogenetically closely clustered but less adaptable to environmental changes than the rare taxa, which were more governed by stochastic assembly. Soil pH regulated the assembly of both abundant and rare sub-communities. A neutral pH led to an increase in stochastic assembly in the rare sub-communities, while it led to a decrease in abundant taxa. These results are beneficial to the understanding of the maintenance of microbial diversity and their related ecosystem functions in lemon farmlands, and should be considered for the development of strategies for ecosystem management.

Author Contributions: Conceptualization, supervision, project administration, writing-review and editing, H.-Q.L.; investigation and methodology, S.-C.L.; data curation and resources; H.-J.L.; validation and methodology, Z.-C.P. All authors have read and agreed to the published version of the manuscript.

Funding: This research was supported by the National Key R & D Program of China (2021YFD1400800, 2020YFD1000102, and 2018YFD0201500); the National Key R & D Program of China (2019YFD1002100); and the Chongqing Scientific Research Project (cstc2021jcyj-bsh0082).

Data Availability Statement: All raw sequences of sediment bacterial communities studied in this study have been submitted to the NCBI Sequence Read Archive (SRA) database under the BioProject number PRJNA1100232.

Conflicts of Interest: The authors declare no conflict of interest.

References

1. Mo, Y.; Zhang, W.; Yang, J.; Lin, Y.; Yu, Z.; Lin, S. Biogeographic patterns of abundant and rare bacterioplankton in three subtropical bays resulting from selective and neutral processes. *ISME J.* **2018**, *12*, 2198–2210. [CrossRef] [PubMed]
2. Zhou, Z.; Zhang, Y.; Zhang, F. Abundant and rare bacteria possess different diversity and function in crop monoculture and rotation systems across regional farmland. *Soil Biol. Biochem.* **2022**, *171*, 108742. [CrossRef]
3. Shao, Q.; Sun, D.; Fang, C.; Feng, Y.; Wang, C. Biodiversity and biogeography of abundant and rare microbial assemblages in the Western subtropical Pacific Ocean. *Front. Microbiol.* **2022**, *13*, 839562. [CrossRef] [PubMed]
4. Dueholm, M.K.D.; Nierychlo, M.; Andersen, K.S.; Rudkjøbing, V.; Knutsson, S.; Albertsen, M.; Nielsen, P.H. MiDAS 4: A global catalogue of full-length 16S rRNA gene sequences and taxonomy for studies of bacterial communities in wastewater treatment plants. *Nat. Commun.* **2022**, *13*, 1908. [CrossRef] [PubMed]
5. Shade, A.; Jones, S.E.; Caporaso, J.G.; Handelsman, J.; Knight, R.; Fierer, N.; Gilbert, J.A. Conditionally rare taxa disproportionately contribute to temporal changes in microbial diversity. *mBio* **2014**, *5*, 10–1128. [CrossRef] [PubMed]
6. Lynch, M.D.; Neufeld, J.D. Ecology and exploration of the rare biosphere. *Nat. Rev. Microbiol.* **2015**, *13*, 217–229. [CrossRef]

7. Jiao, S.; Wang, J.; Wei, G.; Chen, W.; Lu, Y. Dominant role of abundant rather than rare bacterial taxa in maintaining agro-soil microbiomes under environmental disturbances. *Chemosphere* **2019**, *235*, 248–259. [CrossRef] [PubMed]
8. Rivett, D.W.; Bell, T. Abundance determines the functional role of bacterial phylotypes in complex communities. *Nat. Microbiol.* **2018**, *3*, 767–772. [CrossRef]
9. Pascoal, F.; Costa, R.; Magalhães, C. The microbial rare biosphere: Current concepts, methods and ecological principles. *FEMS Microbiol. Ecol.* **2021**, *97*, fiaa227. [CrossRef]
10. Hou, D.; Zhou, R.; Wei, D.; Zeng, S.; Weng, S.; Yan, Q.; He, J.; Huang, Z. Abundant and rare microbial communities respectively contribute to an aquaculture pond ecosystem. *Front. Mar. Sci.* **2022**, *9*, 856126. [CrossRef]
11. Jousset, A.; Bienhold, C.; Chatzinotas, A.; Gallien, L.; Gobet, A.; Kurm, V.; Kusel, K.; Rillg, M.C.; Rivett, D.W.; Salles, J.F.; et al. Where less may be more: How the rare biosphere pulls ecosystems strings. *ISME J.* **2017**, *11*, 853–862. [CrossRef]
12. Kurm, V.; Geisen, S.; Gera Hol, W.H. A low proportion of rare bacterial taxa responds to abiotic changes compared with dominant taxa. *Environ. Microbiol.* **2019**, *21*, 750–758. [CrossRef] [PubMed]
13. Wang, J.; Wang, Y.; Li, M.; Xu, L.; He, N.; Yan, P.; Chen, C.; Lu, Q.; Feng, Y.; Li, J. Differential response of abundant and rare bacterial sub-communities to abiotic and biotic gradients across temperate deserts. *Sci. Total Environ.* **2021**, *763*, 142942. [CrossRef] [PubMed]
14. Ren, Z.; Zhang, C.; Li, X.; Ma, K.; Cui, B. Abundant and rare bacterial taxa structuring differently in sediment and water in thermokarst lakes in the Yellow River Source area, Qinghai-Tibet Plateau. *Front. Microbiol.* **2022**, *13*, 774514. [CrossRef] [PubMed]
15. Smith, C.R.; Blair, P.L.; Boyd, C.; Cody, B.; Hazel, A.; Hedrick, A.; Kathuria, H.; Khurana, P.; Kramer, B.; Muterspaw, K.; et al. Microbial community responses to soil tillage and crop rotation in a corn/soybean agroecosystem. *Ecol. Evol.* **2016**, *6*, 8075–8084. [CrossRef]
16. Zhang, X.; Zhao, W.; Kou, Y.; Fang, K.; Liu, Y.; He, H.; Liu, Q. The contrasting responses of abundant and rare microbial community structures and co-occurrence networks to secondary forest succession in the sub-alpine region. *Front. Microbiol.* **2023**, *14*, 1177239. [CrossRef]
17. Wang, C.; Guo, L.; Shen, R.F. Rare microbial communities drive ecosystem multifunctionality in acidic soils of southern China. *Appl. Soil Ecol.* **2023**, *189*, 104895. [CrossRef]
18. Xu, M.; Huang, Q.; Xiong, Z.; Liao, H.; Lv, Z.; Chen, W.; Luo, X.; Hao, X. Distinct responses of rare and abundant microbial taxa to in situ chemical stabilization of cadmium-contaminated soil. *mSystems* **2021**, *6*, e0104021. [CrossRef]
19. Xue, M.; Guo, Z.; Gu, X.; Gao, H.; Weng, S.; Zhou, J.; Gu, D.; Lu, H.; Zhou, X. Rare rather than abundant microbial communities drive the effects of long-term greenhouse cultivation on ecosystem functions in subtropical agricultural soils. *Sci. Total Environ.* **2020**, *706*, 136004. [CrossRef]
20. Glasl, B.; Webster, N.S.; Bourne, D.G. Microbial indicators as a diagnostic tool for assessing water quality and climate stress in coral reef ecosystems. *Mar. Biol.* **2017**, *164*, 91. [CrossRef]
21. Wan, W.; Liu, S.; Li, X.; Xing, Y.; Chen, W.; Huang, Q. Bridging rare and abundant bacteria with ecosystem multifunctionality in salinized agricultural soils: From community diversity to environmental adaptation. *mSystems* **2021**, *6*, 15. [CrossRef] [PubMed]
22. Xue, Y.; Chen, H.; Yang, J.R.; Liu, M.; Huang, B.; Yang, J. Distinct patterns and processes of abundant and rare eukaryotic plankton communities following a reservoir cyanobacterial bloom. *ISME J.* **2018**, *12*, 2263–2277. [CrossRef] [PubMed]
23. Chen, W.; Ren, K.; Isabwe, A.; Chen, H.; Liu, M.; Yang, J. Stochastic processes shape microeukaryotic community assembly in a subtropical river across wet and dry seasons. *Microbiome* **2019**, *7*, 138.
24. Aguilar, P.; Sommaruga, R. The balance between deterministic and stochastic processes in structuring lake bacterioplankton community over time. *Mol. Ecol.* **2020**, *29*, 3117–3130. [CrossRef]
25. Ortega-Martínez, I.J.; Moreno, C.E.; Rios-Díaz, C.L.; Arellano, L.; Rosas, F.; Castellanos, I. Assembly mechanisms of dung beetles in temperate forests and grazing pastures. *Sci. Rep.* **2020**, *10*, 391. [CrossRef]
26. Wang, J.; Pan, Z.; Yu, J.; Zhang, Z.; Li, Y.Z. Global assembly of microbial communities. *mSystems* **2023**, *8*, e01289-22. [CrossRef]
27. Sun, Y.; Zhang, M.; Duan, C.; Cao, N.; Jia, W.; Zhao, Z.; Ding, C.; Huang, Y.; Wang, J. Contribution of stochastic processes to the microbial community assembly on field-collected microplastics. *Environ. Microbiol.* **2021**, *23*, 6707–6720. [CrossRef] [PubMed]
28. Nemergut, D.R.; Schmidt, S.K.; Fukami, T.; O'Neill, S.P.; Bilinski, T.M.; Stanish, L.F.; Knelman, J.E.; Darcy, J.L.; Lynch, R.C.; Wickey, P.; et al. Patterns and processes of microbial community assembly. *Microbiol. Mol. Biol. Rev.* **2013**, *77*, 342–356. [CrossRef]
29. Dong, Y.; Wu, S.; Deng, Y.; Wang, S.; Fan, H.; Li, X.; Bai, Z.; Zhuang, X. Distinct functions and assembly mechanisms of soil abundant and rare bacterial taxa under increasing pyrene stresses. *Front. Microbiol.* **2021**, *12*, 689762. [CrossRef]
30. Gao, G.F.; Peng, D.; Tripathi, B.M.; Zhang, Y.; Chu, H. Distinct community assembly processes of abundant and rare soil bacteria in coastal wetlands along an inundation gradient. *mSystems* **2020**, *5*, 10–1128. [CrossRef]
31. Su, Y.; Hu, Y.; Zi, H.; Chen, Y.; Deng, X.; Hu, B.; Jiang, Y. Contrasting assembly mechanisms and drivers of soil rare and abundant bacterial communities in 22-year continuous and non-continuous cropping systems. *Sci. Rep.* **2022**, *12*, 3264. [CrossRef] [PubMed]
32. Jiao, S.; Lu, Y. Soil pH and temperature regulate assembly processes of abundant and rare bacterial communities in agricultural ecosystems. *Environ. Microbiol.* **2020**, *22*, 1052–1065. [CrossRef] [PubMed]

33. Yang, Y.; Cheng, K.; Li, K.; Jin, Y.; He, X. Deciphering the diversity patterns and community assembly of rare and abundant bacterial communities in a wetland system. *Sci. Total Environ.* **2022**, *838*, 156334. [CrossRef] [PubMed]
34. Zhao, H.; Brearley, F.Q.; Huang, L.; Tang, J.; Xu, Q.; Li, X.; Huang, Y.; Zou, S.; Chen, X.; Hou, W.; et al. Abundant and rare taxa of planktonic fungal community exhibit distinct assembly patterns along coastal eutrophication gradient. *Microb. Ecol.* **2023**, *85*, 495–507. [CrossRef]
35. Jia, Y.; Zhao, S.; Guo, W.; Peng, L.; Zhao, F.; Wang, L.; Fan, G.; Zhu, Y.; Xu, D.; Liu, G.; et al. Sequencing introduced false positive rare taxa lead to biased microbial community diversity, assembly, and interaction interpretation in amplicon studies. *Environ. Microbiome* **2022**, *17*, 43. [CrossRef]
36. Power, A.G. Ecosystem services and agriculture: Trade-offs and synergies. *Philos. Trans. R. Soc. B Biol. Sci.* **2010**, *365*, 2959–2971. [CrossRef] [PubMed]
37. Wang, Y.; Liao, P.; Zhao, J.F.; Zhang, X.K.; Liu, C.; Xiao, P.A.; Zhou, C.Y.; Zhou, Y. Comparative transcriptome analysis of the Eureka lemon in response to Citrus yellow vein virus infection at different temperatures. *Physiol. Mol. Plant Pathol.* **2022**, *119*, 101832. [CrossRef]
38. Ma, L.; Niu, W.; Li, G.; Du, Y.; Sun, J.; Zhang, Q.; Siddique, K.H. Crucial role of rare taxa in preserving bacterial community stability. *Land. Degrad. Dev.* **2024**, *35*, 1397–1410. [CrossRef]
39. Chen, G.; Wang, W.; Zhang, Y.; Liu, Y.; Gu, X.; Shi, X.; Wang, M. Abundant and rare species may invoke different assembly processes in response to climate extremes: Implications for biodiversity conservation. *Ecol. Indic.* **2020**, *117*, 106716. [CrossRef]
40. Li, Y.; Sun, X.; Zhang, M.; Khan, A.; Sun, W. Dominant role of rare bacterial taxa rather than abundant taxa in driving the tailing primary succession. *J. Hazard. Mater.* **2024**, *462*, 132807. [CrossRef]
41. Bay, S.K.; McGeoch, M.A.; Gillor, O.; Wieler, N.; Palmer, D.J.; Baker, D.J.; Chown, S.L.; Greening, C. Soil bacterial communities exhibit strong biogeographic patterns at fine taxonomic resolution. *mSystems* **2020**, *5*, 10–1128. [CrossRef] [PubMed]
42. Sun, Y.; Deng, X.; Tao, C.; Liu, H.; Shen, Z.; Liu, Y.; Li, R.; Shen, Q. Temporal dynamics of rare and abundant soil bacterial taxa from different fertilization regimes under various environmental disturbances. *mSystems* **2022**, *7*, e00559-22. [CrossRef]
43. Lee, S.A.; Kim, J.M.; Kim, Y.; Joa, J.H.; Kang, S.S.; Ahn, J.H.; Kim, M.; Song, J.; Weon, H.Y. Different types of agricultural land use drive distinct soil bacterial communities. *Sci. Rep.* **2020**, *10*, 17418. [CrossRef] [PubMed]
44. Jansson, J.K.; Hofmockel, K.S. Soil microbiomes and climate change. *Nat. Rev. Microbiol.* **2020**, *18*, 35–46. [CrossRef]
45. Kpalari, D.F.; Mounkaila Hamani, A.K.; Hui, C.; Sogbedji, J.M.; Liu, J.; Le, Y.; Kama, R.; Gao, Y. Soil bacterial community and greenhouse gas emissions as responded to the coupled application of nitrogen fertilizer and microbial decomposing inoculants in wheat (*Triticum aestivum* L.) seedling stage under different water regimes. *Agronomy* **2023**, *13*, 2950. [CrossRef]
46. Wang, T.; Wu, G.; Chen, J.; Cui, P.; Chen, Z.; Yan, Y.; Zhang, Y.; Li, M.; Niu, D.; Li, B.; et al. Integration of solar technology to modern greenhouse in China: Current status, challenges and prospect. *Renew. Sustain. Energy Rev.* **2017**, *70*, 1178–1188. [CrossRef]
47. Eng, A.; Borenstein, E. Taxa-function robustness in microbial communities. *Microbiome* **2018**, *6*, 45. [CrossRef]
48. Astudillo-García, C.; Hermans, S.M.; Stevenson, B.; Buckley, H.L.; Lear, G. Microbial assemblages and bioindicators as proxies for ecosystem health status: Potential and limitations. *Appl. Microbiol. Biotechnol.* **2019**, *103*, 6407–6421. [CrossRef]
49. Wei, H.; Peng, C.; Yang, B.; Song, H.; Li, Q.; Jiang, L.; Wei, G.; Wang, K.; Wang, H.; Liu, S.; et al. Contrasting soil bacterial community, diversity, and function in two forests in China. *Front. Microbiol.* **2018**, *9*, 1693. [CrossRef]
50. Aguilar-Paredes, A.; Valdés, G.; Araneda, N.; Valdebenito, E.; Hansen, F.; Nuti, M. Microbial community in the composting process and its positive impact on the soil biota in sustainable agriculture. *Agronomy* **2023**, *13*, 542. [CrossRef]
51. Islam, Z.F.; Cordero, P.R.F.; Feng, J.; Chen, Y.J.; Bay, S.K.; Jirapanjawat, T.; Gleadow, R.M.; Carere, C.R.; Stott, M.B.; Chiri, E.; et al. Two Chloroflexi classes independently evolved the ability to persist on atmospheric hydrogen and carbon monoxide. *ISME J.* **2019**, *13*, 1801–1813. [CrossRef] [PubMed]
52. Thomas, F.; Hehemann, J.H.; Rebuffet, E.; Czjzek, M.; Michel, G. Environmental and gut Bacteroidetes: The food connection. *Front. Microbiol.* **2011**, *2*, 93. [CrossRef] [PubMed]
53. Wu, Y.; Zeng, J.; Zhu, Q.; Zhang, Z.; Lin, X. pH is the primary determinant of the bacterial community structure in agricultural soils impacted by polycyclic aromatic hydrocarbon pollution. *Sci. Rep.* **2017**, *7*, 40093. [CrossRef]
54. Clark, D.R.; Underwood, G.J.; McGenity, T.J.; Dumbrell, A.J. What drives study-dependent differences in distance–decay relationships of microbial communities? *Glob. Ecol. Biogeogr.* **2021**, *30*, 811–825. [CrossRef]
55. Morlon, H.; Chuyong, G.; Condit, R.; Hubbell, S.; Kenfack, D.; Thomas, D.; Valencia, R.; Green, J.L. A general framework for the distance–decay of similarity in ecological communities. *Ecol. Lett.* **2008**, *11*, 904–917. [CrossRef]
56. Zhang, Y.; Wu, G.; Jiang, H.; Yang, J.; She, W.; Khan, I.; Li, W. Abundant and rare microbial biospheres respond differently to environmental and spatial factors in Tibetan hot springs. *Front. Microbiol.* **2018**, *9*, 345048. [CrossRef] [PubMed]
57. Jiao, S.; Chen, W.; Wei, G. Biogeography and ecological diversity patterns of rare and abundant bacteria in oil-contaminated soils. *Mol. Ecol.* **2017**, *26*, 5305–5317. [CrossRef]
58. Azovsky, A.I.; Chertoprud, E.S.; Saburova, M.A. Small-scale spatiotemporal variability and distance–decay relationships in intertidal micro-and meiobenthic assemblages. *Mar. Ecol.* **2022**, *43*, e12704. [CrossRef]

59. Pompeu, C.R.; Peñas, F.J.; Barquín, J. Large-scale spatial patterns of riverine communities: Niche versus geographical distance. *Biodivers. Conserv.* **2023**, *32*, 589–607. [CrossRef]
60. Zhou, X.; Wu, F. Land-use conversion from open field to greenhouse cultivation differently affected the diversities and assembly processes of soil abundant and rare fungal communities. *Sci. Total Environ.* **2021**, *788*, 147751. [CrossRef]
61. Wang, W.; Ye, Z.; Li, J.; Liu, G.; Wu, Q.; Wang, Z.; He, G.; Yan, W.; Zhang, C. Intermediate irrigation with low fertilization promotes soil nutrient cycling and reduces CO_2 and CH_4 emissions via regulating fungal communities in arid agroecosystems. *J. Environ. Manag.* **2024**, *351*, 119688. [CrossRef] [PubMed]
62. Philippot, L.; Griffiths, B.S.; Langenheder, S. Microbial community resilience across ecosystems and multiple disturbances. *Microbiol. Mol. Biol. Rev.* **2021**, *85*, 10–1128. [CrossRef] [PubMed]
63. Potts, L.D.; Douglas, A.; Perez Calderon, L.J.; Anderson, J.A.; Witte, U.; Prosser, J.I.; Gubry-Rangin, C. Chronic environmental perturbation influences microbial community assembly patterns. *Environ. Sci. Technol.* **2022**, *56*, 2300–2311. [CrossRef]
64. Liang, Y.; Xiao, X.; Nuccio, E.E.; Yuan, M.; Zhang, N.; Xue, K.; Cohan, F.M.; Zhou, J.; Sun, B. Differentiation strategies of soil rare and abundant microbial taxa in response to changing climatic regimes. *Environ. Microbiol.* **2020**, *22*, 1327–1340. [CrossRef]
65. Xiong, C.; He, J.Z.; Singh, B.K.; Zhu, Y.G.; Wang, J.T.; Li, P.P.; Zhang, Q.B.; Han, L.L.; Shen, J.P.; Ge, A.H.; et al. Rare taxa maintain the stability of crop mycobiomes and ecosystem functions. *Environ. Microbiol.* **2021**, *23*, 1907–1924. [CrossRef]
66. Wan, W.; Gadd, G.M.; Yang, Y.; Yuan, W.; Gu, J.; Ye, L.; Liu, W. Environmental adaptation is stronger for abundant rather than rare microorganisms in wetland soils from the Qinghai-Tibet Plateau. *Mol. Ecol.* **2021**, *30*, 2390–2403. [CrossRef]
67. Jiao, S.; Lu, Y. Abundant fungi adapt to broader environmental gradients than rare fungi in agricultural fields. *Glob. Change Biol.* **2020**, *26*, 4506–4520. [CrossRef]
68. Li, C.; Jin, L.; Zhang, C.; Li, S.; Zhou, T.; Hua, Z.; Wang, L.; Ji, S.; Wang, Y.; Gan, Y.; et al. Destabilized microbial networks with distinct performances of abundant and rare biospheres in maintaining networks under increasing salinity stress. *iMeta* **2023**, *2*, e79. [CrossRef] [PubMed]
69. Luan, L.; Shi, G.; Zhu, G.; Zheng, J.; Fan, J.; Dini-Andreote, F.; Sun, B.; Jiang, Y. Biogeographical patterns of abundant and rare bacterial biospheres in paddy soils across East Asia. *Environ. Microbiol.* **2023**, *25*, 294–305. [CrossRef]
70. Amend, A.S.; Martiny, A.C.; Allison, S.D.; Berlemont, R.; Goulden, M.L.; Lu, Y.; Treseder, K.K.; Weihe, C.; Martiny, J.B.H. Microbial response to simulated global change is phylogenetically conserved and linked with functional potential. *ISME J.* **2016**, *10*, 109–118. [CrossRef]
71. Martiny, J.B.; Jones, S.E.; Lennon, J.T.; Martiny, A.C. Microbiomes in light of traits: A phylogenetic perspective. *Science* **2015**, *350*, aac9323. [CrossRef] [PubMed]
72. Chen, Q.L.; Ding, J.; Zhu, D.; Hu, H.W.; Delgado-Baquerizo, M.; Ma, Y.B.; He, J.Z.; Zhu, Y.G. Rare microbial taxa as the major drivers of ecosystem multifunctionality in long-term fertilized soils. *Soil Biol. Biochem.* **2020**, *141*, 107686. [CrossRef]
73. Zhang, Z.; Lu, Y.; Wei, G.; Jiao, S. Rare species-driven diversity–ecosystem multifunctionality relationships are promoted by stochastic community assembly. *mBio* **2022**, *13*, e00449-22. [CrossRef] [PubMed]
74. Dini-Andreote, F.; Stegen, J.C.; van Elsas, J.D.; Salles, J.F. Disentangling mechanisms that mediate the balance between stochastic and deterministic processes in microbial succession. *Proc. Natl. Acad. Sci. USA* **2015**, *112*, E1326–E1332. [CrossRef]
75. Liao, J.; Cao, X.; Zhao, L.; Wang, J.; Gao, Z.; Wang, M.C.; Huang, Y. The importance of neutral and niche processes for bacterial community assembly differs between habitat generalists and specialists. *FEMS Microbiol. Ecol.* **2016**, *92*, fiw174. [CrossRef]
76. Li, X.; Stegen, J.C.; Yu, Y.; Huang, J. Coordination and divergence in community assembly processes across co-occurring microbial groups separated by cell size. *Front. Microbiol.* **2023**, *14*, 1166322. [CrossRef]
77. Zhou, X.; Khashi u Rahman, M.; Liu, J.; Wu, F. Soil acidification mediates changes in soil bacterial community assembly processes in response to agricultural intensification. *Environ. Microbiol.* **2021**, *23*, 4741–4755. [CrossRef]
78. Ning, D.; Yuan, M.; Wu, L.; Zhang, Y.; Guo, X.; Zhou, X.; Yang, Y.; Arkin, A.P.; Firestone, M.K.; Zhou, J. A quantitative framework reveals ecological drivers of grassland microbial community assembly in response to warming. *Nat. Commun.* **2020**, *11*, 4717. [CrossRef]
79. Tripathi, B.M.; Stegen, J.C.; Kim, M.; Dong, K.; Adams, J.M.; Lee, Y.K. Soil pH mediates the balance between stochastic and deterministic assembly of bacteria. *ISME J.* **2018**, *12*, 1072–1083. [CrossRef]
80. Fernández-Calviño, D.; Bååth, E. Growth response of the bacterial community to pH in soils differing in pH. *FEMS Microbiol. Ecol.* **2010**, *73*, 149–156. [CrossRef]
81. Tian, Q.; Jiang, Y.; Tang, Y.; Wu, Y.; Tang, Z.; Liu, F. Soil pH and organic carbon properties drive soil bacterial communities in surface and deep layers along an elevational gradient. *Front. Microbiol.* **2021**, *12*, 646124. [CrossRef]
82. Zhou, X.; Tahvanainen, T.; Malard, L.; Chen, L.; Pérez-Pérez, J.; Berninger, F. Global analysis of soil bacterial genera and diversity in response to pH. *Soil Biol. Biochem.* **2024**, *198*, 109552. [CrossRef]
83. Madegwa, Y.M.; Uchida, Y. Liming improves the stability of soil microbial community structures against the application of digestate made from dairy wastes. *J. Environ. Manag.* **2021**, *297*, 113356. [CrossRef] [PubMed]

84. Yang, Y.; Shi, Y.; Fang, J.; Chu, H.; Adams, J.M. Soil microbial network complexity varies with pH as a continuum, not a threshold, across the North China Plain. *Front. Microbiol.* **2022**, *13*, 895687. [CrossRef] [PubMed]
85. Zhang, S.; Liu, X.; Zhou, L.; Deng, L.; Zhao, W.; Liu, Y.; Ding, W. Alleviating soil acidification could increase disease suppression of bacterial wilt by recruiting potentially beneficial rhizobacteria. *Microbiol. Spectr.* **2022**, *10*, e02333-21. [CrossRef] [PubMed]
86. Xiong, R.; He, X.; Gao, N.; Li, Q.; Qiu, Z.; Hou, Y.; Shen, W. Soil pH amendment alters the abundance, diversity, and composition of microbial communities in two contrasting agricultural soils. *Microbiol. Spectr.* **2024**, *12*, e04165-23. [CrossRef]
87. Li, Z.; Li, W.; Wang, J.; Zhang, J.; Wang, Z. Drip irrigation shapes the soil bacterial communities and enhances jujube yield by regulating the soil moisture content and nutrient levels. *Agric. Water Manag.* **2023**, *289*, 108563. [CrossRef]
88. Tian, G.; Qiu, H.; Wang, Y.; Zhou, X.; Li, D. Short-term legacy effects of rice season irrigation and fertilization on the soil bacterial community of the subsequent wheat season in a rice-wheat rotation system. *Agric. Water Manag.* **2022**, *263*, 107446. [CrossRef]
89. Berg, J.; Brandt, K.K.; Al-Soud, W.A.; Holm, P.E.; Hansen, L.H.; Sørensen, S.J.; Nybroe, O. Selection for Cu-tolerant bacterial communities with altered composition, but unaltered richness, via long-term Cu exposure. *Appl. Environ. Microbiol.* **2012**, *78*, 7438–7446. [CrossRef]
90. Bokulich, N.A.; Kaehler, B.D.; Rideout, J.R.; Dillon, M.; Bolyen, E.; Knight, R.; Huttley, G.A.; Caporaso, J.G. Optimizing taxonomic classification of marker-gene amplicon sequences with qiime 2's q2-featureclassifier plugin. *Microbiome* **2018**, *6*, 90. [CrossRef]
91. Callahan, B.J.; McMurdie, P.J.; Rosen, M.J.; Han, A.W.; Johnson, A.J.A.; Holmes, S.P. DADA2: High-resolution sample inference from Illumina amplicon data. *Nat. Methods* **2016**, *13*, 581–583. [CrossRef]
92. Glöckner, F.O.; Yilmaz, P.; Quast, C.; Gerken, J.; Beccati, A.; Ciuprina, A.; Bruns, G.; Yarza, P.; Peplies, J.; Westram, R.; et al. 25 years of serving the community with ribosomal RNA gene reference databases and tools. *J. Biotechnol.* **2017**, *261*, 169–176. [CrossRef] [PubMed]
93. Dixon, P. VEGAN, a package of R functions for community ecology. *J. Veg. Sci.* **2003**, *14*, 927–930. [CrossRef]
94. Bretz, F.; Hothorn, T.; Westfall, P. On multiple comparisons in R. *R News* **2002**, *2*, 14–17.
95. Kembel, S.W.; Cowan, P.D.; Helmus, M.R.; Cornwell, W.K.; Morlon, H.; Ackerly, D.D.; Blomberg, S.P.; Webb, C.O. Picante: R tools for integrating phylogenies and ecology. *Bioinformatics* **2010**, *26*, 1463–1464. [CrossRef] [PubMed]
96. Virta, L.; Teittinen, A. Threshold effects of climate change on benthic diatom communities: Evaluating impacts of salinity and wind disturbance on functional traits and benthic biomass. *Sci. Total Environ.* **2022**, *826*, 154130. [CrossRef]
97. Stegen, J.C.; Lin, X.; Konopka, A.E.; Fredrickson, J.K. Stochastic and deterministic assembly processes in subsurface microbial communities. *ISME J.* **2012**, *6*, 1653–1664. [CrossRef]
98. Zhou, J.Z.; Ning, D. Stochastic community assembly: Does it matter in microbial ecology? *Microbiol. Mol. Biol. Rev.* **2017**, *81*, e00002-17. [CrossRef]

plants

MDPI

Article

Genome-Wide Analysis of the FNSII Gene Family and the Role of CitFNSII-1 in Flavonoid Synthesis in Citrus

Xinya Liu [1,2,3,†], Beibei Chen [4,†], Ling Luo [1,2,3,†], Qi Zhong [1,2,3], Chee How Teo [4,*] and Shengjia Huang [5,*]

1 Horticulture Research Institute, Sichuan Academy of Agricultural Sciences, Chengdu 610023, China; paul091341@163.com (X.L.); a5363252@163.com (L.L.); zhongqi8707@scsaas.cn (Q.Z.)
2 Key Laboratory of Horticultural Crops Biology and Germplasm Enhancement in Southwest, Ministry of Agriculture and Rural Affairs, Chengdu 611130, China
3 Key Laboratory for Germplasm Innovation & Utilization of Horticultural Crops of Sichuan Province, Chengdu 610066, China
4 University Malaya Centre for Research in Biotechnology for Agriculture, Kuala Lumpur 50603, Malaysia; 22050412@siswa.um.edu.my
5 Institute of Economic Forest Research, Sichuan Academy of Forestry, Chengdu 610081, China
* Correspondence: cheehow.teo@um.edu.my (C.H.T.); huangshengjia0206@163.com (S.H.)
† These authors contributed equally to this work.

Abstract

Flavonoid synthases (FNSs) are key enzymes catalyzing the conversion of flavanones to flavonoids, yet their functions in citrus remain functionally uncharacterized. In this study, we identified three FNSII genes in the citrus genome. Phylogenetic analysis revealed that citrus FNSII genes share the closest evolutionary distance with apple FNSII genes. Chromosomal localization demonstrated that the three FNSII genes are distributed across two out of nine chromosomes. Gene structure analysis indicated that the majority of motifs within these three FNSII genes are highly conserved. We cloned a gene called *CitFNSII-1* from citrus. Transient overexpression of *CitFNSII-1* in citrus leaves significantly increased flavonoid content, while simultaneous virus-induced silencing of *CitFNSII-1* led to synchronously and significantly reduced gene expression levels and flavonoid content in citrus seedlings. Through the *Agrobacterium rhizogenes*-mediated genetic transformation system, overexpression of *CitFNSII-1* was found to markedly enhance flavonoid accumulation in hairy roots, whereas knockout of *CitFNSII-1* resulted in a significant decrease in flavonoid content in hairy roots. Further experiments verified an interaction between CitFNSII-1 and the Chalcone isomerase-1 (CHI-1) protein. The results demonstrated that the flavonoid accumulation patterns of *CHI-1* and *CitFNSII-1* are highly similar. In conclusion, this study advances the understanding of the flavonoid biosynthesis pathway in citrus and provides a theoretical foundation for molecular breeding strategies in citrus.

Keywords: citrus; hairy roots; CitFNSII-1; flavonoid

1. Introduction

Citrus, as one of the most widely cultivated fruits globally, is highly favored by consumers for its unique flavor and rich nutritional components, including carotenoids, vitamin C, folate, dietary fiber, and flavonoids [1]. Functional compounds such as flavonoids are present in citrus fruits, leaves, flowers, and roots [2]. Flavonoids not only influence the color and flavor of citrus fruits but also enhance plant resilience to abiotic and biotic stresses, such as ultraviolet radiation, low temperature, drought, and pathogens [3]. Due to their potent antioxidant properties, flavonoids play significant roles in human health, including anticancer, antiviral, and anti-inflammatory effects [4,5]. Studies have shown that flavonoid accumulation exhibits cultivar specificity; for example, mandarins (*Citrus reticulata*) and sweet oranges (*Citrus sinensis*) accumulate high levels of flavonoids in their peels [6]. As ubiquitous secondary metabolites in plants, extensive in vivo and in vitro studies have confirmed that the unique structures of flavonoids confer biological activity, making them promising candidates for drug development [7]. Consuming more diverse flavonoids can reduce the risk of death and chronic diseases by 6–20%, and the protective effects of intake and diversity on health are independent of each other [8]. Therefore, elucidating the flavonoid biosynthesis pathway in citrus is of great significance for improving the production and application value of these bioactive compounds.

Currently, all identified FNSII proteins in plants almost exclusively belong to the CYP93 subfamily. In monocotyledonous plants, FNSIIs are primarily classified under the CYP93G subfamily, such as CYP93G3 in sorghum (*Sorghum bicolor*) and CYP93G7 in maize (*Zea mays*) [9,10]. In contrast, FNSIIs identified in dicotyledonous plants predominantly belong to the CYP93B subfamily, including CYP93B10 and CYP93B11 in Medicago truncatula, as well as CYP93B16 in soybean (*Glycine max*). Two FNSII genes, *CitFNSII-1* and *CitFNSII-2*, were identified in citrus, belonging to the CYP93B subfamily. These genes catalyze the conversion of flavanones to flavones and are involved in the biosynthesis of polymethoxylated flavones (PMFs) [11]. In Salvia miltiorrhiza, the SmFNSII gene, classified under the CYP93B subfamily, participates in flavonoid biosynthesis [12]. The *Scutellaria baicalensis* genome harbors two FNSII genes (*SbFNSII-1* and *SbFNSII-2*), members of the CYP82D subfamily, which catalyze hydroxylation at the C-6/C-8 positions of flavones, leading to the formation of unique 4′-deoxyflavonoids such as baicalein. In monocots, FNSII enzymes from maize (*Zea mays*) and sorghum (*Sorghum bicolor*) belong to the CYP93G subfamily [13]. These enzymes not only convert flavanones to flavones but also exhibit flavone 2-hydroxylase (F2H) activity, contributing to the synthesis of C-glycosylated flavones [14]. In Arabidopsis thaliana, no typical FNSII genes have been identified; however, members of the CYP93B subfamily may play putative roles in flavonoid metabolism. Research has shown that *SlbHLH95* promotes flavonoid synthesis by directly activating the expression of *SlF3H* and *SlFLS* genes, while inhibiting the expression of the *SlCHS1* gene, thereby regulating flavonoid metabolism [15]. FNSII is an enzyme of the CYP450 family widely present in higher plants, and its identification in citrus species will help improve the synthetic metabolic pathway of citrus flavonoids.

Plant hormones play crucial roles in all stages of plant growth and development. They not only directly regulate plant growth and development but also modulate secondary metabolism, including the biosynthesis of flavonoids. Studies have demonstrated that multiple phytohormones, such as abscisic acid (ABA), gibberellins (GA), methyl jasmonate (MeJA), and methyl salicylate (MeSA), can regulate the synthesis and metabolism of flavonoid compounds [16]. ABA, ethylene, jasmonates, cytokinins, and brassinosteroids promote flavonoid biosynthesis, whereas auxin suppresses this process through negative regulation. Subsequently, transcription factors from the MYB, bHLH, WRKY, NAC, and bZIP families play critical roles in modulating flavonoid biosynthesis [17]. For instance, treatment with MeJA and MeSA has been shown to enhance flavonoid accumulation in *Scutellaria baicalensis* root cultures while simultaneously upregulating the expression of *SbFNSII-2* [18]. Furthermore, MeJA and MeSA have been proven to stimulate flavonoid biosynthesis in tea plants by activating the phenylpropanoid pathway [19]. Previous studies have demonstrated that citrus flavonoids can effectively protect citrus fruits against pathogen attacks. Exogenous SA treatment has been shown to enhance resistance to *Penicillium digitatum* and *Candidatus Liberibacter* asiaticus in citrus fruits, which is associated with the roles of MeSA and MeJA [20]. Therefore, MeSA and MeJA may function by directly modulating flavonoid biosynthesis in citrus, thereby influencing stress resistance responses. However, the specific effects of MeSA and MeJA on citrus flavonoid biosynthesis remain unclear, and whether they regulate the expression of FNSII genes in citrus requires further investigation.

In this study, we conducted the first genome-wide identification and systematic analysis of the FNSII gene family in citrus, encompassing phylogenetic tree construction, gene structure elucidation, chromosomal localization, conserved motif identification, and cis-regulatory element prediction. These findings lay a foundation for elucidating the evolutionary trajectories and biological functions of the citrus FNSII gene family. Through combined validation using transient overexpression, virus-induced gene silencing (VIGS), and *Agrobacterium rhizogenes*-mediated hairy root transformation, we demonstrated that CitFNSII-1 and its interacting protein CHI-1 play critical and functionally similar roles in citrus flavonoid biosynthesis. Furthermore, their expression patterns and regulatory responses to MeSA and MeJA treatments were highly consistent. This study represents the first biological functional validation of citrus FNSII enzymes, significantly advancing our understanding of the flavonoid biosynthetic pathway in citrus.

2. Results

2.1. Identification and Phylogenetic Tree of FNSII Genes in Citrus

Using the HMM profile of the FNSII superfamily as a query, candidate FNSII genes were screened in the citrus genome. A total of three candidate FNSII genes were identified in the citrus genome, and their conserved domains were analyzed via the NCBI database. The results confirmed the presence of three FNSII genes in loquat. A phylogenetic tree of these FNSII gene families from higher plants and citrus was constructed using the maximum likelihood (ML) method, revealing that citrus FNSII genes exhibit the closest evolutionary relationship with apple FNSII genes (Figure 1A).

(A)

(B)

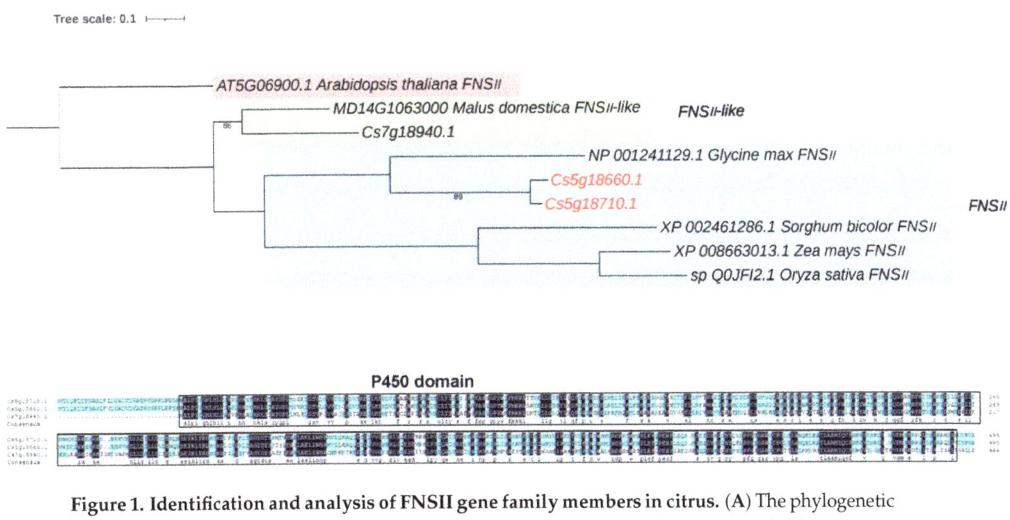

Figure 1. Identification and analysis of FNSII gene family members in citrus. (**A**) The phylogenetic tree including FNSII proteins from *Citrus sinensis* and Arabidopsis thaliana. (**B**) Multiple sequence alignment of FNSII proteins in citrus.

Multiple sequence alignment of the three FNSII amino acid sequences identified through genome-wide analysis was performed using DNAMAN 9.0 software. The results revealed that the P450 domain (spanning residues 30–493) in all three FNSII proteins contains three characteristic conserved motifs, KESFR, PERF, and PFGTGRRGCPG, confirming their classification within the CYP450 superfamily. Highly conserved amino acid residues, including arginine (R), phenylalanine (F), and glycine (G), were observed within these domains (Figure 1B).

2.2. Chromosomal Localization, Synteny Analysis, and Characterizations of the FNSII Genes in Citrus

Physicochemical property analysis revealed that the three FNSII proteins contain 479 to 852 amino acids, with molecular weights ranging from 54,394.97 to 96,714.06 Da. Their isoelectric points (pI) span from 6.18 to 8.34, where *Cs7g18940.1* is an acidic protein, while the others are alkaline. The instability indices ranged from 41.05 to 45.15, classifying all three FNSII proteins as unstable. The aliphatic indices were between 91.63 and 97.65, and the grand average of hydropathicity (GRAVY) values ranged from −0.202 to −0.117 (all < 0), indicating hydrophilic properties (Figure 2A). The corresponding protein three-dimensional structures for *Cs7g18940.1*, *Cs5g18660.1*, and *Cs5g18710.1* were established using the SWISSMODEL website (Figure 2B). Genomic localization analysis showed that two citrus FNSII genes are located on chromosome 5 (chr5). A chromosomal region within 200 kb on chr5 harbors two or more genes, suggesting tandem duplication events. Specifically, *Cs5g18660.1* and *Cs5g18710.1* were identified as tandem duplicates. Additionally, a FNS II-like gene was localized on chromosome 7 (chr7) (Figure 2C).

2.3. Conserved Motif and Gene Structure Analysis of the FNSII Genes in Citrus

Conserved motif analysis revealed that the three FNSII genes exhibit high conservation across most motifs. Notably, with the exception of one FNSII-like gene lacking motif 10, the other two FNSII family members each contain 10 motifs, sharing similar motif distributions but showing slight quantitative variations. The absence of motif 10 in the FNSII-like gene suggests that this motif may play a unique role in specific biological processes mediated by the other two FNSII genes. The loss of motif 10 in the FNSII-like gene might have driven functional divergence or functional constraints, and differences in motif numbers could lead to subtle structural variations between the two FNSII proteins, potentially impacting their biological functions. The domain architectures of the three FNSII genes are largely consistent. Gene structure analysis showed that FNSII genes contain 2–9 exons and 1–8 introns. *Cs5g18710.1* harbors an exceptionally large intron, which may influence its expression, potentially activating or repressing transcription during specific developmental stages (Figure 3A).

All three genes possess a significant number of light-responsive elements. However, *Cs5g18660.1* contains markedly fewer light-responsive elements than the other two genes, suggesting its involvement in light-responsive processes less sensitive to light intensity or photoperiod. *Cs5g18710.1* and *Cs7g18940.1* lack auxin-responsive elements, while *Cs5g18660.1* and *Cs7g18940.1* lack MeJA-responsive elements, indicating divergence in hormonal response regulation among the three genes. *Cs5g18660.1* and *Cs5g18710.1* lack anaerobic induction elements and MYB-binding sites, whereas *Cs7g18940.1* lacks defense- and stress-responsive elements as well as SA-responsive elements, highlighting functional differences in plant defense responses among the three genes (Figure 3B).

(A)

Sequence ID	Number of Amino Acid	Molecular Weight	Theoretical pI	Instability Index	Aliphatic Index	Hydropathicity
Cs5g18710.1	852	96,714.06	8.34	43.53	97.02	−0.117
Cs5g18660.1	503	56,949.97	8.69	45.15	97.65	−0.132
Cs7g18940.1	479	54,394.97	6.18	41.05	91.63	−0.202

(B)

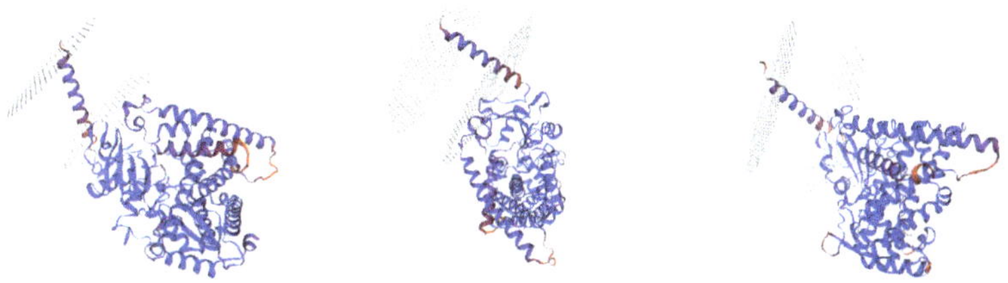

(C)

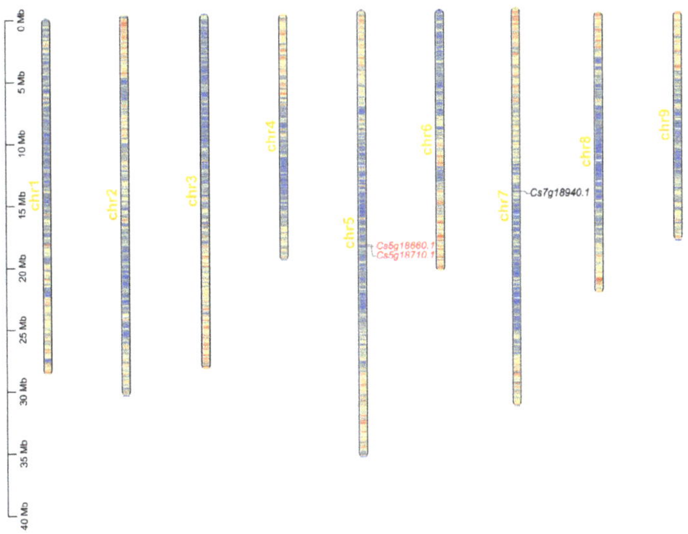

Figure 2. Chromosome localization, homology analysis, and characterizations analysis of citrus FNSII gene. (**A**) The characteristics of FNSII genes in citrus. (**B**) Three-dimensional structural diagram of FNSII protein in citrus. (**C**) The chromosomal localization of the FNSII genes.

2.4. Expression of CitFNSII-1 in Citrus

As shown in Supplementary Figure S1, using the expression level of *CitFNSII-1* in a young leaf as a control, the gene expression level in the mature leaf was 3.7 times higher than that in the young leaf (Supplementary Figure S1A). Using the expression level of *CitFNSII-1* in the young fruit stage as a control, the gene expression level in the full ripe stage was upregulated by 3.9 times (Supplementary Figure S1B). Further analysis of the expression characteristics of *CitFNSII-1* in different tissues showed that the expression level of *CitFNSII-1* in peel was 2.8 times higher than that in leaf veins (Supplementary Figure S1C). The expression level of *CitFNSII-1* in *Carrizo citrange* was used as the control, and the gene expression level in *Citrus grandis* was upregulated by 4.7 times (Supplementary Figure S1D).

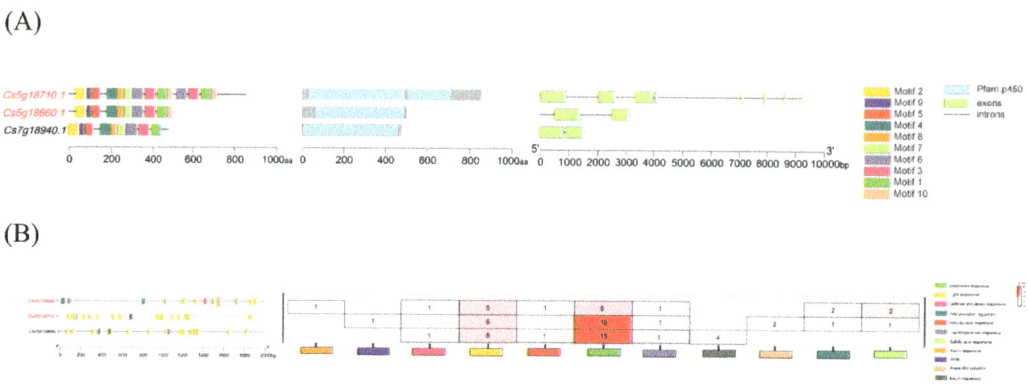

Figure 3. Conserved motif and gene structure analysis of citrus FNSII gene. (**A**) Conserved motifs in FNSII proteins are represented by colored boxes. (**B**) UTRs, exons, and introns are represented by green squares, yellow squares, and gray lines, respectively. Black lines indicate length.

The results showed that, compared to the control group, exogenous MeSA and MeJA treatments led to a decrease in flavonoid content in citrus leaves (Supplementary Figure S1E,G). Subsequently, qPCR was performed to measure the expression levels of *CitFNSII-1* under MeSA and MeJA treatments to determine whether the reduced flavonoid content was associated with its expression. As shown in Supplementary Figure S1F,H, the transcriptional levels of *CitFNSII-1* in MeSA- and MeJA-treated citrus leaves were significantly downregulated. These results suggested that MeSA and MeJA treatments may reduce flavonoid accumulation by suppressing *CitFNSII-1* expression.

2.5. Generation of TRV-CitFNSII-1 Plants

TRV2-*CitFNSII-1* vector was constructed (Figure 4A), and positive plants were identified by PCR, resulting in a total of six TRV2-*CitFNSII-1* plants (Figure 4B). Phenotypic observation revealed that there were no significant differences between TRV2-*CitFNSII-1* plants and WT plants (Figure 4C). The results showed that the gene expression levels of these plants were significantly lower than those of the WT plants (Figure 4D). The significant reduction in *CitFNSII-1* gene expression levels in VIGS plants led to a decrease in flavonoid content (Figure 4E).

2.6. Transient Overexpression Analysis of CitFNSII-1 in Citrus Leaves

To rapidly validate the involvement of *CitFNSII-1* in flavonoid biosynthesis in citrus, transient overexpression analysis of *CitFNSII-1* was performed in citrus leaves, and changes in flavonoid content were analyzed. We constructed the *CitFNSII-1* overexpression vector and introduced it into leaves via *Agrobacterium*-mediated vacuum infiltration (Figure 5A). PCR validation confirmed the successful generation of six *CitFNSII-1* transient overexpression leaves (Figure 5B). Phenotypic observation revealed that there were no significant differences between p1300GMN-*CitFNSII-1* plants and WT plants (Figure 5C). The results showed that, compared to the control group injected with the empty vector p1300GMN, leaves infiltrated with *Agrobacterium* carrying p1300GMN-*CitFNSII-1* exhibited significantly elevated relative expression levels of *CitFNSII-1* (Figure 5D) and a marked increase in flavonoid content (Figure 5E). These results provide compelling evidence that *CitFNSII-1* participates in flavonoid biosynthesis in citrus.

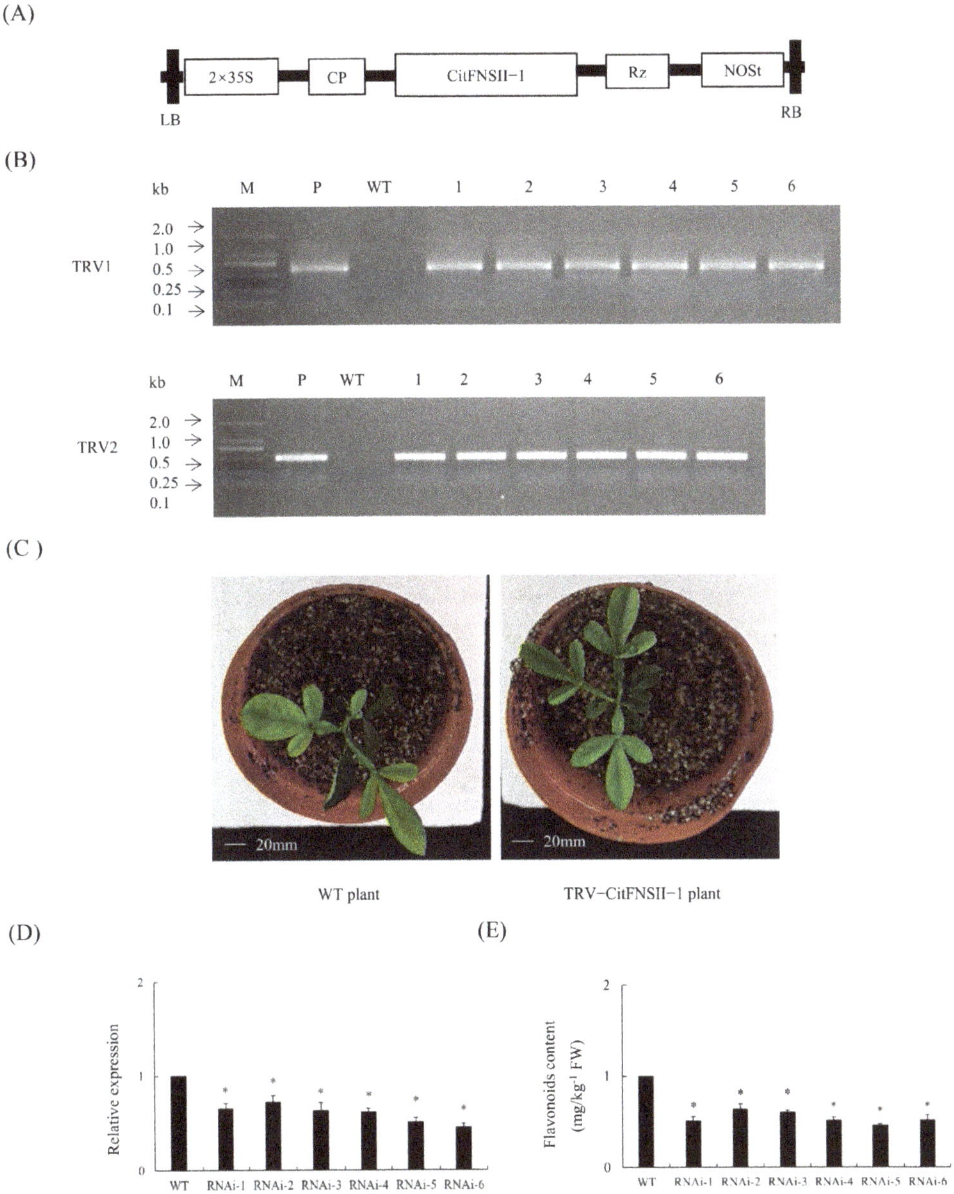

Figure 4. Transgenic plants silencing *CitFNSII-1*. (**A**) A 35S, 35S promoter; NOSt, the nopaline synthase terminator; LB, left border; RB, right border. (**B**) Identification of silencing plants by PCR. M, DNA marker; T, TRV plasmid; WT, wild-type control. (**C**) Phenotypic observation of silencing plants. (**D**) Relative expression levels of *CitFNSII-1* in silencing plants. (**E**) Determination of flavonoid content in citrus seedlings with TRV-induced silencing of *CitFNSII-1*. Values are expressed as means ± standard deviation of three independent tests. * on top of the bars indicates significant differences compared to WT control ($p < 0.05$, Student's t-test).

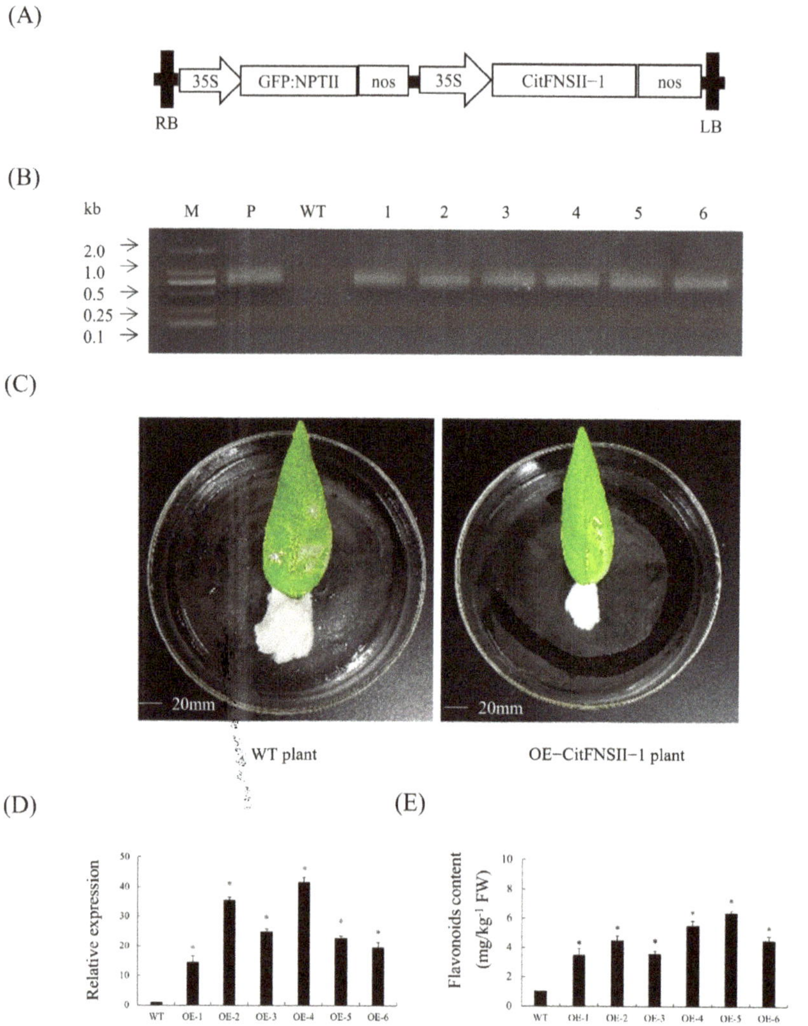

Figure 5. Gene expression and flavonoid content analysis of transient overexpression of *CitFNSII-1* in citrus leaves. (**A**) A 35S, 35S promoter; NOSt, the nopaline synthase terminator; LB, left border; RB, right border. (**B**) Identification of transient citrus leaves by PCR. M, DNA marker; P, p35S: *CitFNSII-1* plasmid; WT, wild-type control. (**C**) Observation of symptoms of transient citrus leaves. (**D**) Relative expression levels of *CitFNSII-1* in transient citrus leaves. **(E)** Comparison of flavonoid content in *CitFNSII-1* transiently overexpressing leaves. Values are expressed as means ± standard deviation of three independent tests. * on top of the bars indicates significant differences compared to WT control ($p < 0.05$, Student's t-test).

2.7. *Overexpression of CitFNSII-1 in Transgenic Hairy Roots*

Nine *CitFNSII-1* transgenic hairy roots were identified by PCR (Figure 6A). The rooting rate of transgenic hairy roots was 30% (Figure 6B). There was no significant difference in phenotype between *CitFNSII-1* transgenic hairy roots and WT hairy roots. *CitFNSII-1* transgenic hairy roots are usually white or yellow, with a length of 7 cm to 12 cm (Figure 6C). The results showed that the gene expression level of *CitFNSII-1* transgenic hairy roots was significantly higher than that of the WT group (Figure 6D). Similarly, the flavonoid content levels in *CitFNSII-1* transgenic hairy roots were significantly higher than those in the WT group (Figure 6E). The results showed that overexpression of *CitFNSII-1* promoted the biosynthesis of flavonoids in citrus hairy roots.

(A)

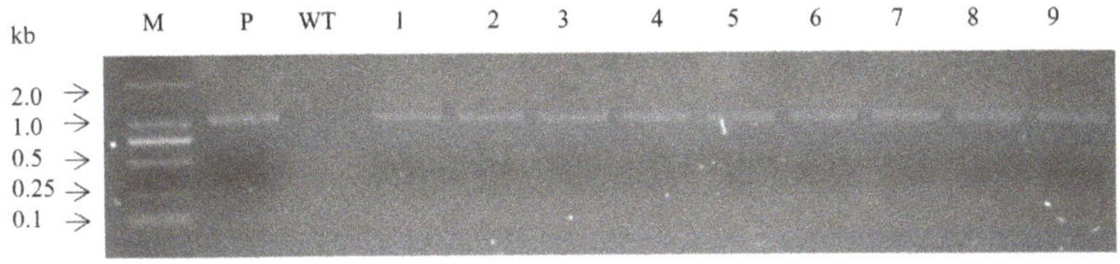

(B)

Material	Number of hairy roots	Hairy root efficiency (%)	Number of positive hairy roots
CitFNSII-1 transgenic hairy roots	30	30	9
Control hairy roots	30	20	6

(C) (D) (E)

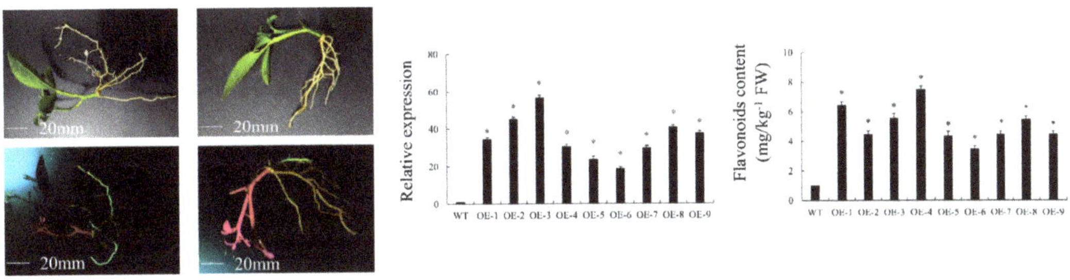

WT hairy root CitFNSII-1 transgenic hairy root

Figure 6. Transgenic hairy roots overexpressing *CitFNSII-1*. (**A**) Identification of transgenic hairy roots by PCR. M, DNA marker; P, p35S: *CitFNSII-1* plasmid; WT, wild-type control. (**B**) Statistics of transgenic hairy roots. (**C**) Observation of symptoms of transgenic hairy roots. (**D**) Relative expression levels of *CitFNSII-1* in transgenic hairy roots. (**E**) Comparison of flavonoid content in *CitFNSII-1* overexpressing hairy roots. Values are expressed as means ± standard deviation of three independent tests. * on top of the bars indicates significant differences compared to WT control ($p < 0.05$, Student's t-test).

2.8. Identification and Editing Efficiency Analysis of CitFNSII-1 in CRISPR/Cas9-Edited Citrus Hairy Roots

We selected the target site (sgRNA1: GCATGGCTAAAGAAAGGCCAGGG) to construct the *CitFNSII-1* gene editing vector (Figure 7A). Preliminary phenotypic observation of hairy roots revealed no significant differences in length or size between the gene-edited and wild-type hairy roots (Figure 7B). DNA extracted from green-positive hairy roots was subjected to PCR amplification targeting the Cas9 sequence in the pKSE401-GFP vector, confirming six *CitFNSII-1*-edited hairy roots (Figure 7C) with a positive rate of 30% (Figure 7D). Editing patterns predominantly included one bp insertion and four bp deletions (Figure 7E), with an overall editing efficiency of 100%. Notably, flavonoid content in *CitFNSII-1*-edited hairy roots was significantly lower than in wild-type controls (Figure 7F).

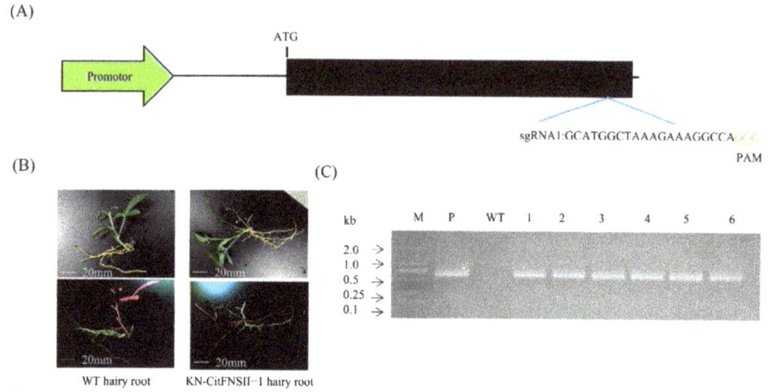

(D)

Material	Number of hairy roots	Hairy root efficiency (%)	Number of positive hairy roots
CitFNSII-1 transgenic hairy roots	20	30	6
Control hairy roots	18	50	9

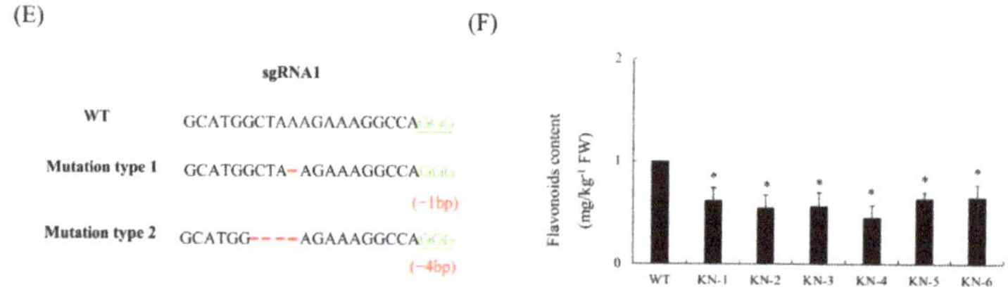

Figure 7. Identification and phenotypic characterization of gene-edited hairy roots. (**A**) Schematic diagram of gene editing vector construction. (**B**) Observation of symptoms of transgenic hairy roots. (**C**) Identification of gene-edited hairy roots by PCR. M, DNA marker; P, pKSE401-GFP plasmid; WT, wild-type control. (**D**) Statistics of transgenic hairy roots. (**E**) The green font represents PAM sequences, the red font represents mutated bases, and the red dashed line represents base deletions. (**F**) Comparison of flavonoid content in *CitFNSII-1* gene-edited hairy roots. Values are expressed as means ± standard deviation of three independent tests. * on top of the bars indicates significant differences compared to WT control ($p < 0.05$, Student's *t*-test).

2.9. Characteristics of Changes in SA, MeSA, JA, and MeJA Content in CitFNSII-1 Transgenic Hairy Roots

To further investigate the potential regulatory mechanisms of *CitFNSII-1* in hormonal signaling, we compared the hormone content differences between *CitFNSII-1* overexpression and gene-edited hairy roots. The results demonstrated that, in *CitFNSII-1* overexpression hairy roots, the levels of SA, MeSA, JA, and MeJA were significantly downregulated compared to control hairy roots (Supplementary Figure S2A–D). In contrast, in *CitFNSII-1* gene-edited hairy roots, the levels of SA, MeSA, JA, and MeJA were significantly upregulated relative to the control hairy roots (Supplementary Figure S2E–H).

2.10. CitFNSII-1 Interacts with CHI-1

Using PPI online prediction of potential interacting proteins of CitFNSII-1 in citrus, it was found that CHI-1 protein may interact with it (Figure 8A). Molecular docking of CitFNSII-1 and CHI-1 was performed using GRAMM software, and the potential complex formed by the two proteins was predicted (Figure 8B). In the Y2H assay, CitFNSII-1 interacted with CHI-1 when CitFNSII-1 was used as the bait (Figure 8C). In the LCA assay, constructs encoding nLUC-CitFNSII-1 and CHI-1-cLUC were infiltrated into *N. benthamiana* leaves, and luminescence signals were observed in the infiltrated regions, confirming their interaction in vivo (Figure 8D). In the BiFC assay, co-expression of YFPn-CitFNSII-1 and YFPc-CHI-1 in *N. benthamiana* leaves resulted in fluorescence signals localized to the nucleus, indicating that these proteins interact within the nucleus (Figure 8E).

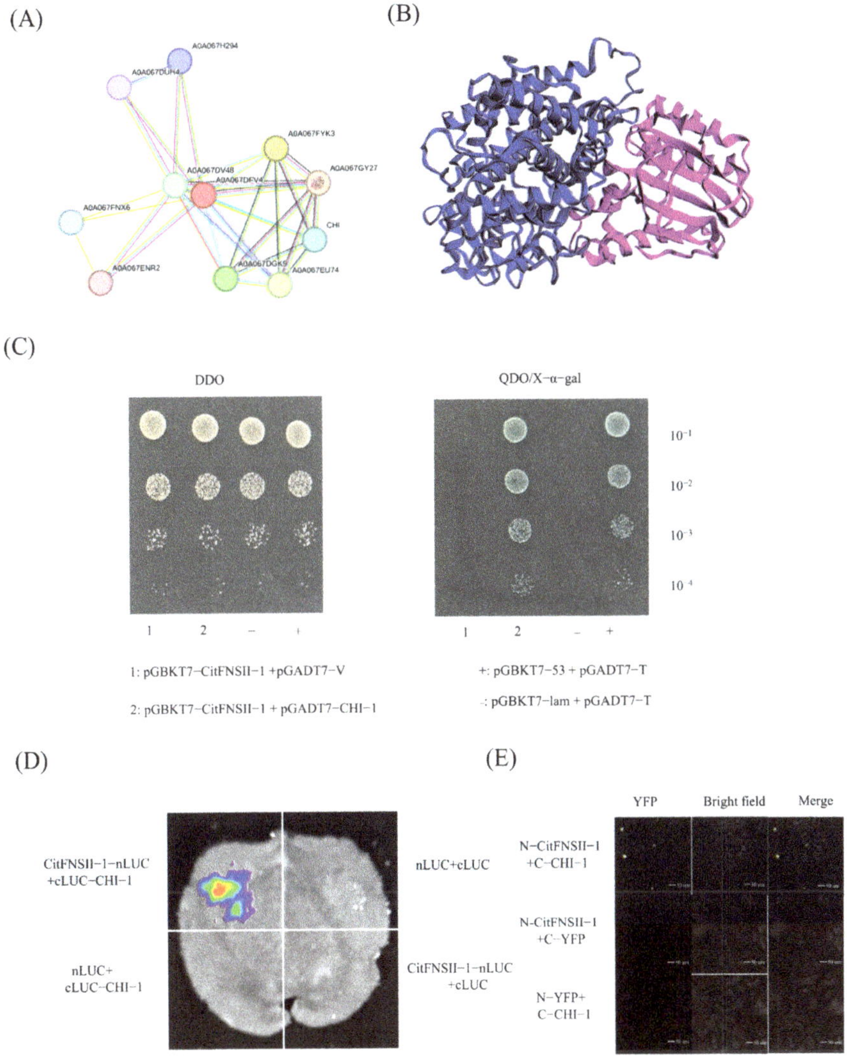

Figure 8. CitFNSII-1 interacts with CHI-1. (**A**) Predicting potential interacting proteins of CitFNSII-1 through PPI analysis. (**B**) Using molecular docking technology to predict the complex structure of the interaction between CitFNSII-1 and CHI-1. (**C**) Y2H assay showing that CitFNSII-1 interacts with CHI-1. Positive transformants were spotted onto synthetic defined (SD) medium lacking Trp and Leu and SD medium lacking Trp, Leu, His, and Ade containing x-α-Gal to test protein–protein interaction. (**D**) LCA assay showing that CitFNSII-1 and CHI-1 interact in planta. *N. benthamiana* leaves were co-infiltrated with mixed *Agrobacterium* cultures, each harboring nLUC-CitFNSII-1 or CHI-1-cLUC. (**E**) BiFC assay showing that nYFP-CitFNSII-1 and cYFP-CHI-1 constructs were co-infiltrated in *N. benthamiana* leaves (nYFP-CitFNSII-1 and cYFP-CHI-1 were used as negative controls). Three independent repetitions were carried out with similar results.

2.11. Expression of CHI-1 in Citrus

Transcript abundance in mature leaves exhibited a 2.4-fold increase relative to juvenile leaves (Supplementary Figure S3A). During fruit development, *CHI-1* expression demonstrated a 2.8-fold elevation at full ripeness compared to the young fruit stage (Supplementary Figure S3B). Expression analysis identified preferential accumulation in peel tissue, with a 3.5-fold higher level than leaf veins (Supplementary Figure S3C). The results showed that *Citrus grandis* displayed 3.3-fold enhanced *CHI-1* mRNA levels compared to *Carrizo citrange* (Supplementary Figure S3D).

The results demonstrated that exogenous MeSA and MeJA treatments induced a concomitant reduction in leaf flavonoid content (Supplementary Figure S3E,G). Transcriptional analysis via RT-qPCR revealed significant downregulation of *CHI-1* under both treatments (Supplementary Figure S3F,H). The results suggested that MeSA/MeJA-mediated suppression of *CHI-1* expression contributes to compromised flavonoid biosynthesis.

2.12. Generation of TRV-CHI-1 Plants

The TRV2-*CHI-1* vector was successfully generated (Figure 9A), with four transgenic plants confirmed through PCR (Figure 9B). Phenotypic observation revealed that there were no significant differences between TRV2-*CHI-1* plants and WT plants (Figure 9C). The results showed that the gene expression levels of these plants were significantly lower than those of the WT plants (Figure 9D). The significant reduction in *CHI-1* gene expression levels in VIGS plants led to a decrease in flavonoid content (Figure 9E).

2.13. Transient Overexpression Analysis of CHI-1 in Citrus Leaves

To functionally investigate CHI-1's role in citrus flavonoid metabolism, we conducted transient expression experiments in foliar tissues. The *CHI-1* coding sequence was cloned into a plant expression vector (Figure 10A) and delivered into leaf tissues through *Agrobacterium*-mediated vacuum infiltration. PCR-based screening generated six overexpressing plants (Figure 10B). Phenotypic observation revealed that there were no significant differences between p1300GMN-*CHI-1* plants and WT plants (Figure 10C). The results showed that leaves expressing the p1300GMN-*CHI-1* construct demonstrated substantially enhanced *CHI-1* transcription levels relative to empty vector-transfected controls (Figure 10D). This transcriptional upregulation corresponded with measurable increases in flavonoid accumulation (Figure 10E). The results indicated CHI-1's functional participation in the flavonoid biosynthetic pathway of citrus plants.

2.14. Overexpression of CHI-1 in Transgenic Hairy Roots

PCR screening successfully identified nine *CHI-1* transgenic hairy roots (Figure 11A), exhibiting a 30% root induction efficiency (Figure 11B). Morphological comparison revealed no observable phenotypic variations between transgenic and WT hairy roots. Both transgenic and WT hairy roots displayed similar coloration (white to yellow) and size characteristics, ranging from 5 to 14 cm in length (Figure 11C). The results showed markedly enhanced *CHI-1* transcript accumulation in transgenic hairy roots relative to WT controls (Figure 11D). Consistent with gene expression patterns, transgenic roots showed corresponding increases in flavonoid content compared to WT controls (Figure 11E). The results indicated that *CHI-1* overexpression enhances flavonoid biosynthesis in citrus hairy roots.

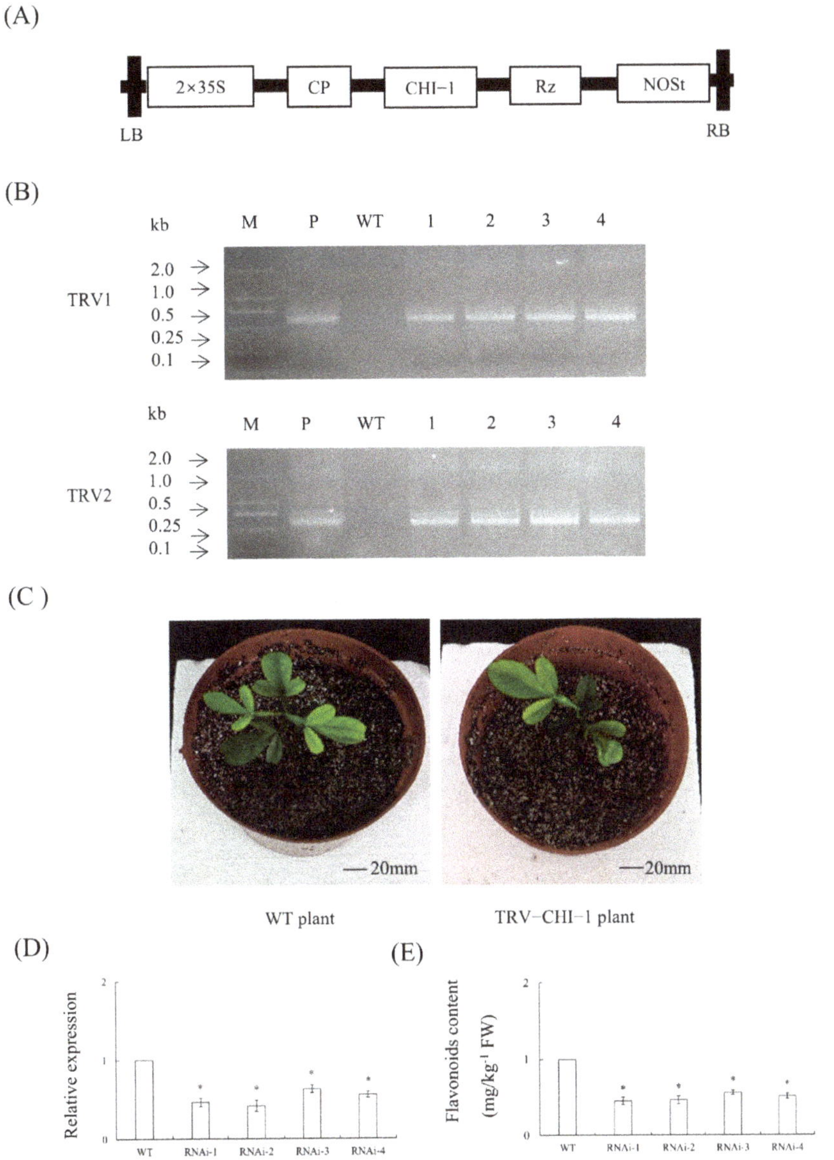

Figure 9. Transgenic plants silencing *CHI-1*. (**A**) A 35S, 35S promoter; NOSt, the nopaline synthase terminator; LB, left border; RB, right border. (**B**) Identification of silencing plants by PCR. M, DNA marker; T, TRV plasmid; WT, wild-type control. (**C**) Phenotypic observation of silencing plants. (**D**) Relative expression levels of *CHI-1* in silencing plants. (**E**) Determination of flavonoid content in citrus seedlings with TRV-induced silencing of *CHI-1*. Values are expressed as means ± standard deviation of three independent tests. * on top of the bars indicates significant differences compared to WT control ($p < 0.05$, Student's *t*-test).

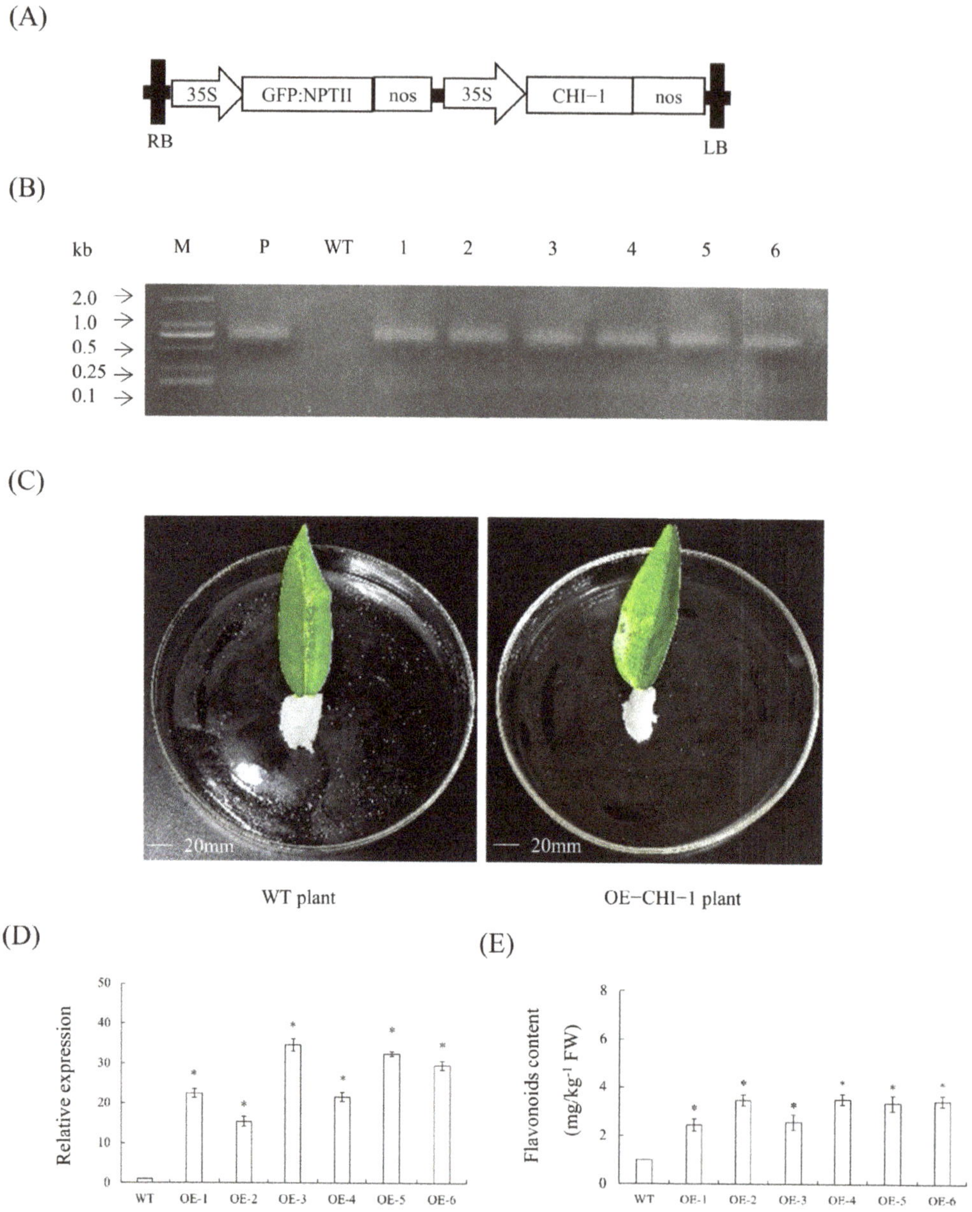

Figure 10. Gene expression and flavonoid content analysis of transient overexpression of *CHI-1* in citrus leaves. (**A**) A 35S, 35S promoter; NOSt, the nopaline synthase terminator; LB, left border; RB, right border. (**B**) Identification of transient citrus leaves by PCR. M, DNA marker; P, p35S: *CHI-1* plasmid; WT, wild-type control. (**C**) Observation of symptoms of transient citrus leaves. (**D**) Relative expression levels of *CHI-1* in transient citrus leaves. (**E**) Comparison of flavonoid content in *CHI-1* transiently overexpressing leaves. Values are expressed as means ± standard deviation of three independent tests. * on top of the bars indicates significant differences compared to WT control ($p < 0.05$, Student's *t*-test).

(B)

Material	Number of hairy roots	Hairy root efficiency (%)	Number of positive hairy roots
CHI-1 transgenic hairy roots	30	30	9
Control hairy roots	30	20	6

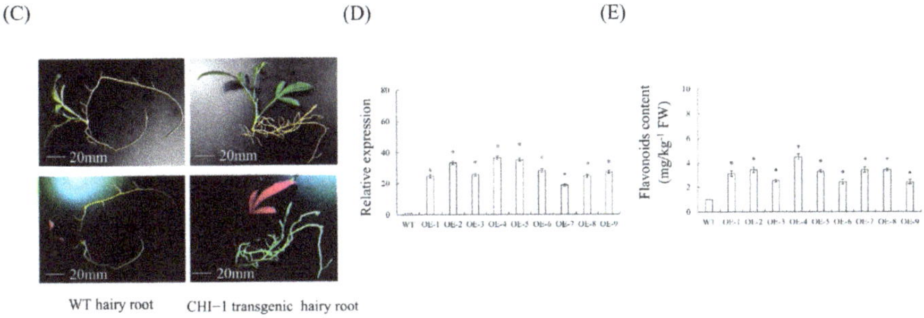

Figure 11. Transgenic hairy roots overexpressing *CHI-1*. (**A**) Identification of transgenic hairy roots by PCR. M, DNA marker; P, p35S: CHI-1 plasmid; WT, wild-type control. (**B**) Statistics of transgenic hairy roots. (**C**) Observation of symptoms of transgenic hairy roots. (**D**) Relative expression levels of *CHI-1* in transgenic hairy roots. (**E**) Comparison of flavonoid content in *CHI-1* overexpressing hairy roots. Values are expressed as means ± standard deviation of three independent tests. * on top of the bars indicates significant differences compared to WT control ($p < 0.05$, Student's t-test).

2.15. Identification and Editing Efficiency Analysis of CHI-1 in CRISPR/Cas9-Edited Citrus Hairy Roots

A CRISPR/Cas9-based editing construct targeting *CHI-1* was engineered using the designed sgRNA2 (GTGAAATTTACCGCGATTGG) (Figure 12A). Preliminary phenotypic observation of hairy roots revealed conserved growth characteristics between edited and WT hairy roots, with comparable length and diameter parameters (Figure 12B). PCR-based screening using Cas9-specific primers confirmed four edited hairy roots derived from GFP-positive root tissues (Figure 12C), representing a 30% transformation efficiency (Figure 12D). Editing patterns predominantly including one bp insertion and two bp deletions (Figure 12E). The results showed reduced flavonoid accumulation in edited hairy roots relative to WT hairy roots (Figure 12F).

2.16. Characteristics of Changes in SA, MeSA, JA, and MeJA Content in CHI-1 Transgenic Hairy Roots

To elucidate CHI-1's regulatory interactions with phytohormonal pathways, we conducted phytohormone profiling in *CHI-1* overexpression and CRISPR-edited hairy roots. The results revealed coordinated downregulation of SA, MeSA, JA, and MeJA in overexpression hairy roots compared to WT controls (Supplementary Figure S4A–D). Conversely, CRISPR-mediated *CHI-1* suppression resulted in increased levels of these hormones relative to WT hairy roots (Supplementary Figure S4E–H).

(A)

(B) (C)

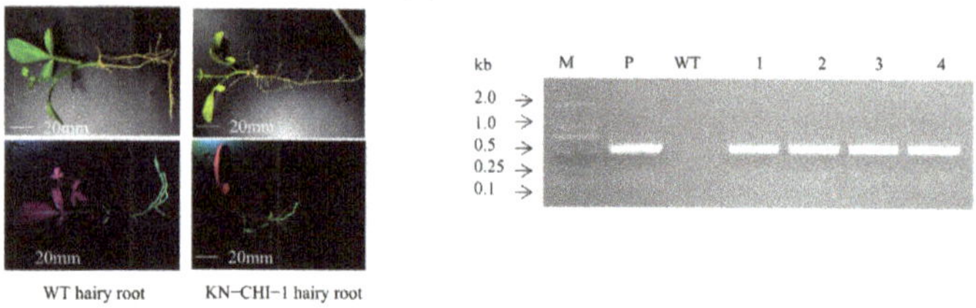

(D)

Material	Number of hairy roots	Hairy root efficiency (%)	Number of positive hairy roots
CHI-1 transgenic hairy roots	20	20	4
Control hairy roots	18	50	9

(E) (F)

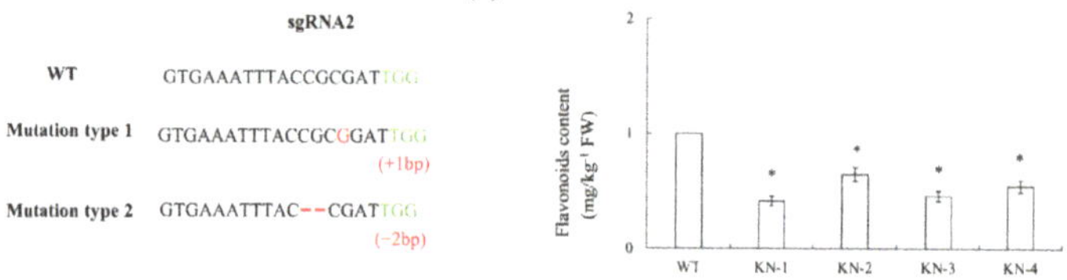

Figure 12. Identification and phenotypic characterization of gene-edited hairy roots. (**A**) Schematic diagram of gene editing vector construction. (**B**) Observation of symptoms of transgenic hairy roots. (**C**) Identification of gene-edited hairy roots by PCR. M, DNA marker; P, pKSE401-GFP plasmid; WT, wild-type control. (**D**) Statistics of transgenic hairy roots. (**E**) The green font represents PAM sequences, the red font represents mutated bases, and the red dashed line represents base deletions. (**F**) Comparison of flavonoid content in *CHI-1* gene-edited hairy roots. Values are expressed as means ± standard deviation of three independent tests. * on top of the bars indicates significant differences compared to WT control ($p < 0.05$, Student's t-test).

3. Discussion

To investigate the functional architecture and regulatory mechanisms of the citrus FNSII gene, this study first performed genome-wide bioinformatics analysis of the FNSII gene family in citrus, followed by functional validation of the target gene *CitFNSII-1*. Transient overexpression and VIGS techniques were employed alongside *Agrobacterium rhizogenes*-mediated genetic transformation to generate *CitFNSII-1* transgenic materials. Quantitative analysis of physiological parameters and flavonoid content in transgenic plants was conducted to elucidate the regulatory role of *CitFNSII-1* in citrus flavonoid biosynthesis.

Functional FNSII genes have been identified across diverse plant systems, including cereal crops, medicinal plants, ornamental plants, and fruit trees [21–23]. All currently characterized plant FNSII proteins belong to the CYP93 subfamily, which also includes flavonoid 2-hydroxylases (F2Hs). These F2Hs catalyze the hydroxylation of flavanones at the C2 position to generate 2-OH flavanones, serving as precursors for flavonoid-C-glycoside biosynthesis [24,25]. Furthermore, FNSII has been demonstrated to function as an essential component of flavonoid metabolic complexes, enhancing flavonoid biosynthesis through biochemical coordination [26]. As a CYP450 family enzyme ubiquitously present in higher plants, the characterization of FNSII in citrus species will contribute to the systematic elucidation of flavonoid biosynthetic pathways in Rutaceae. However, genome-wide identification and functional characterization of FNSII genes remain unreported in citrus. This study completed the whole genome identification of the citrus FNSII gene family for the first time and explored the mechanism of *CitFNSII-1* in citrus flavonoid synthesis.

Phylogenetic analysis revealed a high conservation of *CitFNSII-1* during evolutionary processes, suggesting potential gene duplication events. Previous studies demonstrated a dynamic accumulation pattern of flavonoids in citrus fruit development, characterized by an initial increase followed by a gradual decline [27]. Comparative analysis across cultivars showed the highest flavonoid content in mandarin (*Citrus reticulata*) peels, intermediate levels in sweet orange (*C. sinensis*), and negligible accumulation in pomelo (*C. grandis*). Notably, Satsuma mandarin exhibited significantly elevated flavanone content compared with Ponkan mandarin (*Citrus reticulata 'Ponkan'*) [28]. Further investigation identified *CitFNSII-1* as a duplicated gene whose expression strongly correlated with interspecific flavonoid variation and showed stage-specific associations with flavonoid accumulation during fruit and leaf development.

Flavonoids modulate hormonal signaling pathways, including those involving IAA [29], ABA [30], and SA [31] signaling pathways. The phytohormone MeSA has been documented to regulate flavonoid metabolism [32]. In this study, MeSA and MeJA treatments resulted in significantly reduced flavonoid accumulation. Further analysis revealed concomitant downregulation of *CitFNSII-1* expression, indicating heightened sensitivity of *CitFNSII-1* to both MeSA and MeJA, potentially through similar regulatory mechanisms. The results indicated that flavonoid depletion directly correlates with *CitFNSII-1* suppression, confirming its central role in flavonoid biosynthesis. Analogously, MeJA specifically induces *SbFNSII-2* expression to promote root-specific flavonoid synthesis in *Scutellaria baicalensis*, while MeSA and MeJA likely activate convergent metabolic pathways in *Bidens pilosa* leaves [33]. In tea (*Camellia sinensis*), 1 mmol/L MeSA enhances flavonoid production via phenylpropanoid pathway activation [32], paralleled by SA-induced upregulation of flavonoid biosynthetic genes (F3H, DFR) and increased flavonoid content in wheat (*Triticum aestivum*) leaves [34], aligning with our experimental results. Nevertheless, the precise molecular mechanisms underlying MeSA/MeJA-mediated flavonoid reduction in citrus require further investigation.

Methyl salicylate (MeSA), a volatile derivative of salicylic acid (SA), acts as a key signaling molecule for systemic acquired resistance (SAR) in citrus. However, MeSA concurrently suppresses the biosynthesis of secondary metabolites (including flavonoids) to reallocate metabolic resources toward combating pathogens or abiotic stressors [35]. Experimental evidence demonstrates that MeSA treatment significantly reduces total flavonoid content in citrus and downregulates *CitFNSII-1* expression. This inhibition likely stems from preferential energy allocation to non-flavonoid protective metabolites. MeJA strongly induces lignin monomer biosynthesis genes (e.g., Phenylalanine ammonia-lyase (PAL) and Cinnamoyl-CoA reductase (CCR)), thereby depleting shared precursors like coumaroyl-CoA. Consequently, flavonoid synthase activity (e.g., FNSII and CHI) decreases due to substrate limitations. Additionally, MeJA treatment activates antioxidant enzymes (e.g., Superoxide dismutase (SOD) and Catalase (CAT)), enhancing endogenous antioxidant capacity. This reduces plant reliance on flavonoids (exogenous antioxidants), leading to feedback inhibition of their synthesis [36].

Flavonoids, as a vital class of secondary metabolites, are ubiquitously present in terrestrial plants. CHI, a key rate-limiting enzyme, catalyzes the stereospecific isomerization of chalcones into corresponding flavanones [37]. CHI typically exists as a multigene family and is classified into four functional types (I-IV) based on biochemical activity [38]. Since the first identification of CHI from Phaseolus vulgaris cell cultures [39], CHI genes in higher plants, including *Arabidopsis thaliana* [40], *Lotus japonicus* [41], and *Solanum lycopersicum* [42], have been systematically cloned and functionally characterized, demonstrating their role in promoting flavonoid biosynthesis.

Transcription factors including MYB, bHLH, and AP2/ERF regulate CHI expression by binding to its promoter. For instance, AP2/ERF family members enhance flavonoid accumulation through transcriptional activation of CHI genes [43]. In leguminous plants, CHI-1 interacts with symbiotic-related transcription factors (NSP2) to drive isoflavone biosynthesis, facilitating rhizobial recruitment [44]. Environmental stressors such as low temperature and UV-B radiation upregulate CHI expression to promote flavonoid synthesis. For example, cold stress induces the interaction between apple MdMYB308L and CHI, enhancing anthocyanin accumulation to mitigate chilling injury [45]. CHI competes with FNS for flavanone substrates, thereby directing metabolic flux toward either flavone or anthocyanin biosynthesis pathways [46]. Heterologous overexpression of CHI significantly increases flavonoid content, as demonstrated by a threefold elevation in pericarp flavones observed in tomato expressing petunia CHI [47]. These results collectively validate CHI's role in promoting flavonoid biosynthesis, consistent with the biological function of citrus CHI genes elucidated in our experiment.

This study focused on changes in total flavonoid content but did not resolve specific alterations in flavonoid subtypes. PMFs constitute key bioactive components in citrus peels, yet whether *CitFNSII-1*'s catalytic efficiency toward different substrates affects PMF biosynthesis remains unclear. The lack of targeted metabolomics analysis (LC-MS/MS) precluded definitive determination of *CitFNSII-1*'s functional specialization in synthesizing specific flavonoid subclasses [48]. Experiments utilized only a limited number of citrus varieties (primarily leaves and hairy roots), failing to cover high-flavonoid cultivars or critical fruit developmental stages. Given significant inter-varietal differences in flavonoid accumulation patterns, the generalizability of conclusions requires further validation [49]. We combined targeted metabolomics with single-cell transcriptomics to map *CitFNSII-1* expression profiles and flavonoid subtype distribution in specialized tissues (oil glands in fruit peel, leaf veins). We incorporated high-flavonoid germplasms (Satsuma mandarin) and low-accumulation germplasms (pummelo) to analyze correlations between CitFNSIIs allelic variations and flavonoid phenotypes [50]. In summary, while this work reveals *CitFNSII-1*'s central role in citrus flavonoid biosynthesis, the complexity of its metabolic network (subtype differentiation and environmental interactions) requires systematic resolution. Future studies must adopt integrated strategies spanning diverse cultivars, multi-omics approaches, and gene editing to precisely identify breeding targets and establish an end-to-end theoretical framework for developing high-flavonoid citrus varieties.

4. Materials and Methods

4.1. Plant Materials, Microbial Strains, and Growth Conditions

The experimental materials such as citrus (*Poncirus trifoliata* × *Citrus sinensis* (citrange)) and *Nicotiana benthamiana* were taken from the greenhouse of the Horticulture Research Institute at the Sichuan Academy of Agricultural Sciences. Citrus seedlings subjected to vacuum infiltration treatment, along with leaves and hairy roots, were cultivated in vermiculite alongside *Nicotiana benthamiana* in a controlled-environment growth chamber maintained at 22 °C under long-day conditions (16 h light/8 h dark photoperiod). The *Escherichia coli* strain DH5α was cultured in Luria–Bertani (LB) medium at 37 °C, while *Agrobacterium* strains EHA105, K599, and GV3101 were grown in LB medium supplemented with 50 μg/mL kanamycin at 28 °C.

4.2. Identification of the FNSII Genes in Citrus

The Hidden Markov Model (HMM) of the Homeobox (HOX) superfamily (PF00046) was obtained from the Pfam database (http://pfam.xfam.org/). The FNSII protein sequences of the model plant *Arabidopsis thaliana* were retrieved from the TAIR database (https://www.arabidopsis.org/). Protein sequences of the FNSII gene family from previously reported higher plants were acquired from the NCBI database (https://www.ncbi.nlm.nih.gov/). Candidate FNSII gene family sequences were identified through bidirectional BLAST alignment with published citrus genome sequences. The presence of the conserved P450 domain was verified using the NCBI Conserved Domain Database (CDD) (https://www.ncbi.nlm.nih.gov/cdd/).

4.3. Phylogenetic Tree, Multiple Sequence Alignment, and Characterizations Analysis of the FNSII Proteins

Multiple sequence alignment of the amino acid sequences of all FNSII family members was conducted using MEGA 7.0 [51]. The aligned sequences were trimmed with trimAI, and a phylogenetic tree was constructed using the maximum likelihood (ML) method in IQ-TREE. The resulting tree was visualized and annotated using the ITOL online platform (https://itol.embl.de). Multiple sequence alignment of citrus FNSII proteins was further analyzed using DNAMAN software (v9.0) with default parameters [52]. Protein characteristics, including coding sequence length, theoretical isoelectric point (pI), molecular weight, and amino acid length, were predicted using the ExPASY database (https://web.expasy.org/protparam/) [53].

4.4. Chromosomal Localization, Gene Structure, Conserved Motif, and Synteny Analysis of the FNSII Genes in Citrus

The citrus genome files were downloaded from the NCBI database, and the genome annotation file (GFF format) was obtained. Using TBtools-II, the annotation files of citrus FNSII gene family members were extracted, and chromosomal localization visualization was performed. The intron–exon structures of citrus FNSII genes were analyzed based on genomic sequences and coding sequences. Conserved protein motifs were identified using the MEME suite (https://meme-suite.org/meme/tools/meme) [54]. Structural domains of citrus FNSII proteins were predicted via the NCBI CDD (https://www.ncbi.nlm.nih.gov/Structure/bwrpsb/bwrpsb.cgi). Finally, gene structures, domains, and conserved motifs of citrus FNSII genes were visualized using TBtools-II [55].

4.5. Cloning and Sequence Analysis of Genes and Promoter

The 2000 bp upstream sequences of citrus FNSII genes were extracted using TBtools-II software. Promoter prediction analysis was performed via the PlantCARE online platform (https://bioinformatics.psb.ugent.be/webtools/plantcare/html/). Subsequently, the distribution and abundance of cis-acting elements were visualized using TBtools and the ggplot2 package [56].

4.6. RT-qPCR Analysis

Total RNA was extracted using the EASYspin Plus Plant RNA Extraction Kit (Aidlab, Shanghai, China). cDNA synthesis was performed with the PrimeScript™ RT Reagent Kit with gDNA Eraser (Takara Bio, Dalian, China). qPCR was carried out using NovoStart® SYBR qPCR SuperMix Plus (Novoprotein, Shanghai, China). qPCR primers for *CitFNSII-1* and *CHI-1* genes were designed using Primer Blast in NCBI (Supplementary Table S1). Using a young leaf, the young fruit stage, a vein, and *carrizo citrange* as references, the relative expression level of *CitFNSII-1* and *CHI-1* genes was calculated using the $2^{-\Delta\Delta Ct}$ method. The test was repeated three times.

4.7. Exogenous MeSA and MeJA Treatment of Citrus Leaves and Fruit

Uniform-sized citrus leaves and fruits free of mechanical damage were selected. A 5 mL volume of 1 mM MeSA and MeJA was injected into one side of each sample as the treatment group, while an equal volume of distilled water was injected into the opposite side as the control group [57]. Each treated fruit or leaf constituted one biological replicate, with five biological replicates established. Treated samples were stored at room temperature for one week. The injected regions were subsequently excised, cut into small pieces, snap-frozen in liquid nitrogen, and stored at −80 °C for subsequent flavonoid content quantification and gene expression analysis. The test was repeated three times.

4.8. Vectors Construction

TRV2-*CitFNSII-1* and TRV2-*CHI-1* primers were designed (Supplementary Table S1), and PCR amplification was performed using pGEM-Teasy containing *CitFNSII-1* and *CsCHI-1* interfering fragments as templates. P1300GMN-*CitFNSII-1* and P1300GMN-*CHI-1* primers were designed (Supplementary Table S1), and pGEM-Teasy plasmid containing the CDS sequence of *CitFNSII-1* and *CHI-1* was used as a template for PCR amplification. The recombinant plasmid was transformed into *E. coli* DH5α, and positive clones were screened by sequencing. For *CitFNSII-1* and *CHI-1*-CRISPR, one sgRNA targeting the exon of *CitFNSII-1* (sgRNA1: CCATACGAGCAGTTACGAAG) and one sgRNA targeting the exon of *CHI-1* (sgRNA2: TGTTGAGCCAGCCAAAGGAC) were designed using the web server CRISPR-P [58] and cloned into the binary vector pKSE401G [59] by golden gate assembly.

4.9. Citrus Transformation

The TRV-mediated VIGS experiment in citrus was conducted following a previously established method [60]. *Agrobacterium tumefaciens* cultures carrying the TRV1 with TRV2-*CitFNSII-1* and TRV2-*CHI-1* constructs were grown in liquid LB medium to an optical density (OD600) of 0.8, followed by centrifugation and resuspension in an infiltration buffer containing 10 mmol/L MES, 10 mmol/L $MgCl_2$, and 200 μmol/L acetosyringone (AS). Citrus seedlings were vacuum-infiltrated with a 1:1 mixture of *Agrobacterium* cultures harboring TRV1 with TRV2-*CitFNSII-1* and TRV2-*CHI-1* constructs, while seedlings infiltrated with TRV1 and an empty TRV2 vector served as the control. After infiltration, the seedlings were cultivated in darkness for 3 days and then transferred to a photoperiod of 16 h light/8 h dark until root establishment, followed by an additional month of growth in a greenhouse.

The transient overexpression experiment was conducted following a previously described method [61]. Transformed strains carrying recombinant vectors and the empty vector were first cultured in liquid LB medium at 28 °C, followed by centrifugation and resuspension according to the protocol outlined in the VIGS experiment. Bacterial suspensions containing the target genes and the control were infiltrated into citrus leaves. The suspensions were resuspended in a buffer and infiltrated into the leaves. After infiltration, the leaves were cultivated in darkness for 24 h and then subjected to a 16 h light/8 h dark photoperiod for 3 days.

Agrobacterium rhizogenes-mediated hairy root transformation assay was referenced from a previously established method [62]. Citrus branches with diameters of approximately 0.5 cm were collected and cut into stem segments (~5 cm in length) containing one or more axillary buds using a sterilized blade. Root systems were removed, and hypocotyls were retained. Bacterial suspensions (OD_{600} = 0.6–0.8) were infiltrated into hypocotyl incisions via vacuum infiltration for 30 min. The hypocotyls were then inserted into moist vermiculite and cultivated in a constant temperature incubator at 22 °C under a 16 h light/8 h dark photoperiod.

4.10. Determination of Hormone Content

Three hairy roots from each transgenic line were collected, ground into a fine powder in liquid nitrogen, and homogenized. The contents of SA, MeSA, JA, and MeJA in the supernatant were quantified using plant enzyme-linked immunosorbent assay (ELISA) kits (Jiweibio, Shanghai, China). Absorbance (OD) at 450 nm was measured using a SpectraMax® M2 microplate reader (Molecular Devices Corporation, Menlo Park, CA, USA) [63]. Hormone concentrations per gram of hairy root fresh weight were calculated using Excel 365. All analyses ensure three biological and technical replicates.

4.11. Flavonoids Extraction and Measurement

The extraction and quantification of flavonoids were conducted according to the previously described method, with three biological and technical replicates. Briefly, 0.2 g of leaf or hairy root tissue was ground into a fine powder in liquid nitrogen and transferred to a centrifuge tube. A 700 μL aliquot of extraction solvent (methanol: DMSO = 1:1, *v*/*v*) was added, and the mixture was vortexed thoroughly, followed by ultrasonic extraction for 30 min. After centrifugation at 12,000 rpm for 10 min, the supernatant was collected. The pellet was re-extracted with 700 μL of the same solvent twice, and all supernatants were pooled. The combined supernatant was adjusted to a final volume of 2.5 mL with methanol, filtered through a 0.22 μm membrane, and stored in the dark for subsequent analysis.

Flavonoid separation was performed using a high-performance liquid chromatography (HPLC) system equipped with an Xbridge UPLCC18 column (5 μm particle size, 4.6 × 150 mm). Mobile phase A consisted of water containing 0.2% acetic acid, and mobile phase B was methanol. A gradient elution program was applied with an injection volume of 5 μL, column temperature of 24 °C, and flow rate of 0.8 mL/min. Detection wavelengths were set at 283 nm and 330 nm. The test was repeated three times.

4.12. Identification of Positive Transgenic Plants and Detection of Editing Efficiency

Genomic DNA was extracted from putative transgenic plants. Positive transgenic plants were confirmed at the DNA level by amplifying the sequence of the pKSE401 vector using specific primers (Supplementary Table S1). To assess editing efficiency, the target gene regions were PCR-amplified and subjected to Sanger sequencing to analyze editing patterns. The PCR products were subsequently cloned into the pTOPO-T vector, and single colonies were selected for sequencing to calculate editing efficiency.

4.13. Homology Modeling and Molecular Docking

PPI (https://cn.string-db.org/cgi/network) online prediction of potential interacting proteins of CitFNSII-1 in citrus was used. To establish suitable protein templates for homology modeling, the amino acid sequences of CitFNSII-1 and CHI-1 were individually queried on the SWISS-MODEL platform (https://swissmodel.expasy.org/). Three-dimensional (3D) homology models of CitFNSII-1 and CHI-1 were generated using the SWISS-MODEL server. Molecular docking was performed following a previous report [64] with the GRAMM software (https://gramm.compbio.ku.edu/request) [65].

4.14. Yeast Two-Hybrid (Y2H) Assay

Using the pGEM-T-*CitFNSII-1* plasmid as a template, the *CitFNSII-1* gene was amplified using primer BD-*CitFNSII*-1-F/R (Supplementary Table S1). *CitFNSII-1* was ligated to a pGBKT7 bait vector using the homologous recombination method and transformed into *Escherichia coli* DH5α, positive clones were identified through PCR amplification and sequencing, and the pGBKT7-*CitFNSII*-1 plasmid was constructed. The construction of the pGADT7-*CHI*-1 prey vector is the same as above.

The pGADT7-*CHI*-1 vector was co-transformed with the pGBKT7-*CitFNSII*-1 plasmid into Y2Hgold yeast. pGBKT7-53 was co-transformed with pGADT7-T, pGBKT7-Lam, and pGADT7-T as positive and negative controls, respectively. The bacterial solution was then coated on DDO/X (SD/-Leu/-Trp), TDO (SD/-Leu/-Trp/-His), and QDO/X (SD/-Leu/-Trp/-His/Ade /X-α-gal) media. The test was repeated three times.

4.15. Bimolecular Fluorescence Complementation Assay

The coding sequence of *CitFNSII-1* was cloned into the n-YFP vector, and the coding sequence of *CHI-1* was cloned into the c-YFP vector. Primers used for cloning are listed in Supplementary Table S1. The constructed fusion vectors or empty vectors were transformed into *Agrobacterium tumefaciens* strain GV3101. All vectors were verified by Sanger sequencing prior to transformation into *A. tumefaciens* GV3101. Bacterial suspensions (mixed at a 1:1 v/v ratio and OD = 0.6–0.8) were co-infiltrated into 4-week-old *Nicotiana benthamiana* leaves. Transiently expressed fusion proteins were observed via confocal laser scanning microscopy 72 h post-infiltration. Three independent biological replicates were performed.

4.16. Luciferase Complementation Assay

The amplified coding sequence (CDS) of *CitFNSII-1* (without the stop codon) was ligated into the nLuc vector to construct the fusion protein *CitFNSII-1*-nLUC. The CDS of *CHI-1* was ligated into the cLUC vector to construct the fusion protein *CHI*-1-cLUC. Recombinant plasmids were transformed into *Agrobacterium tumefaciens* strain GV3101. Agrobacterium suspensions carrying the respective constructs were mixed at a 1:1 ratio and co-infiltrated into *Nicotiana benthamiana* leaves [66]. Luminescence signals resulting from reconstituted luciferase activity were detected using the IndiGo™ in vivo molecular imaging system 48 h post-infiltration. Three independent biological replicates were performed.

4.17. Statistics Analysis

All experiments were performed with a minimum of three independent biological replicates, and data were expressed as mean ± standard error (SE). Statistical analyses included one-way analysis of variance (ANOVA) followed by Duncan's multiple-range test for multi-group comparisons, while pairwise comparisons were assessed using Student's *t*-test. Statistical significance was evaluated with SPSS version 26 at predefined thresholds: $p < 0.05$, $p < 0.01$, $p < 0.001$, and $p < 0.0001$. All data analyses were performed with at least three biological replicates and three technical replicates.

5. Conclusions

This study conducted systematic genome-wide identification and analysis of the FNSII gene family in citrus, identifying three FNSII family members. Evolutionary characterization revealed tandem duplication events within this family, suggesting tandem duplication as the primary mechanism driving family expansion. We analyzed the expression patterns of *CitFNSII-1* in citrus and functionally validated its predicted interacting protein CHI-1. Through transient expression techniques and *Agrobacterium rhizogenes*-mediated genetic transformation, we generated *CitFNSII-1*- and *CHI-1*-expressing transient expression leaves and transgenic hairy roots, followed by phytohormone quantification and flavonoid content analysis to investigate their roles in citrus flavonoid biosynthesis (Figure 13). The results revealed a theoretical foundation for the functional characterization of FNSII genes and provided a critical basis for molecular breeding applications in citrus improvement.

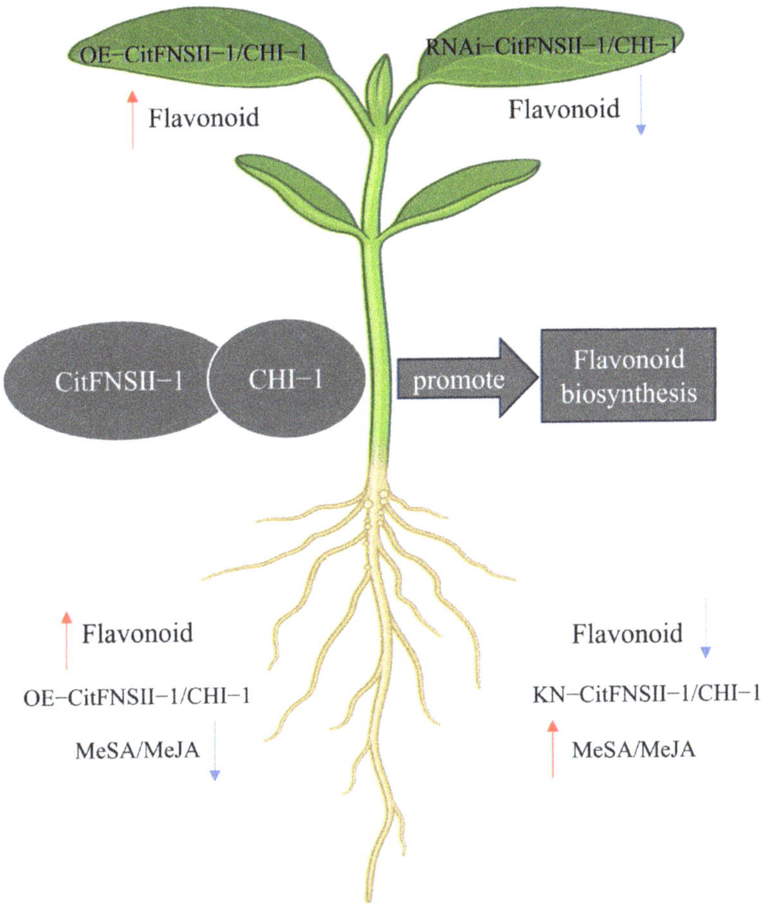

Figure 13. A preliminary model illustrating the role of *CitFNSII-1* in flavonoid synthesis in citrus. The CitFNSII-CHI complex positively regulates flavonoid biosynthesis. Red arrows indicate a significant increase; blue arrows indicate a significant decrease.

Supplementary Materials: The following supporting information can be downloaded at https://www.mdpi.com/article/10.3390/plants14131936/s1. Supplementary Table S1: The primers used in the study; Supplementary Figures file; Supplementary Figure S1: Analysis of the expression characteristics of *CitFNSII-1* gene in citrus; Supplementary Figure S2: Determination of *CitFNSII-1* transgenic and gene-edited hairy roots hormone content; Supplementary Figure S3: Analysis of the expression characteristics of *CHI-1* gene in citrus; Supplementary Figure S4: Determination of *CHI-1* transgenic and gene-edited hairy roots hormone content.

Author Contributions: X.L. designed the experiments. B.C. and L.L. analyzed the data. X.L., B.C., L.L., Q.Z. and C.H.T. performed the experiments. X.L., B.C., L.L. and S.H. wrote and revised the manuscript. C.H.T. and S.H. supervised the research. All authors have read and agreed to the published version of the manuscript.

Funding: This work was funded by The National Key Research and Development Program of China (2021YFD1600802), the Sichuan Science and Technology Support Plan (2025ZNSFSC1099), and the Free Exploration Project by the Horticulture Research Institute of Sichuan Academy of Agricultural Sciences (2025ZYTS01).

Data Availability Statement: The datasets generated during and/or analyzed during the current study are available from the corresponding author upon reasonable request.

Acknowledgments: Thanks to all of the authors who contributed to this paper.

Conflicts of Interest: The authors declare no conflicts of interest.

References

1. Wang, F.; Wang, M.; Liu, X.; Xu, Y.; Zhu, S.; Shen, W.; Zhao, X. Identification of Putative Genes Involved in Limonoids Biosynthesis in Citrus by Comparative Transcriptomic Analysis. *Front. Plant Sci.* **2017**, *8*, 782. [CrossRef] [PubMed]
2. Durand-Hulak, M.; Dugrand, A.; Duval, T.; Bidel, L.P.; Jay-Allemand, C.; Froelicher, Y.; Bourgaud, F.; Fanciullino, A.L. Mapping the genetic and tissular diversity of 64 phenolic compounds in Citrus species using a UPLC-MS approach. *Ann. Bot.* **2015**, *115*, 861–877. [CrossRef]
3. Torregrosa, C.; Cluzet, S.; Fournier, J.; Huguet, T.; Gamas, P.; Prospéri, J.M.; Esquerré-Tugayé, M.T.; Dumas, B.; Jacquet, C. Cytological, genetic, and molecular analysis to characterize compatible and incompatible interactions between Medicago truncatula and Colletotrichum trifolii. *Mol. Plant Microbe Interact.* **2004**, *17*, 909–920. [CrossRef] [PubMed]
4. Sun, W.; Meng, X.; Liang, L.; Jiang, W.; Huang, Y.; He, J.; Hu, H.; Almqvist, J.; Gao, X.; Wang, L. Molecular and Biochemical Analysis of Chalcone Synthase from Freesia hybrid in flavonoid biosynthetic pathway. *PLoS ONE* **2015**, *10*, e0119054. [CrossRef] [PubMed]
5. Bondonno, N.P.; Dalgaard, F.; Kyrø, C.; Murray, K.; Bondonno, C.P.; Lewis, J.R.; Croft, K.D.; Gislason, G.; Scalbert, A.; Cassidy, A.; et al. Flavonoid intake is associated with lower mortality in the Danish Diet Cancer and Health Cohort. *Nat. Commun.* **2019**, *10*, 3651. [CrossRef]
6. Peng, Z.; Zhang, H.; Li, W.; Yuan, Z.; Xie, Z.; Zhang, H.; Cheng, Y.; Chen, J.; Xu, J. Comparative profiling and natural variation of polymethoxylated flavones in various citrus germplasms. *Food Chem.* **2021**, *354*, 129499. [CrossRef]
7. Wang, J.; Jiang, Y.; Sun, T.; Zhang, C.; Liu, X.; Li, Y. Genome-Wide Classification and Evolutionary Analysis Reveal Diverged Patterns of Chalcone Isomerase in Plants. *Biomolecules.* **2022**, *12*, 961. [CrossRef]
8. Parmenter, B.H.; Thompson, A.S.; Bondonno, N.P.; Jennings, A.; Murray, K.; Perez-Cornago, A.; Hodgson, J.M.; Tresserra-Rimbau, A.; Kühn, T.; Cassidy, A. High diversity of dietary flavonoid intake is associated with a lower risk of all-cause mortality and major chronic diseases. *Nat. Food.* [CrossRef]
9. Du, Y.; Chu, H.; Wang, M.; Chu, I.K.; Lo, C. Identification of flavone phytoalexins and a pathogen-inducible flavone synthase II gene (SbFNSII) in sorghum. *J. Exp. Bot.* **2010**, *61*, 983–994. [CrossRef]
10. Righini, S.; Rodriguez, E.J.; Berosich, C.; Grotewold, E.; Casati, P.; Falcone Ferreyra, M.L. Apigenin produced by maize flavone synthase I and II protects plants against UV-B-induced damage. *Plant Cell Environ.* **2019**, *42*, 495–508. [CrossRef]
11. Zheng, J.; Zhao, C.; Liao, Z.; Liu, X.; Gong, Q.; Zhou, C.; Liu, Y.; Wang, Y.; Cao, J.; Liu, L.; et al. Functional characterization of two flavone synthase II members in citrus. *Hortic Res.* **2023**, *10*, uhad113. [CrossRef] [PubMed]
12. Deng, Y.; Li, C.; Li, H.; Lu, S. Identification and Characterization of Flavonoid Biosynthetic Enzyme Genes in Salvia miltiorrhiza (Lamiaceae). *Molecules.* **2018**, *23*, 1467. [CrossRef] [PubMed]
13. Qiu, S.; Wang, J.; Pei, T.; Gao, R.; Xiang, C.; Chen, J.; Zhang, C.; Xiao, Y.; Li, Q.; Wu, Z.; et al. Functional evolution and diversification of CYP82D subfamily members have shaped flavonoid diversification in the genus Scutellaria. *Plant Commun.* **2025**, *6*, 101134. [CrossRef]
14. Zhao, Q.; Yang, J.; Cui, M.Y.; Liu, J.; Fang, Y.; Yan, M.; Qiu, W.; Shang, H.; Xu, Z.; Yidiresi, R.; et al. The Reference Genome Sequence of Scutellaria baicalensis Provides Insights into the Evolution of Wogonin Biosynthesis. *Mol. Plant.* **2019**, *12*, 935–950. [CrossRef]
15. Su, D.; Wu, M.; Wang, H.; Shu, P.; Song, H.; Deng, H.; Yu, S.; Garcia-Caparros, P.; Bouzayen, M.; Zhang, Y.; et al. Bi-functional transcription factor SlbHLH95 regulates fruits flavonoid metabolism and grey mould resistance in tomato. *Plant Biotechnol. J.* **2025**, *23*, 2083–2094. [CrossRef]
16. Wu, Z.; Singh, S.K.; Lyu, R.; Pattanaik, S.; Wang, Y.; Li, Y.; Yuan, L.; Liu, Y. Metabolic engineering to enhance the accumulation of bioactive flavonoids licochalcone A and echinatin in *Glycyrrhiza inflata* (Licorice) hairy roots. *Front. Plant Sci.* **2022**, *13*, 932594. [CrossRef] [PubMed]
17. Li, X.; Cao, L.; Jiao, B.; Yang, H.; Ma, C.; Liang, Y. The bHLH transcription factor AcB2 regulates anthocyanin biosynthesis in onion (*Allium cepa* L.). *Hortic. Res.* **2022**, *9*, uhac128. [CrossRef]
18. Zhao, Q.; Zhang, Y.; Wang, G.; Hill, L.; Weng, J.K.; Chen, X.Y.; Xue, H.; Martin, C. A specialized flavone biosynthetic pathway has evolved in the medicinal plant, Scutellaria baicalensis. *Sci. Adv.* **2016**, *2*, e1501780. [CrossRef]
19. Zou, X.; Zhao, K.; Liu, Y.; Du, M.; Zheng, L.; Wang, S.; Xu, L.; Peng, A.; He, Y.; Long, Q.; et al. Overexpression of Salicylic Acid Carboxyl Methyltransferase (CsSAMT1) Enhances Tolerance to Huanglongbing Disease in Wanjincheng Orange (*Citrus sinensis* (L.) Osbeck). *Int. J. Mol. Sci.* **2021**, *22*, 2803. [CrossRef]

20. Wang, W.; Li, T.; Chen, Q.; Yao, S.; Deng, L.; Zeng, K. CsWRKY25 Improves Resistance of Citrus Fruit to Penicillium digitatum via Modulating Reactive Oxygen Species Production. *Front. Plant Sci.* **2022**, *12*, 818198. [CrossRef]
21. Mizuno, H.; Yazawa, T.; Kasuga, S.; Sawada, Y.; Kanamori, H.; Ogo, Y.; Hirai, M.Y.; Matsumoto, T.; Kawahigashi, H. Expression of Flavone Synthase II and Flavonoid 3′-Hydroxylase Is Associated with Color Variation in Tan-Colored Injured Leaves of Sorghum. *Front. Plant Sci.* **2016**, *7*, 1718. [CrossRef]
22. Tian, S.; Yang, Y.; Wu, T.; Luo, C.; Li, X.; Zhao, X.; Xi, W.; Liu, X.; Zeng, M. Functional Characterization of a Flavone Synthase That Participates in a Kumquat Flavone Metabolon. *Front. Plant Sci.* **2022**, *13*, 826780. [CrossRef]
23. Luo, C.; Liu, L.; Zhao, J.; Xu, Y.; Liu, H.; Chen, D.; Cheng, X.; Gao, J.; Hong, B.; Huang, C.; et al. CmHY5 functions in apigenin biosynthesis by regulating flavone synthase II expression in chrysanthemum flowers. *Planta* **2022**, *257*, 7. [CrossRef]
24. Zhao, Y.; Wu, Z.; Li, J.; Qi, Y.; Zhang, X.; Shen, C. The key role of cytochrome P450s in the biosynthesis of plant derived natural products. *Plant Physiol Biochem.* **2025**, *222*, 109695. [CrossRef]
25. Lam, P.Y.; Zhu, F.Y.; Chan, W.L.; Liu, H.; Lo, C. Cytochrome P450 93G1 Is a Flavone Synthase II That Channels Flavanones to the Biosynthesis of Tricin O-Linked Conjugates in Rice. *Plant Physiol.* **2014**, *165*, 1315–1327. [CrossRef] [PubMed]
26. Fujino, N.; Tenma, N.; Waki, T.; Ito, K.; Komatsuzaki, Y.; Sugiyama, K.; Yamazaki, T.; Yoshida, S.; Hatayama, M.; Yamashita, S.; et al. Physical interactions among flavonoid enzymes in snapdragon and torenia reveal the diversity in the flavonoid metabolon organization of different plant species. *Plant J.* **2018**, *94*, 372–392. [CrossRef] [PubMed]
27. Ledesma-Escobar, C.A.; Priego-Capote, F.; Robles Olvera, V.J.; Luque de Castro, M.D. Targeted Analysis of the Concentration Changes of Phenolic Compounds in Persian Lime (*Citrus latifolia*) during Fruit Growth. *J. Agric. Food Chem.* **2018**, *66*, 1813–1820. [CrossRef] [PubMed]
28. Seoka, M.; Ma, G.; Zhang, L.; Yahata, M.; Yamawaki, K.; Kan, T.; Kato, M. Expression and functional analysis of the nobiletin biosynthesis-related gene CitOMT in citrus fruit. *Sci. Rep.* **2020**, *10*, 15288. [CrossRef]
29. Peer, W.A.; Murphy, A.S. Flavonoids and auxin transport: Modulators or regulators? *Trends Plant Sci.* **2007**, *12*, 556–563. [CrossRef]
30. Luo, P.; Shen, Y.; Jin, S.; Huang, S.; Cheng, X.; Wang, Z.; Li, P.; Zhao, J.; Bao, M.; Ning, G. Overexpression of Rosa rugosa anthocyanidin reductase enhances tobacco tolerance to abiotic stress through increased ROS scavenging and modulation of ABA signaling. *Plant Sci.* **2016**, *245*, 35–49. [CrossRef]
31. Yang, W.; Xu, X.; Li, Y.; Wang, Y.; Li, M.; Wang, Y.; Ding, X.; Chu, Z. Rutin-Mediated Priming of Plant Resistance to Three Bacterial Pathogens Initiating the Early SA Signal Pathway. *PLoS ONE.* **2016**, *11*, e0146910. [CrossRef] [PubMed]
32. Li, X.; Zhang, L.P.; Zhang, L.; Yan, P.; Ahammed, G.J.; Han, W.Y. Methyl Salicylate Enhances Flavonoid Biosynthesis in Tea Leaves by Stimulating the Phenylpropanoid Pathway. *Molecules* **2019**, *24*, 362. [CrossRef]
33. Ramabulana, A.T.; Steenkamp, P.A.; Madala, N.E.; Dubery, I.A. Profiling of Altered Metabolomic States in Bidens pilosa Leaves in Response to Treatment by Methyl Jasmonate and Methyl Salicylate. *Plants* **2020**, *9*, 1275. [CrossRef]
34. Gondor, O.K.; Janda, T.; Soós, V.; Pál, M.; Majláth, I.; Adak, M.K.; Balázs, E.; Szalai, G. Salicylic Acid Induction of Flavonoid Biosynthesis Pathways in Wheat Varies by Treatment. *Front. Plant Sci.* **2016**, *7*, 1447. [CrossRef]
35. Cheng, B.; Xu, L.; Bilal, M.S.; Huang, Q.; Niu, D.; Ma, H.; Zhou, S.; Peng, A.; Wei, G.; Chen, F.; et al. Small RNAs contribute to citrus Huanglongbing tolerance by manipulating methyl salicylate signaling and exogenous methyl salicylate primes citrus groves from emerging infection. *Plant J.* **2023**, *116*, 1309–1324. [CrossRef]
36. Pauwels, L.; Morreel, K.; De Witte, E.; Lammertyn, F.; Van Montagu, M.; Boerjan, W.; Inzé, D.; Goossens, A. Mapping methyl jasmonate-mediated transcriptional reprogramming of metabolism and cell cycle progression in cultured Arabidopsis cells. *Proc. Natl. Acad. Sci. USA* **2008**, *105*, 1380–1385. [CrossRef]
37. Wang, Y.; Liu, X.J.; Chen, J.B.; Cao, J.P.; Li, X.; Sun, C.D. Citrus flavonoids and their antioxidant evaluation. *Crit. Rev. Food Sci. Nutr.* **2022**, *62*, 3833–3854. [CrossRef] [PubMed]
38. Ngaki, M.N.; Louie, G.V.; Philippe, R.N.; Manning, G.; Pojer, F.; Bowman, M.E.; Li, L.; Larsen, E.; Wurtele, E.S.; Noel, J.P. Evolution of the chalcone-isomerase fold from fatty-acid binding to stereospecific catalysis. *Nature* **2012**, *485*, 530–533. [CrossRef] [PubMed]
39. Mehdy, M.C.; Lamb, C.J. Chalcone isomerase cDNA cloning and mRNA induction by fungal elicitor, wounding and infection. *EMBO J.* **1987**, *6*, 1527–1533. [CrossRef]
40. Jiang, W.; Yin, Q.; Wu, R.; Zheng, G.; Liu, J.; Dixon, R.A.; Pang, Y. Role of a chalcone isomerase-like protein in flavonoid biosynthesis in Arabidopsis thaliana. *J. Exp. Bot.* **2015**, *66*, 7165–7179. [CrossRef]
41. Shimada, N.; Aoki, T.; Sato, S.; Nakamura, Y.; Tabata, S.; Ayabe, S. A cluster of genes encodes the two types of chalcone isomerase involved in the biosynthesis of general flavonoids and legume-specific 5-deoxy(iso)flavonoids in Lotus japonicus. *Plant Physiol.* **2003**, *131*, 941–951. [CrossRef] [PubMed]
42. Kang, J.H.; McRoberts, J.; Shi, F.; Moreno, J.E.; Jones, A.D.; Howe, G.A. The flavonoid biosynthetic enzyme chalcone isomerase modulates terpenoid production in glandular trichomes of tomato. *Plant Physiol.* **2014**, *164*, 1161–1174. [CrossRef] [PubMed]

43. Wheeler, L.C.; Walker, J.F.; Ng, J.; Deanna, R.; Dunbar-Wallis, A.; Backes, A.; Pezzi, P.H.; Palchetti, M.V.; Robertson, H.M.; Monaghan, A.; et al. Transcription Factors Evolve Faster Than Their Structural Gene Targets in the Flavonoid Pigment Pathway. *Mol. Biol. Evol.* **2022**, *39*, msac044. [CrossRef] [PubMed]
44. Longevity, O.M.A.C. Retracted: Two Myricetin-Derived Flavonols from Morella rubra Leaves as Potent α-Glucosidase Inhibitors and Structure-Activity Relationship Study by Computational Chemistry. *Oxid. Med. Cell Longev.* **2024**, *2024*, 9849172. [CrossRef]
45. Li, J.; Yu, Q.; Liu, C.; Zhang, N.; Xu, W. Flavonoids as key players in cold tolerance: Molecular insights and applications in horticultural crops. *Hortic. Res.* **2025**, *12*, uhae366. [CrossRef]
46. Tian, X.; Hu, M.; Yang, J.; Yin, Y.; Fang, W. Ultraviolet-B Radiation Stimulates Flavonoid Biosynthesis and Antioxidant Systems in Buckwheat Sprouts. *Foods.* **2024**, *13*, 3650. [CrossRef]
47. Deng, H.; Wu, M.; Wu, Y.; Xiao, X.; Gao, Z.; Li, H.; Hu, N.; Gao, Y.; Grierson, D.; Liu, M. SlMYC2-SlMYB12 module orchestrates a hierarchical transcriptional cascade that regulates fruit flavonoid metabolism in tomato. *Plant Biotechnol. J.* **2025**, *23*, 477. [CrossRef]
48. Zhang, L.; Zheng, J.; Johnson, M.; Mandal, R.; Cruz, M.; Martínez-Huélamo, M.; Andres-Lacueva, C.; Wishart, D.S. A Comprehensive LC-MS Metabolomics Assay for Quantitative Analysis of Serum and Plasma. *Metabolites.* **2024**, *14*, 622. [CrossRef]
49. Mou, J.; Zhang, Z.; Qiu, H.; Lu, Y.; Zhu, X.; Fan, Z.; Zhang, Q.; Ye, J.; Fernie, A.R.; Cheng, Y.; et al. Multiomics-based dissection of citrus flavonoid metabolism using a Citrus reticulata × Poncirus trifoliata population. *Hortic. Res.* **2021**, *8*, 56. [CrossRef]
50. Baldassari, S.; Klingler, E.; Teijeiro, L.G.; Doladilhe, M.; Raoux, C.; Roig-Puiggros, S.; Bizzotto, S.; Couturier, J.; Gilbert, A.; Sami, L.; et al. Single-cell genotyping and transcriptomic profiling of mosaic focal cortical dysplasia. *Nat. Neurosci.* **2025**, *28*, 964–972. [CrossRef]
51. Munir, S.; Khan, M.R.; Song, J.; Munir, S.; Zhang, Y.; Ye, Z.; Wang, T. Genome-wide identification, characterization and expression analysis of calmodulin-like (CML) proteins in tomato (Solanum lycopersicum). *Plant Physiol. Biochem.* **2016**, *102*, 167–179. [CrossRef]
52. Yu, Y.; Yang, M.; Liu, X.; Xia, Y.; Hu, R.; Xia, Q.; Jing, D.; Guo, Q. Genome-wide analysis of the WOX gene family and the role of EjWUSa in regulating flowering in loquat (*Eriobotrya japonica*). *Front. Plant Sci.* **2022**, *13*, 1024515. [CrossRef]
53. Wilkins, M.R.; Gasteiger, E.; Bairoch, A.; Sanchez, J.C.; Williams, K.L.; Appel, R.D.; Hochstrasser, D.F. Protein identification and analysis tools in the ExPASy server. *Methods Mol. Biol.* **1999**, *112*, 531–552. [CrossRef]
54. Bailey, T.L.; Boden, M.; Buske, F.A.; Frith, M.; Grant, C.E.; Clementi, L.; Ren, J.; Li, W.W.; Noble, W.S. MEME SUITE: Tools for motif discovery and searching. *Nucleic Acids Res.* **2009**, *37*, W202–W208. [CrossRef]
55. Chen, W.; Wang, P.; Wang, D.; Shi, M.; Xia, Y.; He, Q.; Dang, J.; Guo, Q.; Jing, D.; Liang, G. EjFRI, FRIGIDA (FRI) Ortholog from Eriobotrya japonica, Delays Flowering in Arabidopsis. *Int. J. Mol. Sci.* **2020**, *21*, 1087. [CrossRef] [PubMed]
56. Zhao, M.; Haxim, Y.; Liang, Y.; Qiao, S.; Gao, B.; Zhang, D.; Li, X. Genome-wide investigation of AP2/ERF gene family in the desert legume Eremosparton songoricum: Identification, classification, evolution, and expression profiling under drought stress. *Front. Plant Sci.* **2022**, *13*, 885694. [CrossRef] [PubMed]
57. Zhang, Y.; Ren, Y.; Yang, D.; Liu, H.; Zhang, Y.; Wang, X.; Bai, F.; Cheng, S. Foliar methyl jasmonate (MeJA) application increased 2-acetyl-1-Pyrroline (2-AP) content and modulated antioxidant attributes and yield formation in fragrant rice. *J. Plant Physiol.* **2023**, *282*, 153946. [CrossRef] [PubMed]
58. Lei, Y.; Lu, L.; Liu, H.Y.; Li, S.; Xing, F.; Chen, L.L. CRISPR-P: A web tool for synthetic single-guide RNA design of CRISPR-system in plants. *Mol. Plant.* **2014**, *7*, 1494–1496. [CrossRef]
59. Tang, T.; Yu, X.; Yang, H.; Gao, Q.; Ji, H.; Wang, Y.; Yan, G.; Peng, Y.; Luo, H.; Liu, K.; et al. Development and Validation of an Effective CRISPR/Cas9 Vector for Efficiently Isolating Positive Transformants and Transgene-Free Mutants in a Wide Range of Plant Species. *Front. Plant Sci.* **2018**, *9*, 1533. [CrossRef]
60. Wang, S.; Yang, C.; Tu, H.; Zhou, J.; Liu, X.; Cheng, Y.; Luo, J.; Deng, X.; Zhang, H.; Xu, J. Characterization and Metabolic Diversity of Flavonoids in Citrus Species. *Sci. Rep.* **2017**, *7*, 10549. [CrossRef]
61. Zhao, C.; Liu, X.; Gong, Q.; Cao, J.; Shen, W.; Yin, X.; Grierson, D.; Zhang, B.; Xu, C.; Li, X.; et al. Three AP2/ERF family members modulate flavonoid synthesis by regulating type IV chalcone isomerase in citrus. *Plant Biotechnol. J.* **2021**, *19*, 671–688. [CrossRef] [PubMed]
62. He, X.; Wang, H.; Wei, W.; Han, Z.; Zuo, J.; He, Q. Expression characteristics of CsESA1 in citrus and analysis of its interacting protein. *Plant Signal Behav.* **2025**, *20*, 2439249. [CrossRef] [PubMed]
63. Du, M.; Wang, S.; Dong, L.; Qu, R.; Zheng, L.; He, Y.; Chen, S.; Zou, X. Overexpression of a "Candidatus Liberibacter Asiaticus" Effector Gene CaLasSDE115 Contributes to Early Colonization in Citrus sinensis. *Front. Microbiol.* **2022**, *12*, 797841. [CrossRef] [PubMed]
64. Liu, T.; Liu, H.; Xian, W.; Liu, Z.; Yuan, Y.; Fan, J.; Xiang, S.; Yang, X.; Liu, Y.; Liu, S.; et al. Duplication and sub-functionalization of flavonoid biosynthesis genes plays important role in Leguminosae root nodule symbiosis evolution. *J. Integr. Plant Biol.* **2024**, *66*, 2191–2207. [CrossRef]

65. Tovchigrechko, A.; Vakser, I.A. GRAMM-X public web server for protein-protein docking. *Nucleic Acids Res.* **2006**, *34*, W310–W314. [CrossRef]
66. Chen, C.; Wang, Y.; Wu, K.; Ding, Y.; Tang, M.; Zhang, X.; Pan, Y.; Wu, L.; Su, C.; Hong, Z.; et al. The DnaJ1 heat shock protein interacts with the flavanone 3-hydroxylase-like protein F3HL to synergistically enhance drought tolerance by scavenging reactive oxygen species in tomato. *Plant J.* **2025**, *121*, e70097. [CrossRef]

plants

MDPI

Article

Effects of Cultivation Modes on Soil Protistan Communities and Its Associations with Production Quality in Lemon Farmlands

Haoqiang Liu [1,2,*], Hongjun Li [1,2], Zhuchun Peng [1,2], Sichen Li [1,2] and Chun Ran [1,2]

1 Citrus Research Institute, Southwest University, Beipei District, Chongqing 400715, China
2 National Engineering Research Center for Citrus, Chinese Academy of Agricultural Sciences, Beipei District, Chongqing 400712, China
* Correspondence: 113883823816@126.com; Tel./Fax: +86-023-6834-9005

Abstract

Citrus is one of the most widely consumed fruits in the world, and its cultivation industry continues to develop rapidly. However, the roles of soil protistan communities during citrus growth are not yet fully understood, despite the potential significance of these communities to the health and quality of citrus. In this study, we examined the soil properties and protistan communities in Eureka lemon farmlands located in Chongqing, China, during the flowering and fruiting stages of cultivation, both in greenhouse and open-field settings. In general, the majority of the measured soil properties (including nutrients and enzyme activities) exhibited higher values in open-field farmlands in comparison to those observed in greenhouse counterparts. According to the results of high-throughput sequencing based on the V9 region of eukaryotic 18S rRNA gene, the diversity of soil protistan communities was also higher in open-field farmlands, and both lemon growth stage and cultivation modes showed significant effects on soil protistan compositions. The transition from traditional agricultural practices to greenhouse farming resulted in a significant transformation of the soil protistan community. This transformation manifested as a shift towards a state characterized by diminished nutrient cycling capabilities. This decline was evidenced by an increase in phototrophs (Archaeplastida) and a concomitant decrease in consumers (Stramenopiles and Alveolata). Community assembly analysis revealed deterministic processes that controlled the succession of soil protistan communities in lemon farmlands. It has been established that environmental associations have the capacity to recognize nitrogen in soils, thereby providing a deterministic selection process for protistan community assembly. Furthermore, a production index was calculated based on 12 quality parameters of lemons, and the results indicated that lemons from greenhouse farms exhibited a lower quality compared to those from open fields. The structure equation model revealed a direct correlation between the quality of lemons and the cultivation methods employed, as well as the composition of soil protists. The present study offers insights into the mechanisms underlying the correlations between the soil protistan community and lemon quality in response to changes in the cultivation modes.

Keywords: greenhouse; open-field cultivation; nutrient cycling; protists; community assembly mechanism; biochemical properties of the soil; fruit quality

1. Introduction

Agricultural ecosystems have provided humans with food, forage, bioenergy, and pharmaceuticals, which are essential for human well-being [1]. It is important to note that there are usually multiple planting modes for the same crop, such as greenhouse vs. natural growth [2]. Furthermore, agricultural land designated for the cultivation of the same crop may exhibit divergent geographical characteristics due to limitations imposed by the topography, such as plains as opposed to hills [3]. Furthermore, the practice of rotation is a prevalent cultivation method, and the conditions experienced by corps in rotation can vary significantly, including the contrast between paddy and dry fields [4]. It has been demonstrated that these variations within agricultural ecosystems have the capacity to influence the soil properties and regulate the crop quality. A case in point is that of rice-wheat rotation, a highly intensive agricultural planting mode in China. It has been demonstrated that this practice enhances soil fertility and productivity, a consequence of the effects of straw return [5]. A study of rice paddy fields employing diverse planting techniques also identified significant variations in soil bulk density, tiller number, spikelet number per panicle, panicle length, panicle weight, and grain yields [6]. Furthermore, analysis of pear orchards within the river basin revealed a decline in soil nutrient levels, while concurrently demonstrating enhanced yield in comparison to conventional natural orchards [7]. Consequently, farmlands that cultivate the same crop but employ diverse cultivation techniques or land backgrounds should be the focus of extensive investigation to develop optimized management.

In addition to the soil properties, it should be noted that soil microbial communities of agricultural ecosystems will also change under different planting practices [8]. However, the majority of previous studies have primarily focused on the effects of different planting modes on the soil bacterial communities in agricultural ecosystems [9–11]. In recent years, the significant roles of protists in the promotion of plant growth and enhancement of plant health have come to the forefront of research [12]. A study of agricultural ecosystems across Europe indicated that land use and management practices are significant factors influencing soil protistan communities [13]. Another study demonstrated that soil protists can function as indicators and determinants of plant performance, thereby stimulating plant performance through microbiome interactions [14]. Furthermore, the potential of protists in biotechnological applications aimed at reducing pesticide and fertilizer usage in sustainable agriculture has been explored [15]. Despite the existence of some studies on the subject, there remains a paucity of precise information regarding soil protistan communities and their relationships with the crop quality in agricultural ecosystems.

Citrus fruits are among the most widely traded commodities on the global market, owing to their refreshing flavor and high vitamin C content [16]. Among the diverse range of citrus species, the Eureka lemon is a popular citrus tree that year-round produces an abundance of fine, market-quality fruit with tart juice and flavourful zesty peel [17]. China is a major producer of lemons on a global scale, with Eureka lemons being a prominent variety cultivated in the country. The production of these lemons is consistent throughout the year in regions such as Chongqing and Sichuan Province. It is hypothesized that the cultivation modes and their variations may have an effect on the quality of lemons, based on the ecological functions of soil protistan communities. In order to further expand the scale of the Eureka lemon industry and increase economic benefits, our study investigated the soil protistan communities on farmlands with different cultivation modes in lemon planting areas of Chongqing Province, China. The objectives of this work were: (i) to portray the soil protistan profiles of lemon farmlands with different cultivation modes; (ii) to dissect the assembly mechanisms of soil protistan communities and their associations with soil properties; (iii) to explore the relationships of lemon quality with cultivation modes, soil properties, and soil protistan communities.

2. Materials and Methods

2.1. Experience Scheme and Sample Collection

Soil samples were collected from the agricultural terrain of a lemon farm at the Chongqing Tongnan Seedling Breeding Center located in the Guopo Village, Baizi Town, Chongqing, China. The area under consideration encompasses 16 hectares of citrus farmland, with a history of agricultural activity that spans a period exceeding five years. The collection of soil samples was undertaken in nine farmlands that cultivated the Eureka lemon tree were selected. Among them, three farmlands are orchards with greenhouse (Greenhouse), three farmlands were transformed into orchards from hilly areas (hilly-converted), and three farmlands were transformed into orchards from paddy fields (paddy-converted). Sampling was conducted on the 4th of May, and the 24th of September, 2023, which corresponded to the flowering and fruiting stages of lemon growth, respectively. In a single orchard at the single stage, six trees (a total of 54 trees) were selected to obtain soil samples. For each tree, a multi-point sampling method was employed, which entailed the collection of surface soils at three random points around the lemon trunks (5–10 cm). Each tree was sampled on three occasions to ensure the representation of the data in the final sample. The soil samples were then divided into two parts for the purpose of determining the soil properties. An additional portion of the fresh soil samples was subjected to treatment with liquid nitrogen for the purpose of rapid cooling. Following this, the samples were cryopreserved at −80 °C for the purpose of determining and analyzing the soil protistan communities. A total of 108 soil samples were collected from nine different lemon farms, with six samples being obtained from each farm at each of the lemon growth stages.

2.2. Measurements of Soil Properties

Soil moisture content was determined by calculating the proportion of lost weight through a process of oven drying for 12 h at 105 °C. Before measuring the content of soil micronutrients, the soil samples were air-dried. The total nitrogen (TN) of the soil sample was measured using the Kjeldahl method [18], and ammonia, nitrate, and nitrite levels were detected using an Auto Analyzer 3 Digital Colorimeter (Bran & Luebbe Inc., Norderstedt, Germany). Soil pH was measured in a 1:5 soil/water suspension using a pH meter. The total phosphorus (TP) was measured using NaOH digestion and the molybdenum blue colorimetric method [19]. The concentration of available phosphorus (AP) in the soil was determined by the molybdenum antimony colormetric method [20] using an UV-visible spectrophotometer (UV-1200, Mapada, Shanghai, China). The total potassium (TK) and available potassium (AK) of the soil sample were measured using a flame photometer with extracted by NaOH and NH4OAc, respectively [21]. For dissolved organic carbon (DOC), soil oscillated were placed for in an Erlenmeyer, and distilled water was added. Samples were oscillated for 30 min, filtered, and left to rest for 10 min to obtain the leaching solution. Then, the DOC was analyzed using a UV-8000 spectrophotometer (Mapada, Shanghai, China) based on the leaching solution after oxidizing organic carbon with potassium dichromate [22]. The enzyme activities of alkaline phosphatase (ALP), beta-glucosidase (beta-Glu), urease, and cellulase were determined using respective ELISA kits (Shanghai Jiwei Biological Technology Co., Ltd., Shanghai, China) using an RT-6100 spectrophotometer (Rayto Life and Analytical Sciences Co., Ltd., Shenzhen, China). For each measurement of each soil samples, three replicates were performed.

2.3. Measurements of Lemon Quality Parameters

A series of lemon quality parameters, including fruit weight, fruit width, fruit height, fruit shape index, pericarp thickness, pericarp weight, valve number, valve weight, seed number, seed weight, solid rate, and edible rate, in each studied farmland were measured after the lemon harvest. All measurements were in accordance with the National Standard of China (GB/T 8210-2011) [23]. Detailed, the weight of whole fruit was measured by an electronic balance. Then, the fruit was disassembled to count its valve number and seed number. Then, weights of pericarp, value, and seed were also measured by an electronic balance. The longitudinal and transverse diameters of lemon were measured by a Vernier caliper, and the fruit shape index was calculated as their ratio. The fruit width, fruit height, and pericarp thickness were also measured by the Vernier caliper. The sold rate was the proportion of the weight of fruit except to the juice with the whole weight of fruit. The edible rate was the proportion of the weight of fruit except seed with the whole weight of fruit.

2.4. DNA Extraction and Protistan Community Sequencing

The isolation of microbial DNA from soil samples was accomplished by means of the FastDNA® SPIN kit for soil (MP Biomedicals, Irvine, CA, USA). Agarose gel electrophoresis (1.5%) was performed in order to ascertain the efficacy of the DNA extraction process. The quality of the successfully extracted DNA was then evaluated by NanoDrop ND-1000 Spectrophotometer (NanoDrop, Wilmington, DE, USA). For investigating the protistan communities, the V9 region of the eukaryotic 18S rRNA gene was amplified from the microbial DNA using the primers of 1380F (TCCCTGCCHTTTGTACACAC) and 1510R (CCTTCYGCAGGTTCACCTAC) [24]. The detail processes of PCR and library construction referred to previous studies [25]. Finally, the libraries were subjected to sequencing using the Illumina NovaSeq 6000 platform at BIOZERN Biotech. Co., Ltd. (Shanghai, China).

The assignment of sequenced reads to the corresponding samples was facilitated by the unique barcode at the end of reverse primer for each gene. Subsequently, quality control was performed using the "dada2" package in R v4.2.2 according to the thresholds of Q score > 20 and no ambiguous base [26]. The remaining paired-end reads were then assembled based on the overlap region between them. Subsequently, the assembled reads were clustered into amplicon sequence variants (ASVs), with the chimera was removed using the QIIME2 (Quantitative Insights Into Microbial Ecology 2) program [27]. Taxonomy was assigned to each ASV based on the PR2 v4.12.0 database [28]. Then, ASVs that were assigned into other taxa (Fungi, Metazoa, and Embryophyta) were excluded. Finally, singletons (the total read number of a specific ASV in all samples is 1) were abandoned, and the AVS abundance table was normalized to 18,907 reads before further data analyses.

2.5. Statistics Analysis

All data analyses were performed using R v4.2.2, and the results were visualized by the "ggplot2" package. Four alpha-diversity indices, Chao1 (richness), Pd_faith (evolution), Shannon (diversity), and Pielou_J (evenness), of protistan communities were, respectively, calculated using the "vegan" package v2.6-10. Differences in the soil properties and alpha diversity indices of protistan communities between different cultivation modes and lemon growth stages were examined by Tukey's HSD test ("multcomp" package v1.4-28). Principal coordinate analysis (PCoA) and PERMANOVA ("vegan" package), based on the Bray–Curtis distance, were performed to demonstrate differences in soil protistan composition from farmlands with different cultivation modes and lemon growth stages. Tukey's HSD test was also employed to assess the variations in Bray–Curtis distance and dominant protist taxa between different cultivation modes and lemon growth stages. Protists were assigned to three major functional groups, based on their taxonomic affiliation according to three main trophic modes: phototrophic, parasitic, and consumer. The detailed matching information between protist taxonomy and functional groups is referred to in the table previously reported by Singer et al. [29]. Variations in these functional groups between different cultivation modes and lemon growth stages were also assessed by Tukey's HSD test. PCoA and PERMANOVA were further performed to evaluate variations in the compositions of these three functional groups.

Redundancy analysis (RDA) was performed to evaluate the correlations of soil environmental variables with protistan communities ("vegan" package). Random forest analysis, one of the machine learning method, was constructed based on the soil environmental variables to predict the variations in soil protistan compositions ("RandomForest" package v4.7-1.2). Based on the random forest results, potential key drivers for soil protistan communities were extracted. To explore the community assembly mechanisms, the null model were applied to the protistan communities [30]. In the null model, the phylogenetic distance (betaMNTD) of protistan communities was first calculated and further turns it into the beta nearest taxon index (betaNTI) to distinguish the contribution of deterministic ($|\text{betaNTI}| > 2$) and stochastic ($|\text{betaNTI}| \leq 2$) processes in community assembly. Differences in the betaMNTD and betaNTI of soil protistan communities between different cultivation modes and lemon growth stages were also tested by Tukey's HSD test. Furthermore, the Raup–Crick (RC) metric among different protistan communities were also calculated. The classification of community assembly was then compared with the betaNTI, as per the findings of a previous study [31].

Finally, a production index was calculated based on the quality parameters of lemons to evaluate the lemon quality according to the method for soil multifunctional index reported by [32]. Linear regression was used to assess the relationships between the lemon production index and the diversity or composition of soil protistan communities. Then, the direct and indirect effects of cultivation modes on lemon quality mediated by the soil properties and protistan communities were quantified using structural equation modeling (SEM, "lavaan" package v0.6-19).

3. Results

3.1. Variations in Soil Properties

Here, a total of 14 soil properties from various lemon farmlands at both the flowering and fruiting stages were measured and compared. In the flowering stage, the soil pH was significantly higher in the hilly- and paddy-converted farmlands (alkalescence) compared to that in the greenhouse group (acidescence), while it was similar in all farmlands at the fruiting stage (alkalescence) (Supplementary Figure S1). In contrast, no obvious variation was observed for the soil moisture among all farmlands at the flowering stage, whereas it was significantly higher in the hilly- and paddy-converted farmlands at the fruiting stage (Supplementary Figure S1). With regard to soil nutrients, the highest concentrations of TP and TK were detected in the hilly-converted farmlands at both the flowering and fruiting stages, followed by the paddy-converted farmlands, and the lowest in the greenhouse soils (Figure 1a). Similar results were also evident in the contents of AP and ammonia during the flowering and fruiting stages, respectively (Figure 1a). In addition, the TN concentration at both the flowering and fruiting stages and the AP content at the fruiting stage were shown to be the highest in the paddy-converted farmlands, followed by the hilly-converted samples, and the lowest levels in the greenhouse (Figure 1a). Moreover, we also observed significantly higher concentrations of DOC and AK in the hilly- and paddy-converted farmland compared to those in the greenhouse at both flowering and fruiting stages (Figure 1a). When paying attention to soil enzyme activity, we found significantly higher activities of ALP in the hilly- and paddy-converted farmlands at both flowering and fruiting stages and beta-Glu and Cellulase in the fruiting stage (Figure 1b). Moreover, the activity of urease was found to be significantly higher in the hilly-converted farmlands in comparison to other regions, both at the flowering and fruiting stages (Figure 1b).

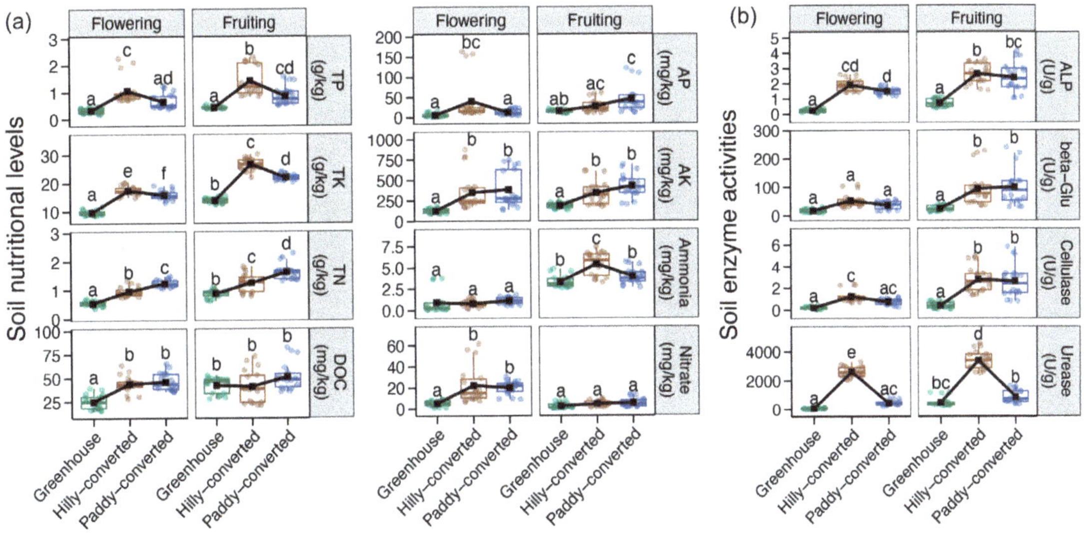

Figure 1. Differences in the soil nutrient levels (**a**) and enzyme activities (**b**) between different cultivation modes and different lemon growth stages. Different lowercases letters in each subfigure represent significant differences between different groups (Tukey's HSD test, $p < 0.05$).

3.2. Variations in Soil Protistan Diversity

In total, we obtained 3,382,668 high-quality sequences from the 108 soil samples, which were then clustered into 3487 ASVs. According to the taxonomy annotation, a total of 10 phyla, 29 classes, 86 orders, 166 families, 300 genera, and 492 species were identified, with more than 90% and 75% annotated rates at the family and species levels, respectively (Supplementary Figure S2). The rarefaction and species accumulation curves of all groups were tended to be horizontal (Supplementary Figure S3), indicating that the sequencing depth and sampling size could reflect the complete soil protistan communities in lemon farmlands. At the flowering stage, we found significantly higher alpha diversity of soil protistan communities (except the Chao index) in the hilly- and paddy-converted farmlands compared to the greenhouse samples (Figure 2a). In contrast, we observed highest alpha diversity indices of soil protistan communities (except the Pielou_J index) in the hilly-converted farmlands, which were significantly higher than those in greenhouse soils (Figure 2a).

The PCoA, based on the Bray–Curtis distance, showed that soil protistan communities from lemon farmlands, which were cultivated in different ways, were distinctively grouped and distinguished by the PC1 axis (Figure 2b). Moreover, soil protistan communities belonging to disparate lemon growth stages were also differentiated by the PC2 axis (Figure 2b). PERMANOVA revealed that both cultivation mode and growth stage significantly affected the soil protistan communities of lemon farmlands, with higher effects of cultivation modes ($p < 0.05$, Figure 2b). Then, we observed highest variations in soil protistan communities in the greenhouse during the flowering stage, followed by the hilly-converted, and the lowest was found in the paddy-converted farmlands (Figure 2c). During the fruiting stage, variations in soil protistan communities in the hilly-converted farmlands were significantly lower than those in the greenhouse and paddy-converted farmlands (Figure 2c).

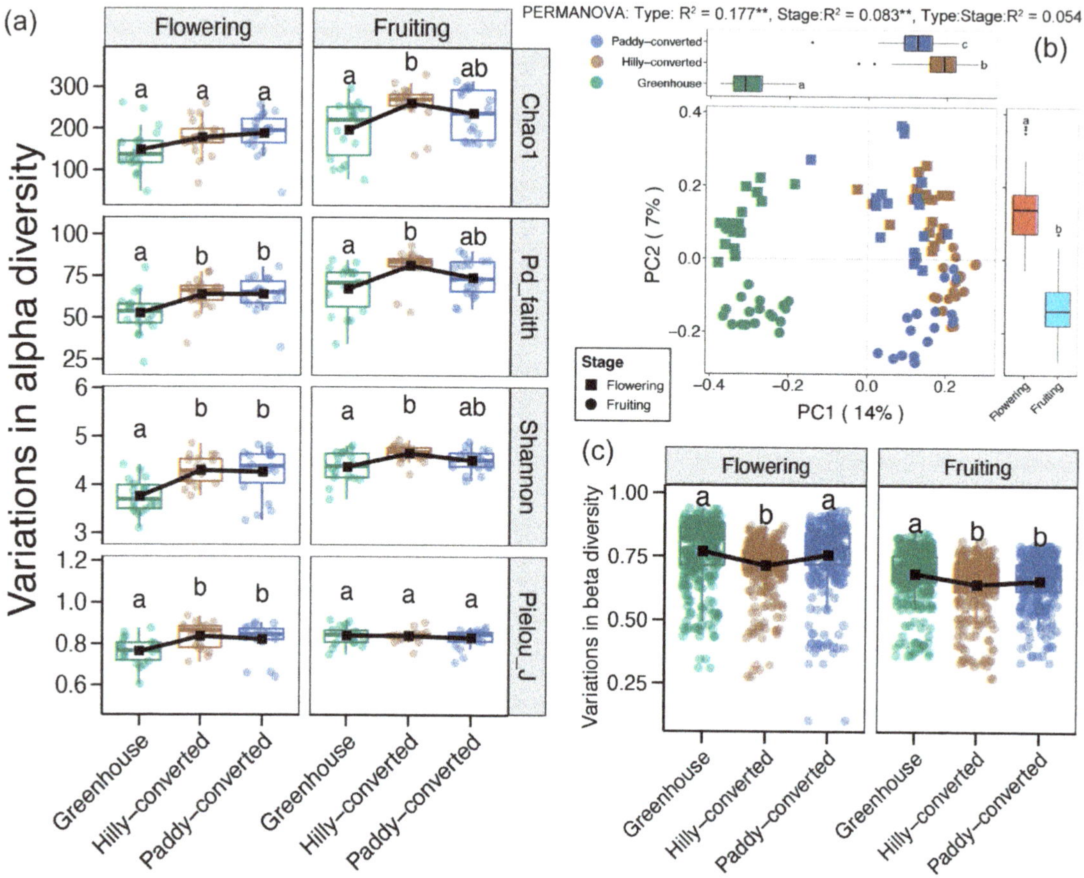

Figure 2. (**a**) Differences in the alpha diversity indices of soil protistan communities between different cultivation modes and different lemon growth stages. (**b**) Principal coordinate analysis (PCoA) and PERMANOVA test based on the Bray–Curtis distance for protistan communities between different cultivation modes and different lemon growth stages. ** represents the significant effects of factors on the variations in protistan communities (adonis test, $p < 0.01$). (**c**) Differences in the Bray–Curtis distance of soil protistan communities between different cultivation modes and different lemon growth stages. Different lowercases letters in each subfigure represent significant differences between different groups (Tukey's HSD test, $p < 0.05$).

3.3. Variations in Soil Protistan Compositions and Functions

Alveolata was found as the dominant protistan phylum in lemon farmlands, followed by Archaeplastida, Rhizaria, and Stramenopiles (Figure 3a). The relative abundance of Alveolata was shown to be significantly higher in hilly- and paddy-converted farmlands than that in greenhouse soils at the fruiting stage (Figure 3b). In addition, Archaeplastida was more abundant in greenhouse soils compared to other farmlands during the both flowering and fruiting stages (Figure 3). Moreover, Stramenoplies was enriched in the hilly-converted farmlands at the flowering stage (Figure 3b). At the species level, Pythium was the most abundant, and other main species included Colpoda, Cercomonas, Oxyricha, and Scenedesmus (Supplementary Figure S4). Among them, Oxytricha was enriched in the hilly-converted farmlands at the fruiting stage, and Scenedesmus was more abundant in greenhouse soils at both flowering and fruiting stages (Supplementary Figure S5).

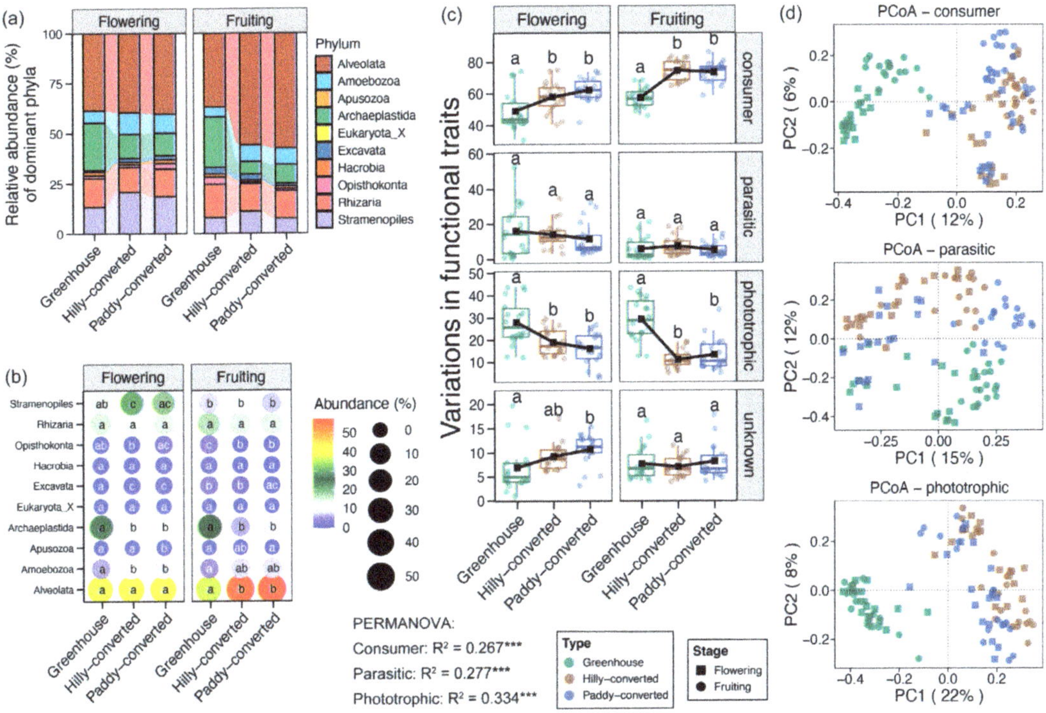

Figure 3. (**a**) Relative abundance (%) of soil protistan phyla among different farmlands. (**b**) Differences in the relative abundance of soil protistan phyla between different cultivation modes and different lemon growth stages. (**c**) Variations in relative abundances of protistan functional traits between different cultivation modes and different lemon growth stages. Different lowercases letters in each subfigure represent significant differences between different groups (Tukey's HSD test, $p < 0.05$). (**d**) Principal coordinate analysis (PCoA) and PERMANOVA test based on the Bray–Curtis distance for different protistan functional traits between different cultivation modes and different lemon growth stages. *** represents the significant effects of factors on the variations in protistan communities (adonis test, $p < 0.001$).

It is evident that protists can be categorized according to their functional traits, which are indicative of their trophic level. The three primary trophic levels of protists are: consumer, parasitic, and phototrophic. In the context of lemon farmlands, a marked increase in the relative abundance of consumers was observed in soils derived from hilly and paddy conversion, both during the flowering and fruiting stages. In contrast, the relative abundance of phototrophic protists exhibited a higher prevalence in greenhouse samples (Figure 3c). In addition, PCoA showed obviously cultivation mode and growth stage were clearly associated with the formation of clusters for consumer and parasitic protists. However, the impact of growth stage on phototrophic protists was found to be comparatively minimal (Figure 3d).

3.4. Environmental Assoications and Assembly Mechanisms of Soil Protistan Communities

The RDA was initially employed to assess the correlations between soil environmental variables and protistan communities in lemon farmlands, and the results are shown in Figure 4a. The majority of measured soil environmental variables exhibited significant correlations with the soil protistan communities. Among the factors under consideration, nitrate was the most significant in terms of discrepancies between the stages, with a substantial correlation to the PC2 axis, the axis differentiated protistan communities from the flowering and fruiting stages. To distinguish protistan communities from farmlands with different cultivation modes, ammonia, TK, and TK positively correlated to the hilly- and paddy-converted samples, whereas nitrate and moisture showed positive effects on the greenhouse soils. Furthermore, a random forest model based on soil environmental variables was constructed to predict variations in soil protistan communities, and a more than 90% predicted accuracy was obtained (Figure 4b). Based on the random forest model, ammonia, TN, TK, and nitrite were recognized as the main drivers for variations in soil protistan communities within lemon farmlands (Figure 4b).

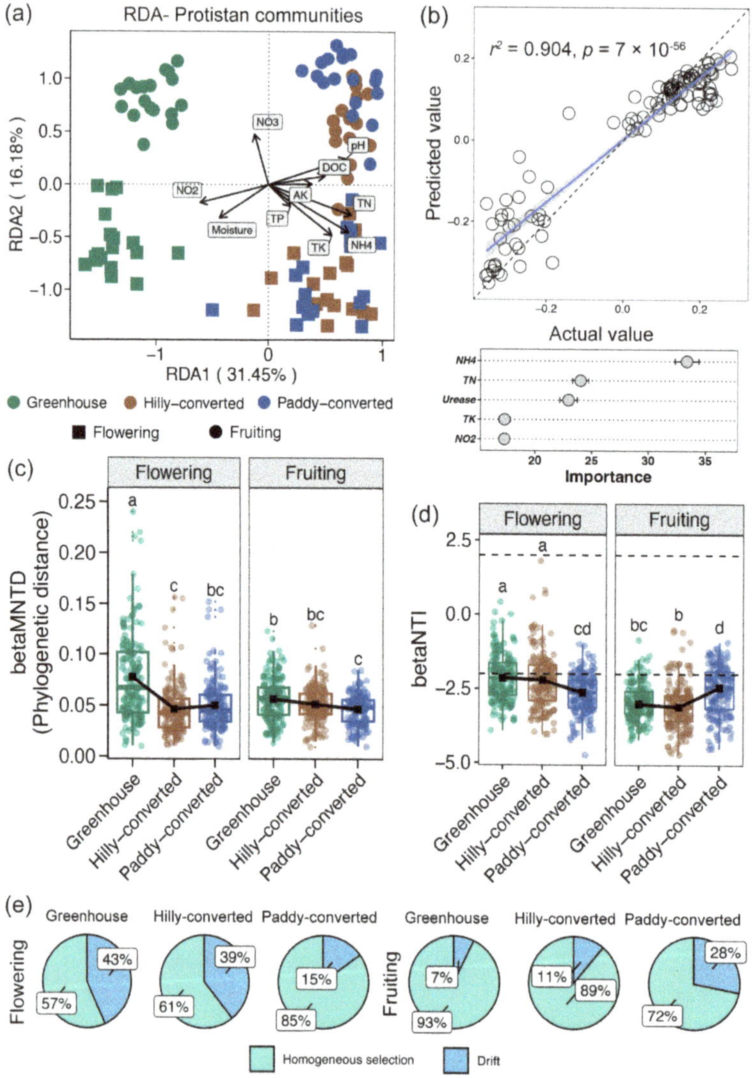

Figure 4. (**a**) Redundancy analysis (RDA) showing the correlations between soil environmental variables with protistan communities in lemon farmlands. (**b**) Linear regression exploring the accuracy of random forest model to predict the variations in soil protistan communities and the key factors for this prediction. Blue line represents the fitted curve and gray shadow represents the 95% confidence interval. Differences in the betaMNTD (**c**) and betaNTI (**d**) of soil protistan communities

between different cultivation modes and different lemon growth stages. Different lowercases letters in each subfigure represent significant differences between different groups (Tukey's HSD test, $p < 0.05$). (**e**) Contributions of different ecological processes for assembly of soil protistan communities.

Based on the comparison of betaMNTD, we observed significantly higher phylogenetic distances for soil protistan communities in greenhouse compared to other farmlands at both flowering and fruiting stages (Figure 4c). For community assembly, we found the median of betaNTI for all groups were lower than −2, indicating determinism-dominated assembly for soil protistan communities in lemon farmlands (Figure 4d). At the flowering stage, the betaNTI of soil protistan communities at the greenhouse and hilly-converted farmlands were more closed to −2, indicating relatively higher contribution of stochastic processes (Figure 4d). In contrast, the betaNTI was significantly decreased in soil protistan communities at the greenhouse and hilly-converted farmlands at the fruiting stage, showing a more deterministic-controlled community assembly (Figure 4d). The processes of homogeneous selection and drift were recognized as the specific ecological processes for deterministic and stochastic processes, respectively, for soil protistan community assembly in lemon farmlands (Figure 4e). In accordance with the findings of betaNTI, the contribution of heterogeneous selection was about 60% in greenhouse and hilly-converted farmlands during the flowering stage, which increased to 84.97% in paddy-converted farmlands (Figure 4e). Conversely, the contribution of heterogeneous selection increased to approximately 90% in greenhouse and hilly-converted farmlands at the fruiting stage, whereas it decreased to 71.9% in the paddy-converted farmlands (Figure 4e).

3.5. Correlations of Soil Protistan Communities and Lemon Quality

Subsequent to the harvesting of lemons, we measured a series of production indices to evaluate the quality of lemons. Although multiple production indices exhibited variations among farmlands with different cultivation modes, only a significantly higher valve number was observed in lemons from the hilly-converted farmlands (Supplementary Figure S6). Furthermore, we calculated a comprehensive production index based on measured quality parameters, and the value of this index was shown to be significantly higher in lemons from the hilly- and paddy-converted farmlands compared to those from greenhouse (Figure 5a). Then, relationships between the diversity and composition of soil protistan communities with the lemon production index were investigated, respectively. Significant correlations between the soil protistan diversity and the lemon production index were only found in paddy-converted farmlands (Linear regression, $p < 0.05$, Figure 5b). For the relationships between the protistan composition and the lemon production index, significant associations were obtained in hilly- and paddy-converted farmlands (Linear regression, $p < 0.05$, Figure 5c). Finally, an SEM was constructed to explore the mechanisms of variations in lemon quality of farmland type mediated by soil protistan communities (Figure 5d). The production index of lemons was found to be directly influenced by the type of farmland and the soil protistan compositions. The findings of the study demonstrated a direct correlation between variations in soil protistan composition and the farm cultivation modes, soil nutrient levels, and protistan diversity. Soil protistan diversity was directly changed by soil enzyme activities (mainly by ALP), which was directly altered by soil nutrient levels (mainly by TN and TK).

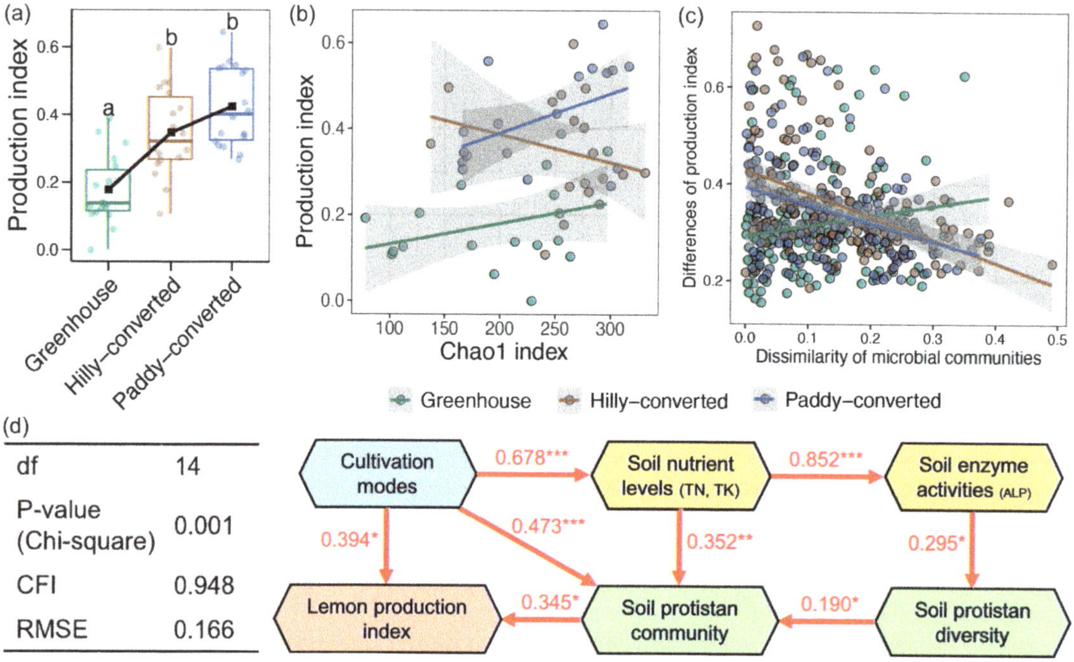

Figure 5. (**a**) Differences in the production index of lemons obtained from different farmcultivation modes. Different lowercases letters represent significant differences between different groups (Tukey's HSD test, $p < 0.05$). Relationships between lemon production index with soil protistan diversity (**b**) and composition (**c**) in different farmlands. (**d**) Structure equation model showing the direct and indirect effects of farmcultivation modes on lemon quality mediated by soil protistan communities. Only significant relationships were shown in the path diagram, and a red color represents a positive effect. *, **, and *** represent the p-value lower than 0.05, 0.01, and 0.001, respectively.

4. Discussion

Protists represent a significant group of eukaryotic microorganisms in soils; however, their specific communities and related functions in lemon farmlands have not been extensively documented. The present study was based on a comprehensive sampling of soil protistan communities across three distinct cultivation modes of lemons in Chongqing, China. To the best of our knowledge, the present study is for the first time to provide the soil protistan profiles and explore their variations in lemon farmlands. We observed significant variations in the diversity, composition, and function of soil protistan communities in lemon farmlands with different cultivation modes and distinct lemon growth stages. Moreover, our results showed stronger effects of cultivation modes on soil protistan communities than those of lemon growth stages. Both seasonality and land use have been found to affect the soil protistan communities with different strengths among studies [33–35]. In this study, the flowering and fruiting stages of lemons were in spring and autumn, respectively, and the climatic conditions were relatively similar. Conversely, the management of greenhouses is subject to a greater degree of regulation as a consequence of human activities when compared with open-field cultivation. This may be the reason for why this study found that cultivation modes have a stronger impact on soil protistan communities in lemon farmlands.

Evidence has demonstrated that biodiversity exerts a significant influence on a variety of ecosystem functions, including, but not limited to productivity, decomposition, nutrient turnover, and stability [36]. A plethora of studies have demonstrated that an increase in species diversity can enhance ecosystem functions by promoting complementary resource use and influencing particular species traits [37]. Furthermore, biodiversity is essential for providing ecosystem services that support human life, such as food, clean water, disease regulation, climate regulation, and pollination [38]. Therefore, maintaining biodiversity is crucial for sustaining ecosystem functions and services that are vital for both natural systems and human well-being [39]. In our results, we observed significantly lower diversity of soil protistan communities in lemon farmlands with greenhouses compared those in open-air environments (Figure 2a), suggesting the degeneration of soil functions in lands with the greenhouses. Previous studies have revealed the land degeneration in greenhouses due to the soil erosion, salinization, and fertility loss induced by the intensive agricultural practices [40,41]. Moreover, the site-to-site variations in soil protistan communities were also found to be largest in the greenhouse fields (Figure 2c), indicating a weak stability of soil protistan communities in greenhouse farming. It is imperative to acknowledge the significance of a stable soil microbial community in ensuring the consistent execution of ecosystem functions [42]. Thus, our results implied the relatively weak and unstable ecosystem functions of lemon farmlands with greenhouses compared to those in open-air environments.

In contrast to taxonomic composition, the application of functional traits is cumulative in nature, facilitating the evaluation of the roles of microbial communities in ecosystem functions and services [43]. According to the lifestyle, protists can be divided into three functional groups, which are consumers, phototrophs, and parasitic taxa [29]. Among them, the consumer group was the dominant group, followed by parasitic taxa and phototrophs [44], consistent results were also found in our study. More specifically, we found significantly higher relative abundance of phototrophs in lemon fields with greenhouse farming (Figure 3c). This result was consistent with more effective lighting and relatively high temperatures in greenhouses. Most of the phototrophs in lemon farmlands belonged to the Archaeplastida phyla, a group of photoautotrophic organisms encompassed by red algae, green algae, and land plants [45]. These organisms are primary producers, responsible for converting light energy into organic compounds through photosynthesis [46]. However, unlike the importance of phototrophy in marine systems for global carbon fixation and nutrient cycling [47], the standing biomass of phototrophs in soils is negligible as compared with plants, although their turnover is much faster [48]. In contrast to the abundant phototrophs, lemon farmlands with greenhouses showed lower relative abundance of consumers. At the flowering stage, the decreased consumers in greenhouses were mainly Stramenopiles (Figure 3b). Non-photosynthetic Stramenopiles are widespread consumers of bacteria, which play a major role in recycling carbon and nutrients within microbial food webs to maintain the nutrient cycles in the ecosystem [49]. At the fruiting stage, the decreased consumers in greenhouses mainly belonged to the Alveolata phyla, especially Hypotrichia (Figure 3b and Supplementary Figure S5). Hypotrichia are important in nutrient cycling and for the maintenance of soil structure, thus influencing the overall health and productivity of these ecosystems [50]. Based on these information, we could deduce that greenhouses might inhibit nutrient cycling in soil and then damage the ecosystem functions of lemon farmlands.

A determinism (heterogeneous selection)-dominant assembly of soil protistan communities was found in our studied lemon farmlands (Figure 4d,e). This finding suggested that the soil protistan community was mainly governed by environmental filtering [51]. Our results further revealed nitrogen, including TN, nitrate, nitrite, and ammonia, as the key factors to drive the variations in the soil protistan communities of lemon farmlands (Figure 4a,b). Nitrogen has also been found to be a dominant driver for soil protistan community in other agricultural ecosystems, expressed by changes in the protistan community due to nitrogen inputs [52,53]. Moreover, research has shown that protists are more sensitive to nitrogen fertilization than other microorganisms in agricultural soils [54]. These evidence suggest that nitrogen can exert a notable influence on soil protist communities in agricultural systems, potentially leading to changes in their diversity and composition. Given that the crucial roles of soil protists in agricultural ecosystems and changes in soil protist communities due to factors such as nitrogen fertilization can impact nutrient cycling, microbial interactions, and plant health, they ultimately influence the stability and functioning of ecosystems [55,56]. These findings further highlight the importance of understanding and monitoring soil protist communities to assess and manage their implications for ecosystem functioning in the context of environmental changes, including nitrogen inputs.

The primary function of agricultural ecosystems is to produce higher quality agricultural products, and it is the considered opinion of the present author that the ultimate goal of other related research should be related to this fact. Thus, we evaluated the lemon quality of the farmlands involved in our study and its associations with soil properties and protistan communities. Greenhouses have been believed to produce crops with higher yields and consistent seasonality compared to the open-field cultivation as they offer climate control systems [57]. Moreover, greenhouses also help lower the risk of pests and disease transmission, which can lead to increased yields and improved overall product quality [58]. However, it is possible to unknowingly follow practices that can contribute to or directly cause a reduction in crop quality in a greenhouse, such as insufficient venting or poorly maintained exhaust fans [59]. In our results, we observed a substantial reduction in lemon quality in greenhouse farming compared to that in open fields (Figure 5a). We further found the direct effects of cultivation modes and soil protistan compositions on lemon quality (Figure 5d). For the greenhouse practices, several potential reasons could lead to the reduction in lemon quality. Greenhouses are more suitable for planting short-term small plants; in contrast, the lemon tree is a perennial large plant and long-term exposure to greenhouse conditions can lead to a decrease in its vitality [60,61]. In addition, greenhouses generally possess a higher plant density, which increases the competition for light and then influences their growth [62]. For the effects of soil protistan community on lemon quality, our results indicated relatively weak nutrient cycling of protists in greenhouses, which could be driven by the low level on nitrogen in greenhouse farmlands (Figure 1a). In conclusion, open-field cultivation could be considered a more suitable agricultural mode for lemon cultivation.

5. Conclusions

The findings of this study demonstrated that the cultivation mode exerted an influence on the soil properties, instigated succession in the soil protistan community, and regulated the production quality in lemon farmlands. In contrast to open-field cultivation, the use of a greenhouses has been demonstrated to engender a decline in the nutrient content and enzyme activity of the soil. This has been shown to result in a reduction in diversity and alteration to the composition of protistan communities. During the growth phase of the lemon, the process of greenhouse farming resulted in an enrichment of Archaeplastida in farmland soils, while Stramenoplies and Alveolata were reduced at the flowering and fruiting stages, respectively. The analysis of functional traits indicated that the greenhouse environment led to an increase in the relative abundance of phototrophs while concurrently resulting in a decrease in consumers within the farmland soils when compared to open-field cultivation. Furthermore, deterministic assembly was demonstrated to govern the succession of soil protistan communities in lemon farmlands, with nitrogen level identified as the dominant driving factor. Finally, the lowest quality of lemons was observed in greenhouse farmlands, which resulted in the direct effects of greenhouse farming and indirect effects mediated by soil protistan communities. The findings of this study have enhanced our comprehension of soil protists in agricultural ecosystems and provide a foundation for the future enhancement of the management of lemon cultivation. The improvement of planting strategies and operational methods in greenhouses to enhance lemon quality represent a future research direction.

Supplementary Materials: The following supporting information can be downloaded at: https://www.mdpi.com/article/10.3390/plants14132024/s1, Figure S1: Differences in the soil properties between different cultivation modes and different lemon growth stages. Different lowercases letters in each subfigure represent significant differences between different groups (Tukey's HSD test, $p < 0.05$); Figure S2: Annotation ratio of protistan ASVs at different taxonomy levels; Figure S3: Rarefaction (a) and species accumulation (b) curves of all groups; Figure S4: Relative abundance (%) of soil protistan species among different farmlands; Figure S5: Differences in the relative abundance of soil protistan species between different cultivation modes and different lemon growth stages. Different lowercases letters in each subfigure represent significant differences between different groups (Tukey's HSD test, $p < 0.05$). Figure S6: Differences in the lemon quality parameters between different cultivation modes and different lemon growth stages. Different lowercases letters in each subfigure represent significant differences between different groups (Tukey's HSD test, $p < 0.05$).

Author Contributions: H.L. (Haoqiang Liu): the funding and Project administration, methodology, ideas, data curation, statistical analysis. H.L. (Hongjun Li), S.L., C.R. and Z.P. participated in this work. All authors have read and agreed to the published version of the manuscript.

Funding: This research was supported by the National Key R & D Program of China (2021YFD1400800, 2020YFD1000102 and 2018YFD0201500); the National Key R & D Program of China (2019YFD1002100), and the Chongqing Scientific Research Project (cstc2021jcyj-bsh0082).

Data Availability Statement: The raw sequencing data for this study can be found in the NCBI SRA database with accessible number PRJNA1063860.

Conflicts of Interest: The authors declare no conflicts of interest.

References

1. Power, A.G. Ecosystem services and agriculture: Tradeoffs and synergies. *Philos. Trans. R. Soc. B* **2010**, *365*, 2959–2971. [CrossRef] [PubMed]
2. Xue, M.; Guo, Z.; Gu, X.; Gao, H.; Weng, S.; Zhou, J.; Gu, D.; Lu, H.; Zhou, X. Rare rather than abundant microbial communities drive the effects of long-term greenhouse cultivation on ecosystem functions in subtropical agricultural soils. *Sci. Total Environ.* **2020**, *706*, 136004. [CrossRef]
3. Zhang, L.; Tan, S.; Liu, C.; Wang, S. Influence of labor transfer on farmland sustainable development: A regional comparison of plain and hilly areas. *Qual. Quant.* **2018**, *52*, 431–443. [CrossRef]
4. Jin, Z.; Shah, T.; Zhang, L.; Liu, H.; Peng, S.; Nie, L. Effect of straw returning on soil organic carbon in rice–wheat rotation system: A review. *Food Energy Secur.* **2020**, *9*, e200. [CrossRef]
5. Song, S.L.; Luo, X.; Wu, H.; Lu, X.L.; Xu, F.J.; Zhang, Z.H.; Guan, Y.X.; Dai, C.C. Combined System of Organic Substrate and Straw-Degrading Microbial Agents Improved Soil Organic Matter Levels and Microbial Abundance in a Rice–Wheat Rotation. *Curr. Microbiol.* **2020**, *79*, 172. [CrossRef] [PubMed]
6. Ebrahimi, M.; Majidian, M.; Alizadeh, M.R. Effect of different planting techniques and puddling methods on soil properties, growth, yield, and grain quality characteristics of rice (*Oryza sativa* L.). *Commun. Soil Sci. Plant Anal.* **2022**, *53*, 2543–2557. [CrossRef]
7. Fu, H.; Chen, H.; Ma, Q.; Chen, B.; Wang, F.; Wu, L. Planting and mowing cover crops as livestock feed to synergistically optimize soil properties, economic profit, and environmental burden on pear orchards in the Yangtze River Basin. *J. Sci. Food Agric.* **2023**, *103*, 6680–6688. [CrossRef]
8. Yang, T.; Lupwayi, N.; Marc, S.A.; Siddique, K.H.; Bainard, L.D. Anthropogenic drivers of soil microbial communities and impacts on soil biological functions in agroecosystems. *Glob. Ecol. Conserv.* **2021**, *27*, e01521. [CrossRef]

9. Dincă, L.C.; Grenni, P.; Onet, C.; Onet, A. Fertilization and soil microbial community: A review. *Appl. Sci.* **2022**, *12*, 1198. [CrossRef]
10. Gupta, A.; Singh, U.B.; Sahu, P.K.; Paul, S.; Kumar, A.; Malviya, D.; Singh, S.; Kuppusamy, P.; Singh, P.; Paul, D.; et al. Linking soil microbial diversity to modern agriculture practices: A review. *Int. J. Environ. Res. Public Health* **2022**, *19*, 3141. [CrossRef]
11. Philippot, L.; Chenu, C.; Kappler, A.; Rillig, M.C.; Fierer, N. The interplay between microbial communities and soil properties. *Nat. Rev. Microbiol.* **2023**, *22*, 226–239. [CrossRef]
12. Chandarana, K.A.; Amaresan, N. Soil protists: An untapped microbial resource of agriculture and environmental importance. *Pedosphere* **2022**, *32*, 184–197. [CrossRef]
13. Santos, S.S.; Schoeler, A.; Nielsen, T.K.; Hansen, L.H.; Schloter, M.; Winding, A. Land use as a driver for protist community structure in soils under agricultural use across Europe. *Sci. Total Environ.* **2020**, *717*, 137228. [CrossRef] [PubMed]
14. Guo, S.; Xiong, W.; Hang, X.; Gao, Z.; Jiao, Z.; Liu, H.; Mo, Y.; Zhang, N.; Kowalhuk, G.A.; Li, R.; et al. Protists as main indicators and determinants of plant performance. *Microbiome* **2021**, *9*, 64. [CrossRef]
15. Jousset, A. Application of protists to improve plant growth in sustainable agriculture. In *Rhizotrophs: Plant Growth Promotion to Bioremediation*; Mehnaz, S., Ed.; Springer: Singapore, 2017; Volume 2, pp. 263–273.
16. Mditshwa, A.; Magwaza, L.S.; Tesfay, S.Z.; Opara, U.L. Postharvest factors affecting vitamin C content of citrus fruits: A review. *Sci. Hortic.* **2017**, *218*, 95–104. [CrossRef]
17. Zhong, S.; Ren, J.; Chen, D.; Pan, S.; Wang, K.; Yang, S.; Fan, G. Free and bound volatile compounds in juice and peel of Eureka lemon. *Food Sci. Technol. Res.* **2014**, *20*, 167–174. [CrossRef]
18. Lee, S.B.; Sung, J.K.; Lee, Y.J.; Lim, J.E.; Song, Y.S.; Lee, D.B.; Hong, S.Y. Analysis of soil total nitrogen and inorganic nitrogen content for evaluating nitrogen dynamics. *Korean J. Soil Sci. Ferti.* **2017**, *50*, 100–105. [CrossRef]
19. Zhao, G.; Sheng, Y.; Wang, J.; Li, Z.; Yang, J. Optimized digestion methods: Organic phosphorus sequential extraction, total phosphorus, and nitrogen simultaneous determination in sediments. *J. Soils Sediments* **2018**, *18*, 2072–2080. [CrossRef]
20. Moonrungsee, N.; Pencharee, S.; Jakmunee, J. Colorimetric analyzer based on mobile phone camera for determination of available phosphorus in soil. *Talanta* **2015**, *136*, 204–209. [CrossRef]
21. Almu, H.; Abdulkadir, N.A.; Sani, A.; Adamu, U.K.; Aminu, M.A. Potassium Distribution in Surface Soils of Kust Teaching, Research and Commercial Farm, Gaya, Kano State. *Afr. J. Agric. Food Sci.* **2021**, *4*, 26–35.
22. Kang, J.; Liu, Z.; Yu, C.; Wang, Y.; Wang, X. Degradation performance of high-concentration coking wastewater by manganese oxide ore acidic oxidation. *Water Sci. Technol.* **2022**, *86*, 367–379. [CrossRef]
23. *GB/T 8210-2011*; Method of inspection for fresh citrus fruit. AQSIQ and SAC. China National Standardization Administration: Beijing, China, 2011.
24. Stoeck, T.; Behnke, A.; Christen, R.; Amaral-Zettler, L.; Rodriguez-Mora, M.J.; Chistoserdov, A.; Orsi, W.; Edgcomb, V.P. Massively parallel tag sequencing reveals the complexity of anaerobic marine protistan communities. *BMC Biol.* **2009**, *7*, 72. [CrossRef] [PubMed]
25. Zhao, Z.; Li, H.; Sun, Y.; Shao, K.; Wang, X.; Ma, X.; Hu, A.; Zhang, H.; Fan, J. How habitat heterogeneity shapes bacterial and protistan communities in temperate coastal areas near estuaries. *Environ. Microbiol.* **2022**, *24*, 1775–1789. [CrossRef] [PubMed]
26. Callahan, B.J.; McMurdie, P.J.; Rosen, M.J.; Han, A.W.; Johnson, A.J.A.; Holmes, S.P. DADA2: High-resolution sample inference from Illumina amplicon data. *Nat. Methods* **2016**, *13*, 581–583. [CrossRef]
27. Bokulich, N.A.; Kaehler, B.D.; Rideout, J.R.; Dillon, M.; Bolyen, E.; Knight, R.; Huttley, G.A.; Caporaso, J.G. Optimizing taxonomic classification of marker-gene amplicon sequences with qiime 2's q2-featureclassifier plugin. *Microbiome* **2018**, *6*, 90. [CrossRef] [PubMed]
28. Vaulot, D.; Sim, C.W.H.; Ong, D.; Teo, B.; Biwer, C.; Jamy, M.; Lopes dos Santos, A. metaPR2: A database of eukaryotic 18S rRNA metabarcodes with an emphasis on protists. *Mol. Ecol. Resour.* **2022**, *22*, 3188–3201. [CrossRef]
29. Singer, D.; Seppey, C.V.; Lentendu, G.; Dunthorn, M.; Bass, D.; Belbahri, L.; Blandenier, Q.; Debroas, D.; de Groot, G.A.; de Vargas, C.; et al. Protist taxonomic and functional diversity in soil, freshwater and marine ecosystems. *Environ. Int.* **2021**, *146*, 106262. [CrossRef]
30. Stegen, J.C.; Lin, X.; Fredrickson, J.K.; Chen, X.; Kennedy, D.W.; Murray, C.J.; Rockhold, M.L.; Konopka, A. Quantifying community assembly processes and identifying features that impose them. *ISME J.* **2013**, *7*, 2069–2079. [CrossRef]
31. Stegen, J.C.; Lin, X.; Fredrickson, J.K.; Konopka, A.E. Estimating and mapping ecological processes influencing microbial community assembly. *Front. Microbiol.* **2015**, *6*, 370. [CrossRef]
32. Garland, P.E.; Pereira, O.; Hochart, C.; Auguet, J.C.; Debroas, D. A strong link between marine microbial community composition and function challenges the idea of functional redundancy. *ISME J.* **2018**, *12*, 2470–2478. [CrossRef]
33. Pérez-Juárez, H.; Serrano-Vázquez, A.; Lara, E.; Ximénez, C.; Godínez-Alvarez, H.; Rodríguez-Zaragoza, S.; Eguiarte, L.E.; Hernandez Moreno, M.M.; Fernandez, L.D.; Rojas-Velazquez, L.; et al. Population dynamics of amoeboid protists in a tropical desert: Seasonal changes and effects of vegetation and soil conditions. *Acta Protozool.* **2018**, *57*, 231–242. [CrossRef]

34. De Gruyter, J.; Weedon, J.T.; Bazot, S.; Dauwe, S.; Fernandez-Garberí, P.R.; Geisen, S.; De La Motte, L.G.; Heinesch, B.; Janssens, I.A.; Leblans, N.; et al. Patterns of local, intercontinental and interseasonal variation of soil bacterial and eukaryotic microbial communities. *FEMS Microbiol. Ecol.* **2020**, *96*, fiaa018. [CrossRef] [PubMed]
35. Fournier, B.; Samaritani, E.; Frey, B.; Seppey, C.V.; Lara, E.; Heger, T.J.; Mitchell, E.A. Higher spatial than seasonal variation in floodplain soil eukaryotic microbial communities. *Soil Biol. Biochem.* **2020**, *147*, 107842. [CrossRef]
36. Pennekamp, F.; Pontarp, M.; Tabi, A.; Altermatt, F.; Alther, R.; Choffat, Y.; Fronhofer, E.A.; Ganesanandamoorthy, P.; Garnier, A.; Griffiths, J.I.; et al. Biodiversity increases and decreases ecosystem stability. *Nature* **2018**, *563*, 109–112. [CrossRef]
37. Tilman, D.; Isbell, F.; Cowles, J.M. Biodiversity and ecosystem functioning. *Annu. Rev. Ecol. Evol. Syst.* **2014**, *45*, 471–493. [CrossRef]
38. Costanza, R.; de Groot, R.; Sutton, P.; van der Ploeg, S.; Anderson, S.J.; Kubiszewski, I.; Farber, S.; Turner, R.K. Changes in the global value of ecosystem services. *Glob. Environ. Change* **2014**, *26*, 152–158. [CrossRef]
39. Ceballos, G.; Ehrlich, P.R.; Barnosky, A.D.; García, A.; Pringle, R.M.; Palmer, T.M. Accelerated modern human-induced species losses: Entering the sixth mass extinction. *Sci. Adv.* **2015**, *1*, e1400253. [CrossRef]
40. Akrami, M.; Salah, A.H.; Javadi, A.A.; Fath, H.E.; Hassanein, M.J.; Farmani, R.; Dibaj, M.; Negm, A. Towards a sustainable greenhouse: Review of trends and emerging practices in analysing greenhouse ventilation requirements to sustain maximum agricultural yield. *Sustainability* **2020**, *12*, 2794. [CrossRef]
41. Ishtiaq, M.; Waqas Mazhar, M.; Maqbool, M.; Alataway, A.; Dewidar, A.Z.; Elansary, H.O.; Yessoufou, K. Application of smart agricultural practices in wheat crop to increase yield and mitigate emission of greenhouse gases for sustainable ecofriendly environment. *Sustainability* **2022**, *14*, 10453. [CrossRef]
42. De Vries, F.T.; Shade, A. Controls on soil microbial community stability under climate change. *Front. Microbiol.* **2013**, *4*, 265. [CrossRef]
43. Fiore-Donno, A.M.; Richter-Heitmann, T.; Bonkowski, M. Contrasting responses of protistan plant parasites and phagotrophs to ecosystems, land management and soil properties. *Front. Microbiol.* **2020**, *11*, 1823. [CrossRef]
44. Nguyen, B.A.T.; Chen, Q.L.; Yan, Z.Z.; Li, C.; He, J.Z.; Hu, H.W. Distinct factors drive the diversity and composition of protistan consumers and phototrophs in natural soil ecosystems. *Soil Biol. Biochem.* **2021**, *160*, 108317. [CrossRef]
45. Kliman, R.M. (Ed.) *Encyclopedia of Evolutionary Biology*; Academic Press: Cambridge, MA, USA, 2016; pp. 344–360.
46. Li, Y.; Gao, P.; Sun, X.; Li, B.; Guo, L.; Yang, R.; Su, X.; Gao, W.; Xu, Z.; Yan, G.; et al. Primary succession changes the composition and functioning of the protist community on mine tailings, especially phototrophic protists. *ACS Environ. Au* **2022**, *2*, 396–408. [CrossRef]
47. Worden, A.Z.; Follows, M.J.; Giovannoni, S.J.; Wilken, S.; Zimmerman, A.E.; Keeling, P.J. Rethinking the marine carbon cycle: Factoring in the multifarious lifestyles of microbes. *Science* **2015**, *347*, 1257594. [CrossRef] [PubMed]
48. Mann, D.G.; Vanormelingen, P. An inordinate fondness? The number, distributions, and origins of diatom species. *J. Eukaryot. Microbiol.* **2013**, *60*, 414–420. [CrossRef] [PubMed]
49. Corliss, J.O. Biodiversity and biocomplexity of the protists and an overview of their significant roles in maintenance of our biosphere. *Acta Protozool.* **2002**, *41*, 199–220.
50. Bamforth, S.S. Interpreting soil ciliate biodiversity. *Plant Soil* **1995**, *170*, 159–164. [CrossRef]
51. Glaser, K.; Kuppardt, A.; Boenigk, J.; Harms, H.; Fetzer, I.; Chatzinotas, A. The influence of environmental factors on protistan microorganisms in grassland soils along a land-use gradient. *Sci. Total Environ.* **2015**, *537*, 33–42. [CrossRef]
52. Du, S.; Li, X.Q.; Hao, X.; Hu, H.W.; Feng, J.; Huang, Q.; Liu, Y.R. Stronger responses of soil protistan communities to legacy mercury pollution than bacterial and fungal communities in agricultural systems. *ISME Commun.* **2022**, *2*, 69. [CrossRef]
53. Liu, S.; Zhang, X.; Wang, H.; Dungait, J.A.; Pan, J.; Lidbury, I.D.; Ma, Z.; Chen, F.; Tang, Y. Seven-year N and P inputs regulate soil microbial communities via bottom-up effects on carbon and nutrient supply and top-down effects on protist relative abundance. *For. Ecol. Manag.* **2024**, *552*, 121582. [CrossRef]
54. Zhao, Z.B.; He, J.Z.; Geisen, S.; Han, L.L.; Wang, J.T.; Shen, J.P.; Wei, W.X.; Fang, Y.T.; Li, P.P.; Zhang, L.M. Protist communities are more sensitive to nitrogen fertilization than other microorganisms in diverse agricultural soils. *Microbiome* **2019**, *7*, 33. [CrossRef] [PubMed]
55. Geisen, S.; Mitchell, E.A.; Adl, S.; Bonkowski, M.; Dunthorn, M.; Ekelund, F.; Fernandez, L.D.; Jousset, A.; Krashevska, V.; Singer, D.; et al. Soil protists: A fertile frontier in soil biology research. *FEMS Microbiol. Rev.* **2018**, *42*, 293–323. [CrossRef] [PubMed]
56. Wang, J.; Zheng, C.; Lucas-Borja, M.E.; Shi, X. Soil protist functional composition shifts with atmospheric nitrogen deposition in subtropical forests. *J. Appl. Ecol.* **2023**, *60*, 1161–1169. [CrossRef]
57. Gruda, N.; Bisbis, M.; Tanny, J. Influence of climate change on protected cultivation: Impacts and sustainable adaptation strategies-A review. *J. Clean. Prod.* **2019**, *225*, 481–495. [CrossRef]
58. Gamliel, A.; Van Bruggen, A.H.C. Maintaining soil health for crop production in organic greenhouses. *Sci. Hortic.* **2016**, *208*, 120–130. [CrossRef]

59. Ganguly, A.; Ghosh, S. A review of ventilation and cooling technologies in agricultural greenhouse application. *Iran. J. Energy Environ.* **2011**, *2*, 32–46.
60. Chakraborty, T.; Saha, S.; Reif, A. Decrease in available soil water storage capacity reduces vitality of young understorey European beeches (*Fagus sylvatica* L.)—A case study from the Black Forest, Germany. *Plants* **2013**, *2*, 676–698. [CrossRef]
61. Jaeger, A.C.; Hartmann, M.; Conz, R.F.; Six, J.; Solly, E.F. Drought-induced tree mortality in Scots pine mesocosms promotes changes in soil microbial communities and trophic groups. *Appl. Soil Ecol.* **2024**, *194*, 105198. [CrossRef]
62. Postma, J.A.; Hecht, V.L.; Hikosaka, K.; Nord, E.A.; Pons, T.L.; Poorter, H. Dividing the pie: A quantitative review on plant density responses. *Plant Cell Environ.* **2021**, *44*, 1072–1094. [CrossRef]

MDPI AG
Grosspeteranlage 5
4052 Basel
Switzerland
Tel.: +41 61 683 77 34

Plants Editorial Office
E-mail: plants@mdpi.com
www.mdpi.com/journal/plants

www.ingramcontent.com/pod-product-compliance
Lightning Source LLC
Chambersburg PA
CBHW061512210926
51850CB00008B/1888
9783725856213